David Braine, Gifford Fellow and Lecturer in Philosophy in the University of Aberdeen, studied physics, history and philosophy at Oxford from 1958 to 1965. Despite spinal injury in 1977, he has continued to publish in the theory of knowledge, ethics and theology. He plans the present work as part of a trilogy, successively reshaping metaphysics, the theory of man and of language, and epistemology: this first about the world of Time in relation to God; the second, 'Man. Animal open to the infinite'; and the third, 'The Structure of Knowledge'.

THE REALITY OF TIME
AND
THE EXISTENCE OF GOD

TO

KEITH WHITEFIELD

THE REALITY OF TIME AND THE EXISTENCE OF GOD

The Project of Proving God's Existence

David Braine

CLARENDON PRESS · OXFORD
1988

Oxford University Press, Walton Street, Oxford OX2 6DP
Oxford New York Toronto
Delhi Bombay Calcutta Madras Karachi
Petaling Jaya Singapore Hong Kong Tokyo
Nairobi Dar es Salaam Cape Town
Melbourne Auckland
and associated companies in
Beirut Berlin Ibadan Nicosia

Oxford is a trade mark of Oxford University Press

Published in the United States
by Oxford University Press, New York

British Library Cataloguing in Publication
Braine, David
The reality of time and the existence of God : the project of proving God's existence.
1. God—Proof
I. Title
212'.1'01 BT102
ISBN 0-19-824459-2

Library of Congress Cataloging in Publication Data
Braine, David.
The reality of time and the existence of God.
Bibliography: p.
Includes index.
1. God—Proof. 2. Time. I. Title.
BT102.B67 1988 212'.1 87–24880
ISBN 0-19-824459-2

Set by Pentacor Limited
High Wycombe, Bucks
Printed in Great Britain
by Biddles Ltd.,
Guildford and King's Lynn

Preface

In the autumn 1983 and spring 1984 as Gifford Fellow at the University of Aberdeen I delivered two sets of addresses as twin series of Gifford Investigations. In the first, *Gifford Investigations: I*, I presented the argument from the reality of time to God's existence which is developed in this book. In the second, *Gifford Investigations: II*, under the heading 'Man: Animal Open to the Infinite', I dealt with the nature of man and the nature of language. In the mean time I have been occupied also with a third project, the problems of the restructuring of epistemology which I had taken up in 1971. There has been continuous interplay between these co-ordinate investigations, the results of which should represent a kind of philosophical testament.

Such interplay has been vital because to have an idea is one thing, but to develop it and uphold it against objection quite another. In this first project I found myself drawing upon the ideas about time, causation and possibility, logical theory and substance which had been with me since the 1960s, the conception of the primacy of the first order which I had made clear in papers in 1971 and 1972, and the view of human understanding borne in upon me by reflection on Dummett's arguments in the philosophy of language. For me, everything in philosophy has fructified thought in philosophical theology and conversely it has often been problems in philosophical theology which have given edge to questionings in general philosophy. Thus it has in fact been only in the course of tussling with the the first project that I feel that I have broken through to clarity at points where there seemed to be an impasse in the other two: the relative roles of concept and proposition in judgement, the standing of general causal principles, and the means whereby philosophy in general, not just theology, can remain critical.

This book was born of the realization that it was absurd to suppose that God could delegate or give to another the power to give existence. Whether the idea come from philosophy or from Revelation, it was an insight vital to any sound theology that that in virtue of which God is able to create is not communicable. Let this principle once be sullied, then the theology concerned is doomed.

And this realization married naturally with the conceptions of Aquinas. 'Nor can the being of a thing continue after the action of the agent has ceased, if the agent is the cause of the effect not only in respect of becoming but also in respect of being. This is why hot water retains heat after the cessation of the fire's action, whereas the air does not continue to be lit up even for a moment when the sun ceases to act upon it, because water is a matter susceptive of the fire's heat in the same way as it exists in the fire . . . whereas the air is not of such a nature as to receive light in the same way as it exists in the sun, which is the principle of light.' (*Summa Theologica*, Ia, Q. 104, Art. 1.) So, somehow in this conception God is to be principle of all existence, so that He alone exists of Himself, and everything else only in virtue of His action.

Indeed, it would seem, a vital conception theologically. The God who has existence of Himself transcends in nature all other things which wait for their existence upon Him, but He is immediate to all other things, alone able to give existence: transcendence and immanence inseparable.

And likewise vital for philosophy: temporality being allowed to disambiguate the idea of contingency, and the contingency of the temporal being recovered as a truth of metaphysics, not the academic concern only of logic or epistemology; and a way being found between the opposing traditions described by Gilson, a long naturalistic tradition within which the substance of the world is necessary, and the rival tradition which recognizes contingency only by making the 'subjects' of existence in some way abstract.

But how to reach or explain this conception philosophically? How to see that ordinary temporal things cannot stand without a creator sustaining them, that dependence on a creator is but the obverse of a full recognition of contingency? How to grasp in one act both the impotence of creatures in respect of their existence and their need of a cause in this respect, a cause not thus impotent?

I could see no approach to a solution proceeding a priori. Clearly, God's existence would have to be extrinsically necessary, *de facto* inescapable because He has no background leaving it open that He might not exist. But the statement that God exists of intrinsic necessity or of His nature remained quite mysterious to me: it had been given no viable explanation.

Yet, proceeding a posteriori, was it not "obvious" that temporal things are impotent to ground their own continuance, and that anything with any kind of parity with temporal things would be equally impotent? And, further, was it not "obvious" that the existence of temporal things involves continuance, and "obvious" that this 'reliable' or 'non-chance' continuance must have a cause? So would it not be "obvious" that there must be something, cause of temporal continuance, which possesses existence in such a different way from temporal things as not to be open to dependence on something else? And then, might we not have both the germ of a proof of God's existence and an indication of what might be *meant* by saying that He exists 'of Himself'?

Moreover, would not this now conform to the universal pattern in philosophy whereby, whereas proof of incoherence or impossibility was often easy, knowledge of coherence or possibility always depended upon knowledge of the actual?—quite contrary to fashionable assumption in the philosophy of religion.

But how was I to give edge to these ideas, with me since 1976, and make what was "obvious" evident?

I have first to thank the Gifford Trust for giving me the opportunity as Gifford Fellow from 1982 to do this.

The thought in this book has been so long in developing, the first relevant ideas arising in Oxford in 1964–5 while preparing my B.Phil. thesis under Professor Ryle and the last key development being while in mid-flood of revision in November 1986, that it is difficult to trace particular influences. I am above all grateful to Professor Elizabeth Anscombe and Professor David Hamlyn whose suggestions were enormously helpful in detail and vital in sparking off the large-scale revision of the book in late 1985 and 1986. I wish also to thank Graham Hall, Guy Stock, and Roger Clark for their advice on Chapter II.

For a person with my disabilities, achieving practically everything by dictation and instruction as to the handling of papers, producing a work of this nature is a fight and for that reason it is also a fight for those who have worked with me secretarially. First, Mrs Jenny Albertson transcribed almost incomprehensible tapes prior to the delivery of the original papers. Then Mrs Patsy Henderson deciphered the handwriting of many scribes so as to produce the bulk of the first draft of the book and Mrs Barbara Rae at electric speed put the main mass of material into the computer. But, above all, Mr Keith Whitefield mastered the computer itself and welded together the efforts of a diverse team of helpers, himself with his wife Anna filling every gap. Among these helpers, I have most especially to thank Christopher Fremantle, Ashley Gaskin, Teresa McCaffery, Gerard MacFarlane, and Anthony Schmitz, but also a great army of others, David, Karl, Sean, and Simon, with Julie, Mandy, Marion, Colin, Alan, Hugh, Michael, Rosalind, Nigel, Mrs Tough, Mrs Smith, and Miss Fraser also playing their part. There is no substitute for the determination and devotion of friends.

David Braine.
Department of Philosophy, University of Aberdeen.
February 1987.

Contents

Analytical Contents

I

The Strategy of my Argument: Its Basic Statement and Background

What is the project of this book?

To show that God exists from the sheer existence of things as this is exhibited in their temporality, in the real contingency of the future, the temporality whereby their continuance is not yet fixed or settled and whereby at an earlier time they might never have come to be. Not to argue from the orderedness or the special character of anything in the supposed 'inner' or 'outer' world of man, since this may appear dark. Nor to presume the intelligibility of anything, or any right to demand explanation, or to consider the conditions of such intelligibility or explanation. But only to consider the conditions of temporal existence itself.

To do this *from within philosophy*, with all that this takes: with the analyses necessary to re-establish a realist understanding of time, causal agency, and substance; with the necessary rewriting of logical theory, of the metaphysics of thing, situation, and activity, and of the theory of analogous reasoning; and above all with the justification of the according of a structured priority to metaphysics and the first-order over epistemology and the second-order within critical thought.

To establish God as indeed God: not as a demiurge who need not be simple, eternal, or infinite, and need not even be unique, but as transcendent in nature; yet not distant, but immediate to each thing, having a role giving Him an intimacy to each thing which none can share. To establish Him as non-temporal yet in such a way as to establish Him precisely as the one who, living and active, constitutes time in the character it has—a character whereby for His action, as for man's, the future is not yet. To establish these things in a way compatible with God's being indeed personal, good, and holy, no other than the God of Revelation.

To exhibit that all this can be done without metaphor.

Such then is our project.

1. STRUCTURE OF HISTORICAL BACKGROUND

Today, as always, irreligion is not so much a doctrine as a way of life. Among those cultures that have tended to raise causal questions about the Universe and the things in it, irreligion as a doctrine has been rare. Until the eighteenth century we can find in the West only Lucretius and a scattering of others to represent atheism or even agnosticism. Moreover, until the seventeenth century it was almost universally taken as obvious that the things in the Universe had behind them, underlying them, or acting within them, either one or many divine powers or beings. That this no longer appeared thus obvious is signalled in different ways by the writings of Pascal and Descartes. But it took time for religious scepticism to take root and then indeed even to become predominant so as to be not just reflective of a culture and way of life but formative of it.

We need pay no attention to those arguments to God's existence which do not bring in causal considerations. They have no bearing on the understanding of the thought of ordinary men or of the mass of men through the ages. True, there has existed a sense that God's existence was somehow, in itself, to the man of faith, indubitable: the very first certainty, the fixed point on which all other thought should turn. And, among philosophers, there has frequently recurred the theory that God's existence is evident from notions innate to the intellect as such, either from the existence of questions to which one or other answer must be true so that truth must exist; or from the idea of God as being, as such, 'something than which nothing greater can be conceived'; or from the idea of God as being as such, 'Ultimate Reality'; or from the idea of God as being, as such, *qua* God, infinite or possessed of all perfections, including existence or necessary existence.[1] These instincts, hunches, or 'thoughts' have provided the inspiration for attempts to develop a

[1] I refer here to arguments to be found in, or attributed to, St Augustine and a long line of successors reaching to St Bonaventure (see Étienne Gilson, *The Philosophy of St Bonaventure*, pp. 120 ff.), from the existence of truth, arguments from the consideration of God as 'something than which nothing greater can be conceived' to be found in St Anselm, or as 'Ultimate Reality' to be found in Tillich, and arguments from God's infinity or from His possession of all perfections, including existence or necessary existence. This last species of argument is to be found in its purest form in Leibniz and has been resurrected in recent times by Norman Malcolm and Charles Hartshorne; it is perhaps also traceable in Anselm, Descartes, and many others.

so-called 'ontological' argument, i.e. an extended (often short) piece of reasoning from the very definition of the 'essence' or 'idea' of God to His existence. But this type of argument has found its existence and continues its life almost solely among philosophers, and they have come to use it as a kind of plaything, interested only in diagnosing the fallacies which its different versions contain. All other arguments turn out to have a causal element: e.g. moral ideas or experience, the idea of God or religious experience, or the general moral or religious agreement of men, are supposed to need some cause or explanation of existence.

Rather was it the case that either the very existence of the Universe or the things in it, or the order and regularity which these things exhibit, or alternatively the particular features or elements within the Universe, were thought to require for their explanation an active power or cause, external to or underlying the Universe. Thus the notion that the existence of God is shown by, or is demonstrable from, the Universe and the things in it is not so much something imagined to need argument as something normally assumed, a part of the backcloth of the belief of the writers of the Old and New Testaments. For them the province of reason was not primarily to prove divine existence, for this was so evident from the Universe and the things in it (although not 'self-evident' in creaturely modes of thinking). Instead, it was for reason to ridicule and disprove polytheism, i.e. to ridicule the belief in many supreme divine powers, rather than in just one. And this continued to be the situation through all the centuries within Christian theology up until 1800 or after. There is no doubt in the minds of the Fathers of the Church, the Western mediaeval theologians, or the theologians of the Eastern Churches, that the existence of God can be known from the things that are visible, i.e. known even to man's intellect or mind apart from any Revelation—even, that is, to his natural reason.[2]

[2] The Psalms and the Book of Isaiah make it particularly plain that the heavens and the whole universe declare God and His glory, and that the worship of anything man-made is peculiarly ridiculous and contemptible. The same attitude to any form of polytheism, nature worship, or idolatry is evident in the Book of Wisdom in the Greek Old Testament and echoed by St Paul in Romans, ch. 1, and in the whole apologetic of early Christians to the Gentiles (of which St Athanasius's youthful work *Contra Gentes* is entirely representative). When Tertullian spoke of believing things because they were impossible, he was not speaking of the existence of God but of the peculiar doctrines of Christianity such as that God became a man and died (besides that, in the manner of one remarking that truth is stranger than fiction, he was giving absurdity here as even a *reason* for belief).

The role of extended discursive proof and analysis was to rebut objections to this or that particular formulation of the argument from the Universe or from the things in it to God and to choose a supposedly satisfactory formulation, rather than to be preoccupied with establishing the very possibility of such argument itself.

The role of philosophers has seemed to be that of questioning the apparently obvious. Until modern times the existence of God was not thereby brought more into doubt than the existence of the material world, of causal connection, or moral obligation, or of other things which the philosophers might bring into doubt amongst 'intellectuals'. For instance, in the Aristotelian conception, the method prescribed for the philosopher was to take the more obvious seeming truths, the more common opinions, and more evident parts of experience, and to reject the things thus apparently given, or to modify them only to the extent required to bring them into coherence with one another.[3] It was not, as since the time of Descartes it has in stages become, a matter of starting from scratch and, for instance, of looking for proof of the very existence of causal connection, querying its nature as well as its existence, and leaving the existence of God as cause the most doubtful rather than among the most evident of propositions.

Thus it may be that in fact the true task of the philosopher is to draw back the veil that other philosophers have drawn over the obvious, thus raising a dust so that men can no longer see. On this view, we have a very simple and direct line of reasoning or reflection to the effect that what has substantial existence or what has temporal existence, or else the Universe as such as a whole, must have a cause and that this cause can only be God. Then what

[3] Thus Aristotle, in *Nicomachean Ethics*, VII, carefully follows the order of discussion which he describes at the start: 'We must, as in all other cases, set the *phainomena* before us, and, after first discussing the difficulties, go on to prove, if possible, the truth of all the common opinions about these affections of the mind, or, failing this, of the greater number and the most authoritative; for if we both refute the objections and leave the common opinions undisturbed, we shall have proved the case sufficiently' (1145b 2 ff.). After enumerating some 'common opinions', he then concludes (1145b 20–1), 'These, then, are the things that are said (*legomena*)', exhibiting that 'the common opinions (*ta endoxa*)' are typically 'the things that are said', i.e. typically 'matters of linguistic usage, or, if you prefer, of the conceptual structure revealed by language' (G. E. L. Owen, '*Tithenai ta phainomena*', pp. 169–77). He follows this pattern of presentation often elsewhere, e.g. in *Physics*, IV (he shows his self-consciousness about his method in 210b 32–211a 11). But even when he is not thus self-conscious, it is still his most characteristic method.

seems required is simply the removal or rebuttal of the contrived, the specious, or academic queries, or veils, which at a second stage of inquiry sundry earlier philosophers have introduced.

However, this is not how it appears either to philosophers or to modern man today.

Today it has come to seem folly even to suppose the existence of God might be proved. To attempt to prove God's existence has come to seem a wild enterprise, indicating some irrationality of judgement on the part of the person who undertakes it, a turning aside from any safe or reasonable course, outside the pale of shared assumptions or discourse among philosophers. And the view of philosophers, amongst whom the reign of Kant and Hume has become thus triumphant, is shared by Western man in general—not indeed by all men since there still seem to be some scientists and others to whom it appears in some way evident that the world, by its existence and order or wonder, must have a transcendent cause.

And this change among the philosophers reflects a deep spiritual change in feelings or 'sentiments'. There appears in Pascal an attitude whereby nature is dark, not pointing to God but leaving Him hidden: this is part of the human predicament. This feeling deepening its hold on modern educated man has sapped the energy behind arguments from the world, especially from its order, to God, and this process has been assisted by doubts as to the reality of this order in face of evil and in face of the supposed capacity of man to construct order and project it on to the world. At that stage 'inner' phenomena of moral and religious experience still seemed to witness to the existence of God as cause: but since the late nineteenth century the 'inner' world has also come to seem dark. Everything about the 'outer' world and about the 'inner' world having thus become dark and silent, the significance of mere existence, contingent or precarious, has not been attended to.

As a result, not just modern philosophers, but educated men in general, have come in most cases to see belief in God as merely a matter of faith in the sense of opinion, not as a matter in respect of which there could be certainty or even clear reason: a matter for subjective opinion, an area for dispute over which reason no longer holds sway.

In the meantime, the theologians also have become ambivalent.

Thomas Aquinas merely gave expression to a long tradition that the existence of God could be known from the things of nature (although the things relating to God's friendship with man, the direction and goal of history, the existence of relationship or personality within the Godhead, all these, were for him known only by faith). According to this view, the existence, goodness, and livingness of God might be known naturally, although these concepts were all vague and indeterminate, leaving room for some further determination to be given them in Revelation, through faith. (Yet neither faith nor reason reached so far as to bring God within our grasp so as to make Him comprehensible, i.e. as to bring His nature as a whole within our grasp, let alone to make it imaginable.) In this view, the view also of the Fathers of the Church earlier, there was no comprehension,[4] no clear and distinct or non-vague concept in the sense that the rationalist or Duns Scotus would imply, which we could use which would apply only to God, and there was no natural theology or natural science of God. And, even though some theological conclusions might emerge as the end-conclusions of metaphysics, this did not constitute any sort of science of God; the only science of God possessed by man was a science or knowledge resulting from God's communicating His own self-knowledge by an act of revelation to man, having the status thereby of what an authority tells someone who heeds and accepts that authority.[5] This act of revelation authorized some use of models to give some shape or appreciative sense to man's assent to what Authority revealed, but these models by themselves constituted no authority or warrant for counting theology a science.

Duns Scotus, by contrast, thought that the conception of existence was a clear one, but it remained that the key concept of infinity was still held to be vague and indeterminate, with an openness to be further, but still incompletely, determined by revelation.[6] The supposed indeterminacy of concepts such as the "infinite" or the "wholly other" has played a key role in more

[4] The possession (*tenere* or *habere*) of God by the blessed in Heaven who see and enjoy God securely as the ultimate fulfilment of desire is not a comprehension (cf. *S.Th.*, Ia, Q. 12, Art. 7).

[5] This is the only ground proposed by Aquinas for regarding 'theology' or Sacred Doctrine as a science (*S.Th.*, Ia, Q. 1, Art. 2).

[6] I here state Étienne Gilson's interpretation of Duns Scotus in his book *Jean Duns Scot*, as it appears most clearly in pp. 191–95, presaged by pp. 42–84, especially 50–6 and 78–84, and leading into his discussions of omnipotence, especially pp. 355–65, but also *seriatim*.

recent philosophy of religion. Whereas in mathematics, "the infinite" has some clearly definable meaning, and for Aquinas it means, in effect, 'what is without qualification, rather than merely in this or that way or respect' (for this is the effect of being not limited by form or by relation to this rather than that matter or kind of matter), in Duns Scotus and in modern times it cannot be explained in any such way, and has sometimes degenerated into a device of empty rhetoric.

Meantime, the disciples of St Thomas failed to grasp the role played by "existence", "life", and "relation" in his philosophy and failed to emphasize the distinction he made between concrete and abstract modes of speech. As a result, they incurred the danger of giving a Platonic twist to interpretation of Aquinas, making his God into something of an abstract entity, and making the eternity possessed by God excessively analogous to the lifeless eternity exemplified by natural numbers.

Alternatively, they so emphasized the adequacy of our causal argumentation to God as to open the way to the modern deistic way of thinking according to which God is indeed cause of the Universe and reason may know Him as such, but, knowing Him in this way, it thereby knows Him in a way which makes Him like creatures, the very opposite of Jewish and Christian orthodoxy, and the opposite of Aquinas's opinion. In that older tradition which Aquinas represented, whereas everything positive about creatures enshrines some likeness to God, e.g. their existence, goodness, causal power, life, knowledge, understanding, and will, one could not say, conversely, that everything positive about God enshrines some likeness to creatures.[7] Accordingly likeness in respect (e.g.) of shape, size, or colour is symmetrical, but likeness in nature, i.e. likeness in respect of an object's character and properties considered overall, is not in that older tradition a symmetrical relation.[8]

[7] *S. Th.*, Ia, Q. 4, Art. 3, resp. 4.

[8] Likeness is not a symmetrical relation. If a creature is like God in being *F*, e.g. 'living', then *F* will be a transgeneric property, wider in extension than any proper genus, and it will be found that the modes of the creature's being *F* and of God's being *F* are such that to state that the creature is like God in being *F* is to imply a greater unlikeness, although not in respect of being *F*. The statement of the Fourth Lateran Council in 1215 in reply to 'the errors of Abbot Joachim' is instructive: 'But when the Truth (Jesus) prays to the Father for his disciples, "that they may be one, *as* we also are one" (John 17: 23) the word "one" as applied to the disciples is to be taken in the sense of a union of charity in grace, but in the case of the Divine Persons

Or, thirdly, the conception of God's being as infinite proved misleading to the rationalists who supposed us to have a clear and distinct concept of infinity and to get to know His existence precisely under this concept and on this basis to make deductions as to His nature.

As a result of these tendencies either an idolatry of concepts or an idolatry of creature-likeness was liable to arise. And to this Pascal's reaction, that the God of Abraham, Isaac, and Jacob was not the God of the philosophers, seemed appropriate.[9] Hence the instinct of theologians and spiritual men to be suspicious of philosophical theology intensified as the years and the centuries proceeded.

Next, in the nineteenth century, they were presented with the phenomenon of Hegelianism, whereby it was not just that a cause (of existence) might be known or inferred from the existence of the Cosmos, but that, from the drift of history, the goal of history intended by the Absolute Spirit was to be known, so there was no secret waiting for this goal to be revealed in a divine act of revelation. It was the reaction against this extended natural theology which gave the theologians some reason beyond fashion for embracing the conclusion of the sceptical philosophy that the living God of Revelation was known through an act of revelation only and not in any way through metaphysics. Thus, for instance, one can draw from Kierkegaard[10] the argument that from the finite there is no reasoning to the infinite, and this appears to stand as a valid objection to all forms of the teleological argument and most forms of the cosmological. This reaction reached its extreme in the writings and teachings of Karl Barth, inspired by the rejection of that descendant of Hegelianism wherein it was supposed that the goal of history should be discerned as lying in the triumph of the master race, the Aryan people.

Against this background I revert to the older tradition of the Old Testament and the Fathers of the Church, whereby indeed God may

in the sense of a unity of identity of nature. So the Truth said on another occasion, "Be you therefore perfect, *as* also your heavenly Father is perfect" (Matthew 5: 48) as though he would say more clearly, "Be you perfect in the perfection of grace *as* your heavenly Father is perfect in the perfection of nature", in other words each in his own way. For between Creator and creature no similarity can be expressed without including a greater dissimilarity . . .' (Denzinger, *Enchiridion symbolorum,* 806).

[9] Pascal, *Pensées*, nos. 17, 19, 380 and 381.

[10] 'The Obstacle', in *Training in Christianity*, pp. 26–31 and *seriatim.*

be known from the things of the world, their existence and order, without being like these things, and in this work propose to consider the proper form which arguments from existence should take.

2. OUR KEY CONCERN: TEMPORAL EXISTENCE AS SUCH

Physicists, biologists, and others often argue above all from order and, if the order that they find is within the very nature of the things they are speaking of and not a matter of the accidental co-operation or co-ordination of the entities they speak of, then any explanation of this order will be an explanation of the cause of the existence of that which has the order.[11] Therefore the argument from order does reach to a cause of existence and not merely to an architect. Yet it does not clearly demonstrate any difference in kind between the cause of existence and the things of the kinds whose existence is caused. It is only by looking at the question of what is involved in the causation of existence as such that one can demonstrate this difference in kind and so uncover the significance of certain kinds of teleological argument from the order, regularity, or simplicity built into the very nature of things. Therefore I shall not argue from the contingent order, that is to say, the empirically discovered order, in the world, to God, eloquently and well though the rational force of this, as pointing towards the existence of God, has been made clear by Basil Mitchell in his book *The Justification of Religious Belief*.[12] Rather I shall look not at existence in general, but at temporal existence, i.e. the type of existence exhibited by things which have successive states or different states at different moments or, as we say, different times.

Here I am not concerned to argue that the Universe must have

[11] This point, that it is not the marvellous accidental co-operation and co-ordination of things in nature, but a non-accidental co-operation and coordination intrinsic to the natures of things, is one of which Kant fails to take account in his description of what he calls 'the physico-theological argument' (*Critique of Pure Reason*, p. 521). However, he may have had a sense of the separate point that there has to be a radical difference in kind between the divine cause and the existence caused.

[12] Mitchell does not directly present a teleological argument but exhibits how a theistic interpretation may lend a coherence to one's view, otherwise lacking, of the Universe and of history, inviting of a theistic perspective in a way analogous to the way new paradigms in science recommend themselves.

had a beginning in time; I am not here interested in the fact that what exists is contingent in the sense that it might never have existed or might not have been. So far as my argument is concerned in this book it does not matter whether or not the Universe never had a beginning. My concern is rather to insist that there is the same need of explanation for the continuance in existence of things as there might be supposed to be in regard to the coming into existence of things from nothing.

(a) *Continuance through time*

The continuance of the very stuff of the Universe, the fact that it goes on existing, is not self-explanatory. It is incoherent to say that the very stuff of the Universe continues to exist by its very nature since it has to continue to exist in order for this nature to exist or to be operative. Hence, nature presupposes existence.

If we say that something happens according to nature we generally have in mind that it happens in the way that it regularly does. But thus ascribing things to nature explains nothing, since it leaves the question why things regularly behave as they do and, in the example here concerned, why they regularly and reliably continue to exist quite unexplained, so that one has emptied the notion of nature of all explanatory significance. Accordingly, it remains viciously circular to imagine that the continuance of the Universe in its order can be explained in terms of its nature, since the existence and operation of this nature has to be contemporaneous with this exercise (its operation is its exercise), and, to this, previous existence is not as such relevant.

Therefore what concerns us here is not that the Universe or things are contingent in the sense that they might not have been but in the sense that they might cease to be. There is a certain precariousness involved in the fact that its future does not yet exist; whereas the past is settled and the present is settled, the future does not yet exist and so is open. Of course if the continuance of things is secure so that it is of their nature to continue, then, in terms of this, other things may be explained.

This is the background of all the explanations that we commonly accept, presupposing the 'going-on' of things. The potter makes his pots: we do not anticipate that the potter will disappear or that the clay will disappear into nothing without natural explanation. We do not anticipate that the potter or the clay will suddenly have quite different properties from those that they have exhibited previously. It is perfectly right to say the potter exercises causal efficacy. But in accepting this explanation, one presupposes numerous things, namely the continuance and existence of the potter and his clay and many other things about the stuff of the Universe and the regularities of nature, which are not self-explanatory.

This precariousness is something which Sartre, in his book *La Nausée*, and other existentialists sensed. It is not the kind of contingency referred to when people say that things might have been otherwise, or might never have existed, or might never have been; it is rather the kind of precariousness represented in saying that they might not continue, might not exist in the future; or that in general the regularities of the present and past might not persist from the present into the future.

It is to this precariousness that Hume and his successors have drawn attention in its logical aspect. But the logical possibility of things being otherwise is of no great importance, as there is the same contingency, so far as logic is concerned, in respect of the past and the present, as there is in regard to the future. This precariousness implies the logical contingency of the existence of that which is precarious, but not vice versa: logical contingency in respect of existence does not imply real contingency whether in respect of existing, or of coming to exist, or of ceasing to exist.

Some philosophically technical queries arise.

Differences in mode of statement, quasi-grammatical differences, qualify the extent to which we may accept that this or that sentence is without qualification contingent; sometimes the mode of statement imports a reference to an existing object and, if this existence is taken as a datum or sentential presupposition, then the contingency asserted or implied can only apply to the predicate. This and other like complexities mean that the reign of logical contingency and also of epistemic or epistemological contingency is more restricted in respect of the past and present than in respect of the future. But careful restatement of the case will rescue the future from this appearance of difference of status in relation to logic.

In brief, every state of affairs or event is, in a certain sense, 'nested' in a context, and it is only relative to this context that we judge in prospect that it might or might not happen, or in retrospect that (supposing it did not occur) it might have occurred or (supposing it did occur) it might not have occurred. It is therefore only relative to this context that we judge whether or not logical reasons determine the state of affairs or event as necessary or impossible; if they do not then it is logically contingent—although it may be (e.g.) naturally necessary and not contingent *simpliciter*. We treat the case of 'a priori reasons' in a similar way. (Being logically necessary or impossible and being a priori necessary or impossible are not the only ways of being necessary or impossible; so what is 'logically contingent' or 'a priori contingent' may very well not be contingent *simpliciter*). What is vital is that the questions of the logical or a priori contingency of states of affairs and events are not to be judged relative to the same background or in the same way as the questions of the logical or a priori contingency of propositions identifying these states of affairs and events.

However, granted that for such reasons as these the technical queries mentioned earlier can be set aside, the question remains to us: what is that difference whereby the past is already settled, and indeed the present is also settled because it is now existent and therefore likewise established in reality, whereas the future has no establishment as yet, but has a status entirely provisional?—depending on extrapolation from the present and the past, so far as knowledge is concerned, and depending on the persistence of the nature of the things presently existing for its own not yet actualized, but future, existence, so that it has no existence as it were established in its own right, but remains always contingent and dependent in a more radical way than the past or the present.

Obviously this general metaphysical fact about the future has a close tie with the insecurity which may be felt by the individual who realizes that he has no control over his own continuance. It is not within his power either to secure the continued existence of the things round about him or to secure that they will continue to behave according to nature, or even to secure that he himself will not just disappear into nothing. From this type of insecurity there has, rightly or wrongly, sprung (and it is part of the concern of this book to discover whether it is rightly or wrongly) a sense of the immediacy of the dependence on the part of the individual, in respect of his very existence and continued existence, upon God: just as his continuance in the future depends upon God, so likewise even that continuance which allows him to exist at the present moment is itself an exercise of that divine action, so that, in the words of Aquinas, God as its source of being (or existence or *esse* to each thing) is 'most intimate or innermost to each thing'.[13]

Viewing the writings of the existentialists, it seems indeed true that they may make no adequate distinction between any kind of absolute contingency and the more limited contingency which arises within the context of assuming the continuance of nature, with its own much more limited uncertainties. However, it is a mistake to dismiss their instinct for the contingency of things on the ground that they ignore that reliable continuance in the context of which our knowledge of natural science and of natural necessities (the necessities, that is to say, that are expressive of the natures of things) becomes possible, i.e. the reliable continuance which natural science discovers to us. Any adequate philosophy has to give an adequate account of the genuine necessity set by the natures of things, marked by physical law, and to give an account of physical law, not just as a matter of observed regular association of appearances or phenomena, but as an expression or realization of the natures, albeit temporal, of the things that exist.

Any adequate philosophy has to give an account of the distinction between this and some more absolute necessity, a necessity which physical law does not carry with it, since it does not leave the future settled absolutely or without qualification. Neither Sartre nor Hume give an adequate account of this distinction, because the status of physical necessity and of nature and its reliability even in their physical character cannot be given adequate account of solely in epistemological or psychological terms, but is (and must be given an account of as) a matter of the established nature of things, the nature of things as they exist.

But there is no explanation within physics of the persistence in existence of the things, either individually, or as an ordered set or collection of intimately interrelated things: i.e. of the continuance in existence of the things or constellations of things in virtue of whose natures natural or physical necessity, reliable knowledge of the future, etc., depend. So the contingency we are concerned with is not the contingency whereby things simply might have been

[13]. *S. Th.*, Ia, Q. 8, Art. 1. His phrase is reminiscent of St Augustine's statement that God is 'interior to the innermost' of each person (*Confessions*, III, 6).

otherwise or whereby that which does exist might never have existed, but the contingency whereby that which does exist has no absolute necessity to continue but only a limited necessity, if nature persists and things continue to act according to nature; but, for things to continue to act according to nature, they need first to exist.

The persistence of the nature of a thing (or things), and its operativeness, depends upon, presupposes, and cannot explain, the continuance and existence of the thing itself. (When I use the expression 'limited necessity, if nature persists', I do not have in mind a logical necessity, i.e. I do not have in mind what some people call a "hypothetical necessity",[14] whereby if *p*, where *p* logically implies *q*, then *q*, so that if nature persists then this limited necessity obtains as a hypothetical necessity. Rather I think of this necessity as an expression or actualization of the natures of the things that exist. That is, it is rather the natures of the things with active powers in the world in their relations that physically make this or that necessary (i.e. that makes this or that effect follow) than any logical, let alone epistemological, connection: the necessity asserted is asserted as real, the result of how the world is, not asserted hypothetically).

(b) *The mythology of inertia, akin to the mythology of fate*

It is possible to confuse methodological questions with metaphysical questions. In Aristotelian physics it was the movement of something from place A to place B that needed to be explained, i.e. change of place as such. By contrast, in Newtonian physics this does not need explanation in terms of special active powers and their exercise, but what needs such explanation, in terms of special active powers and their exercise, i.e. forces, is change of velocity, not mere change of place.

Correspondingly, it has come to be supposed that things just continuing to exist does not need explanation, but only changes in the way that things go on. However, this represents a point of methodology rather than metaphysics, a view as to the ways in

[14] This distinction between a non-absolute necessity which is still not hypothetical and a hypothetical or inference-relative necessity is explained clearly in David Braine, 'Varieties of Necessity', pp. 150–53.

which different things are to be explained, rather than as to whether or not explanation as such is needed or to be asked for. Upon examination it turns out that even the continuing of things, in the absence of external force without change of velocity, i.e. this constancy as much as any other, and indeed any constancy in the realm of the temporal, calls for explanation. An explanation has been sought for it even by scientists within science, the law of inertia, as it is called, being explained within different cosmological theories in terms of interrelation with some of the most general cosmological features of the Universe. But whatever goes on in science, i.e. so-called empirical science, it remains that, absolutely speaking, the continuance of things in existence is, if time is real, in need of explanation, if any beginning of existence out of nothing is in need of explanation: it is as much in need of explanation as the continuance of anything else, whether this be an intra- or extra-scientific need.

Wittgenstein remarks in the *Tractatus*:

> The whole modern conception of the world is founded on the illusion that the so-called laws of nature are the explanations of natural phenomena. (6.371.)
>
> Thus people today stop at the laws of nature, treating them as something inviolable, just as God and Fate were treated in past ages.
>
> And in fact both are right and both wrong: though the view of the ancients is clearer in so far as they have a clear and acknowledged terminus, while the modern system tries to make it look as if everything were explained. (6.372.)

I am arguing that he is absolutely right that the continuance in application of the so-called laws of nature provides a purely illusory ultimate explanation of natural phenomena. Even physicists, Newton and Mach being succeeded by modern cosmologists, sometimes seek some explanation within physics of such laws as those of inertia and gravitation. But the validity of any explanation they offer will still presuppose the continuance of nature as such, or will do so if time is real in the sense to be explored later.

(c) *The pre-conditions of proof of God as Self-existent Cause*

But from the point of view of theology or of any conceivable attempt at theological conclusions, the historical transfer of attention from the mere coming into existence of things to the causation of coming into existence of things out of nothing is not enough. What is required is a transfer of attention from the causation of coming into existence to the causation of temporal existence as such, whether coming to be or continuing to be, and its pre-conditions. Everyone is agreed that, if the cause of the Universe be thought of as an architect ordering coeval material, this would leave us with no explanation of any supposed infinity attributable to Him. But it needs to be recognized that, even if God is introduced, not merely as a cause of order, but rather as a cause of the existence out of nothing, this remains compatible with thinking of Him as a particular cause of a particular entity, *viz.* the Universe, with a causal relation to it analogous to other causal relations with which we are acquainted in which something prior in nature, or in time of activity, causes something posterior or later than it in time.

That is, unless ideas of time and causation are so developed as to exclude this, even the conception of God as cause of existence out of nothing is still compatible with thinking of God as a finite being preceding in existence and continuing in existence alongside some other entity called the Universe—compatible with giving no adequate explanations of God's 'infinity' and 'eternity' so that to use these terms remains mere rhetoric. A demiurge does not need to be simple, eternal, or infinite, except in name, nor even unique. By contrast, if we recognize that temporal or successive existence is in itself of such a nature that nothing with temporal existence is capable either of giving rise to it from nothing or of causing its continuance, and recognize that all temporal things are equally impotent in this respect, whether in regard to themselves or in regard to other things, then we shall discover that we have arrived at a valid basis for asserting a genuine infinity and a radical eternity in God.

In the argument which I shall develop later we shall find that what makes the temporal in this way impotent is something (I shall call it 'compositeness') which would make anything which is not intrinsically underivative impotent. (I use the word "derivative" in a causal sense.[15]) Only something intrinsically underivative (which

[15] That is, I do not concern myself with the 'derivation' of conclusions from premisses, of means from ends, or of the Son and the Holy Spirit from the Father whose being consists in being their ground.

therewith would be First because it could have no cause), according to this argument, could conceivably cause temporal existence. And I argue that this latter must have a cause, so that it will follow that the intrinsically underivative exists. And, since the temporal is impotent, obviously this cause must be non-temporal or eternal. In this way, we should expect the intrinsically underivative to be non-temporal. And in later chapters I argue that such non-temporality as well as infinity follows from 'incompositeness' and besides that not only must the *intrinsically* underivative be 'incomposite' but also it must be (more puzzlingly) *intrinsically* unique.

There is here no risk of an infinite regress of causes of existence which themselves cause existence. The question rather is as to whether existence or temporality can in the relevant way be caused at all, and philosophically whether successive or temporal existence does thus absolutely require so to be caused. And the meaning or credentials of this word "intrinsically" need to be inquired into, and with it the meaning of this associated term "incomposite". Any causing of temporal existence as such (whether giving things existence out of nothing or upholding things in existence, as it were, out of nothing, i.e. not relying on their previous existence or nature) will depend upon a radical non-temporality and upon the possession of existence in a way quite different from temporal things in such a way that no question of efficient causation would appropriately arise in respect of the existence, number, nature, or activity of the cause concerned: this radically different way of possessing existence is what I call 'incompositeness', but it is this that needs explanation.

This is the viewpoint which deserves to be given a precision rooted in considerations drawn from the philosophy of science and the philosophy of language, and is the basis of the argument which this book opens out.

However, one ought to note that it is plausible to suppose that, if one allows here for the existence of different sorts of time, all involving transitive and asymmetrical relations of before and after between distinct acts or events, but not all involving continuity, i.e. not all subordinate to physical time, then, in this perspective, temporality and possible derivativeness would seem inseparable. That is, not only must the intrinsically underivative be non-temporal, but also, conversely, the non-temporal must be intrinsically underivative. It has seemed plausible to many that the only modes of real existence that are possible are temporal. The only strong reason for supposing there to be some other mode of real existence would seem to lie in the consideration that there has to be a non-temporal cause of temporal existence as such, whether giving things existence out of nothing or upholding things in existence, as it were, out of nothing, i.e. not relying on their previous existence or nature. There seems to be no particular attractiveness in supposing that there are or could be other distinct modes of real non-temporal existence beyond this. That is, there seems to be no role for a theory of non-temporal modes of existence other than that of the 'incomposite'—no role for a theory of possible things other than God which are not temporal, having natures requiring realization in successive acts or stages. But although this book contains many suggestions germane to proving that there can be no non-temporal being other than God, my argument does not depend on it.

3. APPROACHES FROM THE PHILOSOPHY OF SCIENCE

Philosophically it is well known that there is a problem about the thing which Locke and Hume referred to as real causal connection, while the inadequacies of positivism have seemed plain to most men (once the question is clearly presented) and to most scientists in respect of the explanation of phenomena in history and physics. In sum, it is necessary or required (required, that is, if a satisfying philosophical account of the matter is to be given) that we find an account of the causal relation between natural or mundane things which does not depend on logical interconnection and which does not consist in some relation to human knowledge and modes of knowledge, but rather is such that the exercise of causal power is an exercise of the natures of the things that exist.

Yet it has always been a problem that, if the reality of time is denied, we tend to be forced back into a view of phenomena as a kind of heap distributed through four dimensions, three of space and one of time, without any existence of substance as a possible entity to be logical subject of a nature in virtue of which it might have causal power, and thus to exercise causal power. As it were, once we have turned the field of phenomena into a heap, any so-called 'substance' can be nothing more than a logical construction and is thus debarred from any 'real' causal action.

If on the other hand we say that time is real, the sense of this statement seems to involve a contingency in respect of the future which goes beyond any contingency explained within physical science or historical study, so that we are forbidden to regard any natural necessity or causal connection, whether physical or historical, as involving absolute necessity and reliability. In this way we are forbidden, if time is real, to attempt to explain the continuance of things with their natures individually or globally (the whole of nature) in terms of this stability of nature, individual or global, having been given logically or a priori. And yet we are at the same time forbidden to rest content with a reduction of the efficacy of the natures of things to some relation to human knowledge and modes of knowledge.

Accordingly we wish the things with their natures to be in some way metaphysically secured, but yet not absolutely logically established. This effect is achieved if indeed we regard things with their natures as upheld in existence by God in accordance with the traditional account. In this way 'secondary causation', as it will then be called (i.e. the ordinary causation we are familiar with, which is indeed genuine causation), the exercise, that is, of genuine powers by things possessed of these powers, does not exist as it were self-sufficiently in a vacuum. Rather it exists or arises only within a context of these things with powers being upheld in existence, and, if upheld in existence at all, then, it becomes evident, in existence with their natures, by God. In this way, it becomes apparent that all secondary causation, i.e. all ordinary causation, depends upon the primary causal activity of God in respect of the existence of things in their natures.

So it may be grasped how in the course of refuting of positivism and of a positivistic view of causation, and of setting a doctrine of substance or individual things with active powers capable of causal activity upon a secure basis, and in the exigencies of establishing the reality of the distinction between the past and the future in a way that is not dependent on modes of human knowledge (or is prior to difference in modes of human knowledge), we may be led to an argument for the existence of a being of a non-temporal, non-material kind, giving existence, vigour, and activity to whatever else exists.

Accordingly, if, and I emphasize the 'if', there is some kind of existence (the word "exist" has many senses and applications which need to be distinguished) in which existence is an act or actuality, then one can say that God does all the substantial existing, whether His own existing or the existing of creatures, that is done. In God the logical subject, or grammatical subject, of the existential statement will be the same as the doer of the activity which His existing consists in. In the case of creatures, however, God will be the doer of existence and their existence will be that which is done. G. E. M. Anscombe, in her magisterial work *Intention*,[16] makes out that what the agent does, in the case of intentional action, is what happens (so far as it is intended). What the agent knows in knowing what is going to happen in the case of what is done intentionally is what he knows in knowing what he does. And in this way what is done is what happens. And what is done by God is the existing and the continued existing of the things that exist.

God is, in this perspective, one might say in a metaphor, intimate to the world as an animal is to its breathing (albeit He does not depend upon the world), and as the breathing of the animal is a sign to us of its life so to us the existing which we meet with in the world is a sign and work of His life. He is as immediate to each thing as each thing is to itself, and to its action. It is not that God is all things, but that He is in all things: and it is the capacity of our account to give a non-metaphorical explanation of this "in" which reveals the 'truth' in pantheism. The escape from deism is achieved when, in one move, we at once mark the unlikeness in God's way of existing from the way creatures exist and mark this immediacy. He alone has existence in such a way as to be able to communicate it unconditionally or without presuppositions but thereby He has a role which he cannot pass on making Him immediate to the existence of all other things.

[16] In *Intention*, G. E. M. Anscombe raises a query about this position, that the agent does what happens, to the effect that although tempting it might seem mistaken (pp. 52–3). But on p. 57 she indicates how this query ought to be answered, namely in terms of the concept of practical knowledge, the concept which the rest of her monograph is directed towards elucidating. On p. 91, she shows her continued preoccupation with the sameness of the object of knowledge in knowing what I am going to do and in knowing what is going to happen. The same position appears in the, otherwise quite mysterious, declarations of Peter Geach in *Logic Matters*, pp. 324–7. According to this position, since temporal existing is what happens in the world, it is what God does in the world.

4. PROGRAMME OF INVESTIGATION

Thus there opens out to us a programme of inquiry, a philosophical programme. This inquiry is theologically in accord with the thought of the Old Testament and, within the Greek Old Testament, with the thought of the Book of Wisdom, followed by St Paul and the Fathers of the Church, that God alone has power to give existence; and yet this inquiry is of such a nature as to be within the reach of philosophical examination and question. Thus there opens out before us the fearsome fact that nothing can be got right in the area of philosophical theology if it is not first got right in the area of ordinary philosophy. And indeed there is no mistake in 'ordinary philosophy' which will not, if it would affect philosophical theology, also infect and subvert any proposed system of Revealed theology, since faith that seeks understanding has to draw upon and integrate with all avenues of experience and knowledge appropriate to man, especially those inseparable from linguistic capacity as such.

And so firstly we must establish the reality of time against such authors as McTaggart and Mellor (Chapter II). Then we must restore the notion of causal agent and a spread of types of causal agency within the area of natural science against Hume and his followers (Chapter III), and rewrite logical theory, freeing it from the fallacy that the only existence is that expressed by the existential quantifier (Chapter IV), in order to re-establish the notions of substance and 'real' existence on a firm footing (Chapter V, Part I). In Chapter V, Part II, because causation always involves not only an agent but also activity, I develop a theory of activities and situations ('actualities'), in effect re-mapping the ground covered by Wittgenstein's *Tractatus*, in order to make room for a correct account of the contingency of temporal existence. This puts us in a position to explain the notion of "incompositeness": I explain this in such a way that the incomposite, if it exists, must be intrinsically underivative and unique.

Then, in Chapter VI, I review some of the approaches suggested in the history of philosophy to the questions as to why temporal existence as such cannot have a temporal cause (the temporal being in this way impotent) but still needs a cause. It emerges that any such cause would have to be incomposite, but that the reason for thinking that such a cause is necessary arises from understanding the interdependence of two analogous uses of the notion of "cause", one in regard to ordinary mundane causes and the other in regard to the upholder of temporal existence. This affects one's view of the structure of metaphysics and depends upon a certain view of its relation to epistemology.

Any discussion of God's existence is ostrich-like if it ignores the doubts, raised by Locke and Kant, as to what knowledge is within the reach of human faculties in their condition of dependence on the senses, and doubts as to the extent of human knowledge of general statements raised by Hume. And it is equally ostrich-like if it ignores queries as to whether belief in God's personhood is not a piece of primitive anthropomorphism raised by materialists and by certain mystical traditions alike.

Therefore our programme, as will become repeatedly clear, has to be carried through against the background of epistemological questions as to whether and in what way, if any, philosophy can establish any first-order conclusion whatsoever—rather than being essentially second-order and parasitic upon the positive sciences and upon *mores* and *praxis*. It also has to continue against the background of questions as to the roots of human knowledge, as to whether this lies ultimately in general principles or whether it may lie in the understanding of concepts, some of which may be applied analogously rather than uniformly in different arguments. These questions are dealt with in Chapter VII. The standing doubt as to whether a cause of temporal existence as such (a so-called "Ground of Being") can sensibly be conceived as living and personal, rather than being described only in neutral or impersonal terms, will be resolved in Chapter VIII.

This will still leave problem areas, e.g. in respect of the theory of evaluation and accounts of God's goodness, of the omnipotence of God, and of the relationship of our discussions to Revealed theology and knowledge of the 'holy'. In regard to these we have to get at least this far, namely to indicate the structure of approach called for by our emphases on God's being the upholding cause of all existence, His incompositeness, and the absolute need to maintain an adequately a posteriori approach: this is the task of Chapter IX. These points mark the frontiers of metaphysics.

And all this will have been necessary to our project.

Meanwhile we shall have to be watchful against the theologians' fear that, if philosophy should perchance look like a friend, it might turn out to be a yet more grave source of idolatry. This supposed danger arises, it is imagined, either by making God unlike creatures and by transmuting Him into an abstract or 'conceptual' object, or by making Him like creatures, a cause in the same way as other causes. However, this last fear is one upon which I shall in this work spend little time except in considering personhood, individuality, and goodness in God (I regard the discussion of the unity and uniqueness of God as it reflects upon these as the acid test of the satisfactoriness, from the standpoint of Revealed theology, of any proposed philosophical approach to knowledge of God). I spend little time on this fear, on the one hand, because the insistence on God's causal efficacy and the logical requirement that one needs to be real in order to be causally efficacious appear to remove the danger of God being conceived of in excessively abstract and Platonic terms. And it is, on the other hand, because the very mode of proof here envisaged demonstrates that the unlikeness between God and creatures, if God exists, is indeed greater than any likeness that there may be.

Hume, in his *Dialogues*, may conjecture the weak conclusion that the ultimate cause of the Universe bears some 'remote analogy' to human intelligence, and here, as in most arguments from order in the Universe to God, we meet with the suggestion that God is like creatures. At the same time as we need to escape deism, we also need to escape any kind of anthropomorphism or 'process-theology', any view which would purport to establish God in parallelism or symbiosis with the world. In particular, in terms of recent discussion we need to escape the limited God of Richard Swinburne,[17] and to avoid any such acceptance of ways of thinking about causation and the priority of questions of coherence over questions of existence, ways of thinking drawn from the empiricists. By contrast, according to the argument of this book, what allows God to be immediate to every creature is a mode of existence which makes Him at the same time unlike every creature in nature. This will limit the appositeness of empiricistic ways of arguing: we will find that He needs to be self-existent in order to cause existence

[17] R. Swinburne, in *The Coherence of Theism* and *The Existence of God*. It is very important to my general approach that second-order (commonly supposed to be a priori) questions as to the coherence of concepts cannot be decided in advance of the first-order question of whether they have actual application, as will appear in Chapter VII.

and eternal in order to cause temporal existence, unlike all mundane causes.

In the mode of argument proposed in this book, we establish in one act both the immediacy of God to creatures, with their immediacy in respect of dependence on Him for their existence, and the gulf between creatures and God in respect of their manner of existence inasmuch as God is involved in doing something, *viz.* communicating existence, vigour, life, or actuality, which they are in themselves impotent to do, and it is only a non-temporal being that is thus capable of communicating a temporally realized or successive existence and life.

So the feeling that the individual may have of immediacy in respect of dependence upon God, of precariousness in respect of his or her existence, apart from the activity of God in keeping him or her in existence, may have behind it a genuine intellectual foundation and not be a matter of mere subjective feeling. As mere subjective feeling it would provide the basis of no argument. If on the other hand this feeling has, in a certain sense, a rational contour, is an expression of a genuine intellectual necessity, then what among the supposedly philosophically sophisticated appears complex, elaborate, and discursive, involving many steps of argumentation for its presentation, may be the same as that which in the ordinary man or woman retains rationality despite the absence of any such explicit articulation and elaborate argumentation.

II

First Investigation: The Question of the Reality of Time

Our programme of investigation invites us first to consider the question of the reality of time.

I shall contend that those physicists and philosophers who have in key respects only accorded to time a status co-ordinate with and parallel to the three dimensions of space turn out to be compelled by considerations from within physics itself to concede to time the same quite different status which it has for the biologist and the historian, as well as for the individual and the novelist—the individual as a reflective decision-maker and language-user, and the novelist as reflecting this. Our task is to unravel the deep mistake in thought exemplified in Einstein's feelings expressed in his discussion with Carnap: 'The problem of the Now worried him seriously. He explained that the experience of the Now means something special for men, something different from the past and the future, but that this important difference does not and cannot occur within physics'.[1]

But what is it to make time like space in such 'key respects' as to betray this 'quite different status', and what is this special 'quite different status'?

The key to understanding the reality of time lies, I shall argue, in the appreciation of the way in which, at any juncture at which causal action (in the proper sense)—whether inanimate, teleological, or fully intentional—is to take place, there has to be a distinction between, on the one hand, what is future relative to that juncture, a future open and indeed not yet existent, and, on the other hand, what is present or past relative to that juncture, which, in that it is present or past, is thereby fixed, settled, and established as already existent. But to make this clear in the context of modern scientific and philosophical discussion requires some elaboration.

In order to resist the trivialization of any questions about the reality

[1] *The Philosophy of Rudolf Carnap*, p. 37.

of this or that in philosophy, we have always first to show ('exhibit' or 'demonstrate') the connection of the thing concerned with various patterns of 'causal' explanation, the character of which will be the subject of Chapter III.

The intimate connection of time with causality has become almost a commonplace in discussions in this area. Once accepted, it may have huge consequences. Thus, in Chapter V, we shall establish that there is a privileged sense of the word "exist" whereby what possesses 'real existence' in the primary and unqualified sense is that which is capable of exercising active power or causal agency. This connection of existence with causation inevitably gives 'real existence' a very direct connection also with temporality.

However, whether or not this movement of thought, this gambit tying existence to temporality, is a mere cheat depends upon what account is given of causality and its relation to temporality.

In our investigation of the character of time in this chapter, we shall find that it is not enough to acknowledge the existence of, one might say, just 'some intimate connection or other' between time and causality, but vital to recognize temporal order as *determinative* of possible causal order. And we need to understand this in such a way as takes account of the way in which the distinction of past, present, and future is presupposed in any relations of being earlier than, or later than, amongst past events. McTaggart was right in supposing the truth of 'earlier'–'later' propositions (his B–series) to be parasitic on the truth of tensed propositions involving pastness, presentness, or futurity (his A–series). Hence there can be no severance between a reality of time, cognizable by physicists and anyone with knowledge of objective reality (e.g. God), and the reality expressed in tense: no indication of the facts which make propositions of the B–series true which does not presuppose the A–series.

We shall find ourselves in an Either–Or situation.

Either we take a realist view of both time and tense and, together with it, a realist view of causal action (i.e. a view regarding the causal relation as a physical relation or, if non-physical things exist, a 'metaphysical' one, rather than to be understood in epistemological terms, in the way our later discussion makes clear). It will emerge that this realist view involves the notion of causal agency and, with it, also, the notion of a 'real' contingency not explicable in terms of the concepts of epistemology or logic.

Or we find ourselves in a view that represents tensed statements as not expressive of any aspect of reality, but only, for us who act in time, an inescapable mode of perceiving, thinking and speaking about reality.[2] This will be a view which makes causality a matter of the making of a difference within the field of things which are statistically studiable, a field therefore including suitably similar or comparable and numerous cases, so that contingency ends up as a notion to be understood in epistemological (e.g. probabilistic) terms, if not solely as a name for what is neither logically necessary nor logically contradictory.

The question presents itself as to whether the choice presented is to be decided in regard to cause by reference to considerations concerning time, or in regard to time by reference to considerations concerning cause.

In fact, the key advance lies in the grasping of the notions of agent, i.e. substance with active power, and of a field within which such agents act and interact, as alone allowing a realist view of cause, but thereby bringing with it a realist view of time. The relevant sort of 'reality' in causal relations is the one involved in 'real causal action', and, within the world, this is exercised by things with natures persisting through time (in typical cases, so-called 'substances') which exercise active powers in causal actions. By 'causal actions' I mean actions such as those which G. E. M. Anscombe begins to list as follows 'scrape, push, wet, carry, eat, burn, knock over, keep off, squash, make (e.g. noises, paper boats), hurt'.[3]

Likewise, it is from a consideration of the role of possibility, contingency, and necessity in the context of causal explanation that we reach to conclusions about time, rather than *vice versa*. In fact, a realist view of contingency and possibility inherently involves and logically presupposes a realist view of time, but such a realist view of contingency and possibility can only be sustained within the context of a view wherein the world includes agents, animate or inanimate. Within the succeeding chapters, I note the ways in which accounts of space involve substances as subjects of passive as well

[2] This view reappears in D. H. Mellor, *Real Time*, p. 6.
[3] 'Causality and Determination', p. 137.

as active powers, interacting in ways which are necessary if there is to be spatial localization. But my central contention is that it is in a combination of considerations from the philosophy of language with considerations of the nature of causation that the question is to be decided—in favour of the realist view both of time and of causation. The indispensability of substances as subjects of active and passive powers, if change, time, or causal action are, any of them, to exist, is for me a key thesis. Arguments to this effect begin to appear first in this chapter,[4] are indicated in summary in Chapter III,[5] but most explicitly in Chapter V with its account of categorial structures.[6] Chapter V enshrines my principal objection to Whitehead and the background to my earlier remarks: my suggestion in Chapter V, Section 3, is that 'events' (however internal relations may be to the nature of events) are in the relevant ways as much abstract or quasi-linguistic as 'universals' such as 'wisdom'. (It is not having relations internal to its nature that makes anything abstract, and if relations are internal to the nature of some or all substances that will not of itself make these substances abstract.)

Meantime, the first step in our strategy is to establish the reality of time *in the relevant sense*. In disentangling the more general philosophical aspects of the question in Section 1, we shall find in (*c*) that the key question is whether or not temporal relations individually are prior to and determinative of possibilities of causal relation, rather than consisting in such possibilities (despite that the system of temporal relations does not exist independently of the system of causally related natures). And in 1(*d*) and (*e*) we shall find that the possibility of ascribing such priority to temporal relations individually turns on recognizing the primary locus of temporal language as in describing exercises of agency and the order in history, not in describing experiences and their order. In Section 2 we shall find that consideration of the problems special to physics itself confirms these perspectives.

[4] See the end of Section 2(*b*) and the hint at the end of Section 2(*a*) in this chapter.

[5] See the end of Section 1 and especially Section 2(*b*) of Chapter III.

[6] See Chapter V, Section 4, on the structure of 'the categories' as giving the focal place to 'first substances' precisely *qua* subjects of action and passion. This section crucially depends upon Section 3 where I indicate that P. F. Strawson's reasons for giving secondary status to universals, and by parity of reasoning to events, are along the wrong lines, drawing upon epistemology where the philosophy of language is what is in point.

1. GENERAL PHILOSOPHICAL ISSUES

(a) *Resisting the trivialization of the question*

In resisting the trivialization of the question whether time is 'real', it is necessary to proceed in two stages. Firstly, we need to counter the gambit which arises with all such questions in philosophy, namely the gambit of separating 'reality' from causality. But, secondly, we need to resist certain special forms of trivialization which arise, either from phenomenalizing causality, or from failing to grasp the character of time (albeit within a field of causal powers) as determinative of causal possibilities.

When philosophers ask whether or not time is real, it is possible for the discussion to be reduced to an apparently trivial level, in the way that a parallel trivialization has appeared to arise in discussions of the distinction between the so-called 'primary' and 'secondary' qualities.[7]

In the early treatments, the primary qualities—shape, size, solidity, position, motion, and suchlike—had been supposed to exist independently of human knowledge or indeed of human existence, whereas the secondary qualities—such as colour, loudness, and so forth—had been supposed to exist only dependently on human knowledge and indeed even on human sensation.

It then came to be pointed out that there is a distinction between the real colour of a thing and its apparent colour. In this connection it is also pointed out that the colour of a thing is related to certain of its physical properties in respect of the reflection or emanation of light, that is, in this case, of electromagnetic radiation of a certain kind,[8] in such a way that, although what colour is involves some

[7] This distinction seemed obvious to the philosopher Locke, and to most of his scientific contemporaries and predecessors from the time of Galileo. R. Jackson rescued Locke from some of the criticisms to which his informality, and indeed sloppiness of exposition, laid him open, criticisms made standard as a result of their popularization by Berkeley and Hume ('Locke's Distinction between Primary and Secondary Qualities'). The special connection between the primary qualities and physical causal explanation is something of which the mediaevals showed awareness in only bringing the notion of intentionality into play in explaining the secondary qualities, recognizing the more immediate gearing of the primary qualities to geometrical relation and physical interaction. A. J. Ayer in *The Foundations of Empirical Knowledge* gives some acknowledgement to this special connection of the primary qualities with causal explanation.

[8] Some materialist philosophers have been fond of simply identifying colours and (e.g.) hotness and coldness with physical properties, and have brought their own

indication of relation to possible human sensory observers, colours themselves appear to be realities possessed by things even independently of human observation—at least in this sense, that there are certain true propositions to be uttered, e.g. "The star Rigel is blue", "The star Sirius is white" and "The star Antares is red", which are not true in a way dependent on human knowledge. One can continue such analytical discussion almost indefinitely. For example, John Austin[9] points out that the adjective "real" is often used only secondarily in order to contrast some, as it were, normal cases with other cases so that one says that a duck is real in order to contrast it with a toy duck, a decoy duck, and so forth. And he also raises the question what, after all, is the 'real' shape of a cat.

It is arguable that, in such analytical discussion, although it indeed makes the correct point that there are true propositions describing each kind of quality, the mistake is made of *reducing* the question of *status in reality* to questions of the *possibility of true statements* and of largely *parallel sets of idioms*, in the case of two kinds of thing of allegedly different status.

But perhaps only if some sense of the word "cause" is allowed a meaning which has to do not just with appearances but with things with active powers, acting according to their natures (i.e. a 'transcendental', or not merely phenomenal, meaning[10]), will some metaphysical distinction be possible between types of sensory quality.

arguments into appropriate disrespect by their unwillingness to give serious consideration to the systematic effects or influence of contrast phenomena, mixture effects, and cultural formation upon sensory judgements: matters to which I advert incidentally in Section 2 of Chapter VII.

[9] Austin makes these observations and raises these questions in *Sense and Sensibilia*.

[10] The distinction between the transcendental and the phenomenal or merely empirical is due to Kant. When, in the Prefaces to the successive editions and the beginning of the opening of the Transcendental Aesthetic in the *Critique of Pure Reason*, he leaves room for a non-Berkeleian interpretation of his general view of sensation (science does not know the objects as they are in themselves or *noumena*, but cannot exclude their having an underlying role in sensation) and lets us understand that things as they are in themselves stand as causes to our sensations, he thereby follows Locke in conceiving there to be some transcendental or meta-empirical significance in the word "cause". However, if the only human access to knowledge of causal relation is for Kant, as for Hume, through appreciation of the conformity of appearances to law (i.e. their objectively apparent regularity), then there is nothing in Kant which establishes the notion of cause as having a transcendentally real, rather than merely empirical or ideal, meaning or application.

Obviously the question whether or not time is real, if it is of metaphysical interest to us, is unlikely to be the question whether any statements, or sentences uttered with definite declarative meaning, of such sorts as "The Battle of Bannockburn happened later than the Battle of Hastings", are true. Granted that there are a whole mass of sentences uttered with definite meaning or written down with definite meaning, statements or propositions as they are sometimes called, our inquiry will not be as to whether any of these are true, or whether any kind of relation whatsoever in any sense is expressed by such statements, but rather it will be about the significance of these statements. We shall find in the second half of this chapter that the consideration of issues in recent physics both presupposes and confirms the strong (albeit non-Newtonian) view of time which careful philosophical consideration commends. Where the question at issue is that of the special 'reality' of the so-called primary qualities, what is at issue seems to be the more immediate connection of the primary qualities with 'real causal action'. In the case of the discussion of the reality of time, we shall find that the question is not solely of the internality of certain qualities to certain actions, or (less controversially) of their more direct connection with certain so-called causal explanations, but of the priority within a certain context (*viz.* the Universe) of temporal relations over causal relations. The exercise of causal powers involves time in such a way as to involve not merely it as internal to the nature of its unfolding (i.e. to the nature of the very exercise concerned), but as a pre-condition of the possibility of the exercise concerned.

One might have thought of expressing the point by saying that causal relations presuppose temporal relations, but (quite apart from other important inaccuracies) this way of expressing the point brings into prejudice the very point at issue. Relations are between things conceived of as already existent, actual, or established, so that one can properly speak of causal and temporal relations where things past or present are concerned, but in no way properly speak of either temporal or causal relations between things past or present and things whose existence or occurrence is future. This is because the very point of insisting on the reality of time lies in maintaining in clear view the present non-existence, non-establishedness, and non-fixedness of the future; hence there are as yet no future individuals or events within the range of being referred to or of being (extensionally) quantified over or generalized about. What one has to say is that, at any juncture, the past and present determine or limit what is possible at later junctures, i.e. at junctures which are, at that juncture, future.

But this philosophical consideration, for the sake of the clarification of the logic of the questions involved, has to come first. Against this background of preliminary remarks, we must return to our original question, namely as to the *significance* (as distinct from the *truth*) of temporal statements.

We have, for instance, the view of Leibniz that these statements may indeed have truth and falsity and some genuine foundation in reality, but that nonetheless they are all true only in consequence of something else—proximately in consequence of some other kind of relation (internal or external). And one might with Leibniz[11] hold the view that they obtained in consequence of there being a system of causal relations between things which exist, between (suppose we say) events, such that some events are capable of affecting certain others, whereas yet different other events are not thus capable of affecting these others. According to this view, the *whole system of events* is such that, in certain cases, relations called causal are liable to obtain, but not in certain others, and some of the restrictions on what causal relations could exist between events are regarded as, or labelled as, temporal restrictions.

Alternatively, if we were to take Kant's view of the matter, we would regard time and temporal relations as merely a matter of the way in which we experience the field of things, i.e. of the way events as they appear to us are organised in the field of phenomena. In this view, if you like, the presence or absence of certain other types of relation, e.g. causal, presents itself to our consciousness or experience in a temporal guise, so that time, like the three dimensions of space, can be described as a form of our sensibility, a matter not of the way the things or events are as they exist in

[11] Thus Leibniz's view seems to have been that causal relations determine temporal relations, rather than temporal relations being in any way prior. For instance, he tells us, 'If one of two states which are not simultaneous involves a reason for the other, the former is held to be prior, the latter posterior', and the idea of simultaneity is explained in terms of the idea of compatibility (Leibniz, *Philosophical Papers and Letters*, pp. 666–7). And earlier he says, 'space and time taken together constitute the order of possibilities of the one entire Universe, so that these orders—space and time, that is—relate not only to what actually is but also to anything that could be put in its place, just as numbers are indifferent to the things that can be enumerated' (p. 583).

themselves, but a matter of how they appear to us. Indeed, the first hints of such a view might appear traceable in St Augustine's discussion of time in the *Confessions*, and a little later in Boethius's explanation of divine 'foreknowledge'.

In general, philosophical idealism takes two forms.

The first of these does not essentially involve egocentricity or anthropocentricity. It is exemplified in Leibniz, within whose thought every monad was in some way a living thing, and the 'well-foundedness in reality' of temporal relations did not arise from any relativity of time to a particular form of sensibility or experience, by contrast with the situation in Kant's thought. It is also represented in the thought of many 'Idealists' in the tradition established by Hegel: but among these the final position held is often only reached by means of intermediate arguments which still depend upon representing time as we know it as in some way subjective, experience-relative, or relative to differences in modes of experience or knowledge, as in the case of McTaggart.

The second form of philosophical idealism makes reality (or at least reality as properly knowable) always consist in, or depend upon, some relation to the individual human knower, or at least some relation to this race of human beings we belong to and know, taken in general.

In respect of the first stage of his argument in regard to time, McTaggart might be grouped with Kant as belonging to this second category. He might be represented as arguing *at this first stage* that because the *sense* of statements of temporal relation (statements of his B–series asserting relations of earlierness, laterness, or simultaneity) presupposes—and even can be explained in terms of—the *sense* of statements involving relation to particular human beings or groups at particular times (statements of his A–series asserting pastness, futurity, or presentness), it follows that there is no 'reality' indicated by the former kind of statement prior to the 'reality' indicated by the latter set; and that therefore the former set cannot be conceived, *as thus signified or indicated*, to express or signify any objective reality. Such an argument has found a sympathetic ear in such writers as Michael Dummett.[12] And it is further presumed that

[12] The most convenient succinct presentation of the relevant arguments of McTaggart is to be found in pp. 93–8 and 100–3 of P. T. Geach, *Truth, Love and Immortality*. The earlier 'A Defense of McTaggart's Proof of the Unreality of Time" by Michael Dummett (*Phil. Rev.*, 69, reprinted as Paper 20 in *Truth and Other*

the reality indicated by the latter set (statements of the A–series) essentially involves egocentricity or anthropocentricity.[13]

The fact that, *at a second stage*, McTaggart argues that both the two kinds of statement involved indicate some reality, *viz.* his C–series, allow him to maintain that relations of time are *bene fundata*. But this must not be allowed to obscure the fundamentally subjectivist or anthropocentrist strategy of the *first stage* of his argument: for, unmistakably, *this* stage depends on conceiving pastness, futurity, and presentness as properties whose natures are to be defined in terms of their relations to a certain experiential or cognitive standpoint, rather than, for instance, in terms of their fixing a certain situation or position within the unfolding of the world, unfolding through the activity and interaction of a variety of agents, the unfolding which we call 'history'.

Enigmas) is also notable, and presages many continuing elements in Dummett's thought (see *Truth and Other Enigmas*, Preface and Papers 10, 21, *et al.*). In A. N. Prior, *Time and Tense*, p. 102, we find that McTaggart's explanation of "*p* before *q*" as "*p* is past while *q* is present or *p* is present while *q* is future" is supposedly improved by, in effect, explaining "*r* while *s*" as "At some time, *r* and *s*": this attempt to do away the temporal conjunction "while" in favour of quantification over times, for reasons made plain enough by Geach in 'Some problems about time' (*Logic Matters*, p. 311 f.) represents no advance in thought.

[13] D. H. Mellor in his recent work, *Real Time*, accepts a variant of this view in his making tense signify something essentially to be understood in terms of token-reflexivity—although he rejects McTaggart's argument so as to make out, contrary to McTaggart, that the B–series does signify something independent, something objective in whatever way space is objective, and not, like cause, to be explained in terms of probability. (Mellor thinks that the B–series signifies the same reality as it indicates, i.e. as that which makes the propositions concerned true, whereas McTaggart thinks that the B–series so depends for its intelligibility on the A–series as, no more than the A–series, to signify the reality which it indicates.) Of course, Mellor claims to hold a view of probability as objective, but this view still leaves it tied to various statistical and induction-geared notions, i.e. systems of comparison and correlation which have a second-order standing, logically speaking, relative to a presupposed first-order aggregate of data with some kind of prior objectivity or reality including in its character whatever is objective spatio-temporally. However, unless notions of nature presupposing notions of causal agency are brought in in an adequate way, the claim (at a philosophical, as distinct from a less critical scientific, level) to have justified a view of a certain kind of probability as 'objective', is mere self-deception. Moreover, the character of the 'objectivity', even of invariant functions of space and time, let alone of any supposed 'objective probabilities', remains to be explained and rescued, if such is appropriate, from the conventionalism of Nicholas Rescher as its most sophisticated recent exponent.

(b) Mathematical models in physics

It is vital to avoid confusing this philosophical idealism with what we may call 'methodological idealism', as this is represented within the thought of mathematical physicists over the ages. Only by distinguishing the separate issues raised by these two quite different elements or trends in human thought is it possible for us properly to identify the questions at issue in discussing the reality of time and in exposing the weaknesses of the credentials of philosophical idealism.

At each stage in the development of mathematical physics, systems involving mathematical idealization have been brought into play: firstly, the theory of the continuum, points as divides or ordered pairs of connected parts of lines, developed in Aristotle and presupposed in the development of geometry from Eudoxus and Euclid to Pascal; secondly, a more developed theory depending on the utilization of integration and differentiation, seeming to reduce the continuous to points corresponding to the real numbers, speaking of masses or particles as if they could be located at points, and twinning one set of idealizations labelled 'events' with the already established mathematicians' 'point-instants'; thirdly, a mathematical theory of quanta, allowing of a rejection of the classical (second) theory of the continuum. Yet one must be careful not to be deceived: one can have a pure mathematical theory (e.g. of groups) which has prima facie quite unrelated applications (e.g. to propositions and to sets, in the case of Boolean algebra); and one must discover whether or not in the description of a theory a word such as "point" is being used in a 'pure' sense, or whether to indicate an intended type of application. In any application, the sacrosanctness of some primitive ordinary and informal statements has to be preserved.

If any notion of frames of reference, within which points, or point-instants, are to be distinguished by co-ordinates, is to have application, some other rough, but not false, non-mathematical statements of spatial location and temporal relation, often presupposed by or incorporated within reports of causal action or instructions as to courses of action to be followed, must exist: e.g. "You are now close to Balmoral facing eastwards towards the North Sea; I have placed a table in front of you; on the table are two billiard balls; kindly put the red ball on the left of the white billiard ball at arms length without moving the white ball". Again, as Peter Geach makes plain,[14] any use of clocks in order to make judgements of temporal relation depends on a capacity to make judgements of simultaneity ("*P* happened when clock *A* indicated . . ." or "Clocks *A* and *B* were set going at the same time").

(Of course, when such ordinary and informal statements are thus sacrosanct so that their status as presupposed by 'scientific' statements and incapable of being overthrown by them—if science is consistent with its basis—is indisputable, there may still be problems of elucidation. Thus, a judgement of simultaneity of the kind just referred to involves the symmetrical 'simultaneity' of events *P* and *Q*. But this simultaneity cannot be a sameness of time in the sense of temporal point-instant for reasons associated with relativity theory. Rather it must be a sameness of 'moment' in some more informal sense: this informal sense will be related to what speakers and hearers of the judgement can both be taken to concur in when they agree in a judgement of simultaneity.)

Which ordinary and informal statements are thus sacrosanct is liable to be the subject of scientific as well as philosophical dispute.

For instance, it will be an absolutely key part of our later argument that statements of real causal action are among them, and on this the refutation of idealism will in one way hinge. All of these statements of real causal action will imply some statements or other of temporal relation and some of them will imply bodily interaction at the boundaries of the bodies concerned in such a way as to imply spatial location in the most unsophisticated sense.

But, in the course of the progress of science, it became clear, for instance, that the Sky is not an impenetrable dome and that surface-tension phenomena are not the result of liquids having skins in the sense that custard may have a skin. That is, some pruning may be needed among the candidates offered as primitive sacrosanct statements—paradigmly, "The Sun rose" has come to be regarded as in a certain way false except at the level of so-called 'objective appearance', because in its most primitive interpretation it implies geocentrism. I mention the existence of this 'methodological idealism' because it will be our concern later in this chapter (when we discuss cosmological theories, fields, and space) and in Chapter III (in Section 1 on 'Explanation in Physics') to deal with the issues

[14] Peter Geach, in 'Some Problems about Time' in *Logic Matters*, pp. 311 f.

raised by it. What is important at this stage is merely to note that these issues have primarily to do with the status and character of mathematical models and idealizations, and do not directly or necessarily bring any kind of philosophical idealism in their train.

It is important thus to keep the issues confronting us separate or at least unconfused. In this chapter I shall deal with the general philosophical issues first and turn to the more particular problems in the philosophy of physics only in the second half of the chapter.

Note on the theory of the continuum

Zeno's paradoxes purported to show that continuous motion was impossible. Aristotle's solution of them exhibited the necessity of insisting that lines, and in general *continua*, are not composed of atoms, whether points or instants. It is crucial to the consideration of the continuum to recognize that points enter into our thought, not as entities of such a kind as to have any length or other dimensions, space-wise or time-wise, but solely as idealizations or constructions in thought arising from considering (e.g.) lines as having actual or possible *termini*: i.e. inasmuch as motion along a line can be terminated or reversed, there will lie on the line a potentially existing point of termination or reversal, and, inasmuch as, for mathematical purposes, a line may be conceived of as divisible according to procedures corresponding to the identification of division-points corresponding to fractions, algebraic numbers, and other functions, it may be conceived of as comprising sub-lines joining at such division-points which themselves contribute no extra length to the line.

The consideration of motion and its continuity is what inspired Aristotle's treatment. Newton's conception of a line as the path of a moving point preserved the insight that a line is not composed of points and was again inspired by a concern with dynamics; and the idealization involved in the notion of an infinitesimal (i.e. a length which is not yet a point although less than any finite length) did not yet violate the conception that the continuous cannot be composed of the discontinuous. And in the development of the nineteenth-century theory of limits the conception of continua as sets of the discontinuous appeared initially only as a convenience of presentation, not as necessary to the theory.

The commitment to regarding the continuous as composed of the discontinuous, in the way that a set is composed of its members, comes into mathematics, not with the theory of limits, but with the purported reduction of analysis to set theory. This makes the whole development of the theory of the continuum dependent upon postulates within set theory whose status is highly dubious, as well as upon a controversial and elaborate 'measure' theory. These postulates within set theory are highly dubious inasmuch as, except in so far as one introduces sundry arbitrary or ad hoc restrictions, they are liable to engender firstly the logical and secondly the semantic paradoxes, as well as a variety of implausible theorems at every level and a multiplicity of questions of doubtful significance. The implausible theorems which I have in mind include the very elementary theorem[15] that, given a finite line, it is possible to cover every rational point on the line by a finite length in such a way that the limit of the sum of these lengths is as small as any finite length one cares to name (e.g. every point on the line between 0 and 1 by a series of lines of length 0.1, 0.01, 0.001, etc.). But they reach to very recondite theorems in topology such as the Banach–Tarski paradox (1924), whereby a sphere of fixed radius may be decomposed into a finite number of parts and put together again in such a way as to form two spheres with the given radius[16]—Rosser has shown that five parts suffice. It needs to be recognized that these latter peculiar results arise from very artificial reconstructions of the notion of congruency (a notion only at home when the continuous is being dealt with in its own terms) in a theory of sets of the discontinuous.

(c) Identification of the question at issue

Evidently time and temporal order are set more deeply into our modes of experience than space. This is most obviously the case because time and temporal order are involved, not only in the statements we make about so-called outer appearance, but also in the statements we make about so-called inner experience.

However, the way time and temporal order are built into human judgement is much more intimate than this suggests because they are involved in *any* statements whatsoever that express or report thought or experience, *simply as a result of the very structure of language and with it of articulable thought itself*. This is because

[15] This is instanced by W. and M. Kneale, *The Development of Logic*, p. 400.
[16] This is the instance given by Patrick Suppes in *Axiomatic Set Theory*, p. 250.

they are involved in the distinctions of these which we make and because of our incapacity to deviate from the present tense without our deviation being significant. They are also involved in the way, made plain by Peter Geach, whereby words like "when" and "while", "before" and "after", function; for these function in ways more analogous to "and", "or", "if", and "because", than to conjunctions such as "where", having to do with space.[17] We shall see later how, besides the primitive judgements of simultaneity already mentioned, modern physics preserves inviolate all the customary relationships between causal relations (expressed by "because") and observer-independent temporal relations (expressed by "before" and "after", where no qualification is needed).

If we delve yet further into the matter, we find that there are yet further features involved in our ways of speaking, making temporal distinction ever more deeply embedded. For instance, the inquiry as to whether the same dog would have been conceived and born, that is to say, numerically identically the same dog, would have been born if my Alsatian and my Great Dane were to have come together by mistake next month, as would have been born if they had come together by mistake in two years' time, seems to be absurd unless we think of the identity of dogs as dependent upon, or consisting in, the identity of their souls, and think of the souls of dogs as already there, waiting to be enfleshed, if not on one occasion, then on another occasion. If we do not think that there is any such pre-existence of souls, discussion of the identity of these two entirely hypothetical dogs in the future seems meaningless. It is a general logical characteristic of statements about future individuals that the way in which they refer to these individuals, the sense in which they refer to these individuals, and the status of the identity of the individuals thus referred to, need to be given a different explanation from, and are less well established than, the status of individuals and their identity as referred to in statements about the present and past.

But all this might be supposed to be a consequence of some limitation on the vantage-point from which the human being gains knowledge and experience of the Universe, so that although future and past individuals have this different status *in relation to modes of human knowledge*, reflected in the implications of, and

[17] *Logic Matters*, pp. 312 and 314–16.

significance of, human modes of speech and statements, there is no such difference between future and past individuals *as they are in themselves* (e.g. as they present themselves to divine knowledge, if there is indeed a God to have knowledge of them, or as they would present themselves to divine knowledge if there were a God)—although you will notice it is difficult to make this philosophical point coherently: I was involved in saying that there was no difference in the way in which past and future individuals *present* themselves to God, using the present tense of the verb "to present", as if future individuals were *already now* existing, and as if past individuals were *still now* existing, and both ever-presently exercise the activity or the passivity of presenting themselves to a knower.

The position that makes time a matter not just of *appearance*, but a matter of something consequent on the presence or absence of certain *real possibilities*, within a range of causal relations, involves giving a stronger foundation in reality to temporal relations. It also gives us a back-up for a large range of further distinctions in human ways of thinking and speaking. For instance, we conceive of '*real' possibility*. Here, in using the word "real" we are making a contrast with what some people call *epistemic* possibility, i.e. that in regard to which merely there is an absence of certainty that it is false, as when we say it is possible that there were Saxons in Britain in the third century AD: this idiom ("possible that . . ." preceded by the indicative) represents a quite different use of the word "possible". Rather, in speaking of *real* possibility (typically expressed by "is possible for . . . to do . . ."), we are thinking of the range within which the natures of powers of things may be exercised: I may sensibly say that I can, it is possible for me to, get to New Zealand before tomorrow night; I may speculate that it may be possible for me to get there before midnight tonight; but there is a certain nonsense in the notion that it is possible for me to get there before the present moment or before yesterday midnight.[18] We constantly meet, sometimes with surprise, yet again with a conception of causation as relating events or states of affairs or sets of these, without any essential bringing in of the notion of active power, let alone those of substance or causal agent. We meet this not only in undisguised attempts to present refurbished versions of Hume's

[18] The distinction between real and *epistemic* possibility here indicated is made in more precise terms by David Braine in 'Varieties of Necessity', pp. 159–63, using the terms "factual" and "judgemental" in place of "real" and "epistemic" respectively.

account,[19] but also in the discussions of Michael Dummett[20] and David Mellor,[21] bringing in some notion or other of probability. None of these approaches requires the restriction of a thing's effects to what is future relative to it, unless for some epistemological or methodological reason, because none includes the concept of substance with active powers set within a field of co-acting substances with active powers constituting a system of Nature.[22] The field of exercise within which the active powers of a thing are exercised is future, or perhaps contemporary, with the time of the exercise of these active powers—never previous to it.[23]

However, the kind of insistence on the reality of time which is of interest to the philosopher and which we are here interested in understanding, and the status or validity of which we wish to ascertain, is the following. It is the position whereby it is thought, not that temporal relations *consist in the presence or absence of the possibility of certain causal relations*, but rather that temporal relations *are that in virtue of which these causal relations are or are not possible*, i.e. it is the position according to which temporal relations are *in this respect prior* to causal relations.

In other words, our thesis includes this, that temporal and causal relations hold their reality jointly, the temporal determinative of what causal relations are possible, and must be conceded and understood so to hold their reality, or not at all. And this position will be confirmed by our later consideration of the situation of time in the scientific explanations of physicists and such like.

(d) Steering between Scylla and Charybdis: the requirements for a refutation of idealism; the conception of the 'Dramatic Now'

In my discussion of the question of the reality of time, it is vital to avoid the pitfall of explaining the reality of time in such a way as to absolutize it. True, I am arguing that, in each individual case,

[19] E.g. in John Mackie, *The Cement of the Universe*.

[20] E.g. in 'Bringing About the Past' reprinted as 19 in *Truth and Other Enigmas*.

[21] *Real Time*, pp. 11 f., *et al.* (cf. n. 13 above).

[22] See n. 26 below.

[23] This point, observed by David Braine in 'Varieties of Necessity' in 1972, was at that time already a commonplace in the public lectures and discussions of G. E. M. Anscombe and others, as it had been for Aristotle and others aware of the difference between the future and the past.

temporal relations limit what causal relations are possible. But I am not arguing that the whole system of temporal relations, let alone any system of supposed entities called 'times' or any grand entity called 'Time', exists prior to or independently of the existence of the system of things, including substances with active powers, i.e. the causal system arising from the natures and relations of these things.

That view, commonly associated with Newton, is incompatible not only with the notion of God's eternity but also with relativistic physics. It is in any case a view which should be ruled out a priori on purely conceptual grounds, of the same type as should lead us to rule out universals and numbers from any independent metaphysical existence—reasons we shall meet with in Chapter V. Rather time is something co-constituted, or, if there is such a thing as creation, concreated with the system of things in time: time exists by the very act of the existing of things whose nature is temporal. But within the system of Nature, particular temporal relations are prior to certain possibilities of causal relation.

This view of Time as antecedent to Nature, conveniently called 'Newtonian', was rejected by Aristotle and all his followers, by St Augustine and those influenced by him, and indeed by all those who make eternity different from everlastingness. It has in modern times presented a further example of how the absolutizing of an abstract entity is usually followed after only a short space of time by the offering of an idealist account of this entity. Thus, for Kant, Newtonian Time has to be regarded as 'transcendentally ideal'. Firstly, a mistake in logical theory or the philosophy of language leads to certain abstract subjects of discourse (e.g. universals, numbers, times) being ranked amongst basic objects or individuals. This engenders at a second stage the metaphysical mistake commonly labelled Platonism, whereby all these objects are ascribed the same metaphysical reality. Then at a third stage, the difficulty of understanding the nature of these abstract objects leads to the conceptualist or idealist view, whereby they are conceived to be mental entities or 'ideas' in a modern sense. What we need to avoid is the initial logical mistake.

Accordingly, our task is to cut our way towards a view that time is not just 'empirically real', or real at the level of appearances and description only, but unqualifiedly real in a metaphysical sense, without allowing ourselves to be driven either into idealism or into any Newtonian absolutism.

The arguments of Peter Geach are adequate to establish the existence of true temporal propositions as prior to and presupposed by propositions asserting spatial relations, and their difference in status, structure, and behaviour from such space-concerned propositions. But we need to do more than this to establish that their status is not just that of the 'empirically real'. And in order to do this we have to establish that they *together with* causal relations have a metaphysical status prior to experience. It is my view that it is a condition of achieving this that accounts of causal relations in terms of conformity to law be ruled out as inadequate. That is, it is my view that, so long as causal relations are explained in terms of conformity to law, the distinction between the empirically real and the requisite kind of underlying metaphysical reality cannot be made.

Of course, there is a systematic question as to whether showing that explanation requires an attribution of reality to entities exercising active powers and causal agency in this way does in fact get us to the level of so-called 'metaphysical reality' or 'reality as it is in itself'. The question at issue here turns out to be the question of the order of priority between first-order and second-order statements, i.e. questions of so-called fact and questions of so-called methodology. In Chapter VII, I shall give reason to suppose that first-order statements have a certain priority, so that the view that explanation reaches only to the level of the 'empirically real' and not to the level of the unqualifiedly real is incoherent or without sense.

The same argument in Chapter VII might also *seem* to refute McTaggart. His view that a consideration about the priority of sense between two kinds of statement is determinative of one's view of the priority between the 'realities' referred to by these statements involves the same mistake of making the second-order prior to the first-order. The refutation of McTaggart is of peculiar interest to us since we are launching an inverse argument from the reality of time to the existence of God, whereas it was his daring project to launch an argument from the unreality of time to the non-existence of God.

However, it does indeed seem that *what time is* has been left out of view in any account of temporal relations which limits itself, either simply to the logico-linguistic consideration that they are better conceived of as expressed by grammatical conjunctions (in this, parallel to truth-functional and causal relations), or else to the phenomenalistically conceived distribution of possible objects of appearance in respect of one aspect (called temporal) of this supposed distribution. Moreover, there seems to be no way of guarding the interpretation of "earlier", "simultaneous", and "later" against treatment along these defective lines except by explaining them in a way which connects them essentially with the notions of past, present, and future.

Accordingly, what has to be realized is that the fault in basing an idealist view of time on such arguments as those of McTaggart lies in a radical misconception of what is involved when reference to past, present, and future is implied. The idealists with whom we have been concerned have imagined that when such reference is implied what is meant (or signified) turns upon contrasts in modes of experience or knowledge, e.g. St Augustine's memory, sight, and expectation. However, in fact, the essential meaning does not have to do with something primarily cognitive or psychological, but to what we might label the 'dramatic now', the stage presently reached in the unfolding of a causal or dramatic history, the 'now' relevant to what is next to happen (i.e., in my phraseology, at this 'juncture') in this 'locality'. (An accurate account of the applications, in the spatio-temporal context always interdependent, of the terms "locality" and "juncture" requires more extended discussion,[24] giving an account of the co-action of things at junctures which are contemporary or which have no observer-independent order, as well as of puzzles raised by more antique discussions of the spheres of influence of the angels and the stars.)

The point of view which we are expressing can be highlighted by making the following observation. God can indeed know time for what it is in its true character, and the present as the present, not just as a member of a series of things ordered from the earlier to the later, but only because He knows the 'now' of time precisely as the dramatic now. That is, He knows the 'now' as indeed now, not by a theoretical, factual, or propositional observer's knowledge, but by

[24] The notions of "juncture" and "locality" can be explained in such a way as to make plain that, in their applications within a spatio-temporal context, they are used interdependently: within such a context, the notion of "locality" represents a certain abstraction of spatial aspects, while spatial and temporal considerations enter in jointly to the application of the notion of "juncture", which is in a way the more basic one.

the practical knowledge which an agent has of what he does in his doing of it. The notion of "practical knowledge" is a main part of the inspiration of the anti-dualist account of action drawn on in Chapters III and V and expressed in the adage 'the agent does what happens' in Chapter I.[25] The agent does not correlate a 'now' in an 'outer' world, the arena of his action, with another 'now' in his 'inner' life of thoughts, experiences, intentions, and decisions: rather, he knows his own action, only secondarily by observation, but in its own nature and therefore primarily simply because it is what he does intentionally; and in this knowledge he knows the 'now' of his action as no different from the 'now' of the arena of his action. Likewise there can be no question of God correlating a 'now' in an outer created world with another 'now' in His inner life: rather, He, no less than other agents, knows what time is and knows the dramatic now as indeed 'now' precisely in knowing this dramatic now as the juncture or take-off point of His action in the world.

It is the failure to envisage the primary connections of time as with causal agency,[26] and the mistaken supposition that the primary connections of time are with experience and modes of experience, experience being conceived of as primarily passive, which underlie the remark of Einstein, cited earlier, referring to 'the experience of the Now' as 'meaning something special for men', but as having no such meaning different from the past and from the future in physics.

[25] See Chapter I n. 16, Chapter III, Section 2(*c*) and n. 11, and Chapter V, Sections 7 and 8.

[26] Dummett recognizes that a certain 'connection between something's being a cause and the possibility of using it to bring about its effects plays an essential role in the fundamental account of how we ever come to accept causal laws' ('Bringing about the Past' in *Truth and Other Enigmas*, p. 333). Thus, he makes some concession to what we may call 'an agent's experience of the world'. However, he fails to reach to a realist account either of time or of substance because he always conceives causal relations as obtaining primarily between events, states, or possible experiences of pure observers and never as essentially requiring a subject with active power as their subject. Because Dummett lacks the concept that the prime subjects of causal relations are causal agents, he does not conceive of a knowledge of what time is as comprised within the practical knowledge which a causal agent acting intentionally has of his or her own actions (this is a practical knowledge of the action under any description whatsoever which is intentional, not only descriptions of it as a means to the acheiving of a later end).

(e) The influence of different conceptions of time on the conception of creation

This concept of the dramatic now and the difference between three of the main groups of theory about time may conveniently be made more vivid in terms of the three different kinds of picture or model they would suggest of the creation of the world.

Within the first picture, time and space are conceived of as absolutes, constituting an already pre-established container or context for any operation of God or of a demiurge in bringing into existence substance or 'stuff' or a world of such where previously there was no such substance, 'stuff', or world. This picture leaves it open whether the Creator is coeval with what He creates, Himself in time, or whether He is somehow outside time, but it leaves it doubtful whether there could be any sense in which God creates Time or is above it. This view of time, conveniently called Newtonian (although this is unfair to Newton), seems to be almost ubiquitous in contemporary philosophical theology: the God who cannot have foreknowledge of what is not yet determined must have a 'time' coeval with or interacting with the Time of the physical Universe.[27]

In the second picture, time and space have no reality independently of human experience and knowledge or modes of experience and knowledge. The Creator atemporally causes to exist the Universe as a single Object, an integrated whole, constituted of states and events interrelated in complex ways, none of which is essentially temporal or spatial. The regularities in systematic structural relation (which may or may not be called causal) are so presented to human experience and knowledge and to the possibility of human sense of control as to be experienced by man in temporal and spatial ways, without being known by God as in any way in themselves temporal or spatial. In this second picture, we find a strange marriage of positivistic or regularity conceptions of causality with subjectivist or psychologistic accounts of time. This second picture reappears in many disguised variant forms: for

[27] This primitive conception of God as implicitly finite is to be found, for example, in the work of Nelson Pike (*God and Timelessness*), Richard Swinburne (*The Coherence of Theism* and *The Existence of God*), and Stuart Sutherland ('God, Time and Eternity').

instance, in the recent work of David Mellor, we meet with a combination of a subjectivistic account of what he calls 'tense' (which for many would seem to include things belonging to the essence of time) with a probabilistic account of causation.[28]

In the third picture the Creator gives and upholds existence, not in any one once and for all act, but at each and every time and place or at each and every juncture in the unfolding of the history of the Universe or of its parts. 'The Moving Finger writes, and having writ moves on' (Omar Khayyám). Time is 'concreated', i.e. created not separately but in the act of giving and perpetuating the existence of temporal things. At any particular juncture, the past and present with respect to that juncture are already established in full reality fixed for God as much as for man. The future, with respect to that juncture, does not *at that juncture* exist even for God, so that God is free and so that, within this context, created agents are capable of real agency and men are free also. The point of speaking of junctures, and the future in respect to this or that particular juncture, will become clear in our discussion in Section 2 (*d*) of the significance of relativity theory for our conception of the unity of time and the Universe.

The outlining of these three different pictures may perhaps serve the expository purpose of making clear what I mean by the expression "the dramatic now".

I now turn in Section 2 to the more particular problems for us in the philosophy of science, but peculiarly the philosophy of physics.

2 SPECIAL ISSUES IN THE PHILOSOPHY OF SCIENCE AND OF PHYSICS

(a) *Strategic importance of the refutation of positivism*

At this stage, let us postulate that, independently of any question about time, what we need is:

(i) a non-positivistic account of the aims and achievements of natural science, physical and biological, of psychology, history

[28] Mellor, *Real Time*. Cf. remarks on Mellor's views on probability at the end of n. 13 above.

and the biographical element of history that has to do with individuals; and

(ii) a non-positivistic account of the nature of explanation within these contexts.

Leaving this postulate for later examination, let us proceed simply assuming it, rather than first trying to prove it. The question will then arise as to whether this will force upon us a view of explanation which, among very many other elements, involves the notion of the reality of time, involves, that is, a notion of causal relations, in the Aristotelian sense of efficient cause, in a way in which the cause has some externality or priority to the effect and in which a temporal order is presupposed, underlying the causal order, rather than the temporal order simply consisting in the existence of the causal order.

There is a tendency in discussing necessity within physics to attend only to the necessity of physical laws. In order to understand how this prominence came to be given to the notions of law and regularity over those of substance or agent, nature, and function, it will be useful to take the reader aside into some other by-ways of philosophical history.

The preoccupation with laws is partly a consequence of the rationalism which became dominant contemporaneously with Newtonian physics and which was in part the child of the success story of this type of physics, and partly of the background out of which this whole belief in law sprang, a belief consequent upon belief in the rationality of God.

The philosopher Locke did not attempt to explain the notion of active power in terms of the notion of law, and in general the preoccupation with law and coherence is less in Locke than in his successors. Berkeley, in his anticipation of the possibility of explaining the difference between reality and appearance in terms of the orderedness and regularity of veridical experience by contrast with the fitfulness of imagination, anticipates Hume. Hume, in his turn, is mesmerized by the order discovered by Newton, and indeed hopes to find in the area of psychology, and in the explanation of the ordering of human thoughts and ideas, a set of laws, laws of association of ideas, analogous to the Newtonian laws of mechanics and gravitation in their universal application and in their success in respect of ordering one's understanding of mental phenomena, and in the explanation (whatever explanation be supposed to consist in) of the phenomena of thought.

From Hume's notion, not without its background in Aristotle, that the idea of causal connection presupposes some logical connection between cause and effect, and from his view that no such logical connection exists, the way opens to his theory that the connection to be explained is still a connection between the thought of the cause and the thought of the effect, but that this is to be explained, not in logical, but solely in psychological, associational, terms.

Kant[29] always sought to replace theories of what he called the physiology of knowledge—met with in Locke and Hume—with theories of a properly epistemological character. So Hume's psychological theory of causal reasoning provided the groundwork for Kant's epistemological account. Shaped by Hume's ways of thinking, this Kantian account still explains causal connection in terms of conformity to law, i.e. in terms of regularity in respect of appearances, and does not involve our having any access to the things as they are in themselves. Kant offered an *epistemologically* objectivist alternative to psychologism, but without salvaging things as they are in themselves for science, i.e. without salvaging realism or *metaphysical* objectivism.

The alternative development of Hume's psychologism lay in conventionalism. This conventionalism is exemplified most prominently in Poincaré, the mathematician at the beginning of this century, and amongst more recent writings by the careful arguments of Nicholas Rescher, as well as the more ambitious and impressionistic discussions of Quine. According to these thinkers, both the identity of the primitive data, along with their mode of statement, and the general methods and laws of science, are a matter of the consensus of a scientific community, or of some assessment of coherence not otherwise validated. Thus all the most common views involve either a rejection of the idea of knowledge as such, or else a making of knowledge into, or constitutive of, rather than something posterior to, consequent upon, the objective reality of the things known.

[29] Kant, *Critique of Pure Reason*, pp. 8, 122.

If, as against all these views, one supposes that some non-positivistic account must be given of the knowledge we have of nature and of the things of nature, and, with this, some non-positivistic account of method and discovery in science, then a question arises as to what this rejection of positivism—in favour of supposing a certain access to knowledge of objective reality, or access to knowledge of things as they are in themselves—may involve.

Any such account will need to have shed many of the prejudices inherited by Kant from Leibniz and his predecessors, e.g. the prejudice that knowledge of things in themselves must be knowledge of things possessed of natures specifiable in entirely non-relational terms. Rather, on the contrary, these natures will typically precisely involve relations. Once possessed of such an account, we will then be in a position to proceed with the main thrust of our argument. *The question will now be as to whether or not this rejection of positivism and of a positivistic account of explanation involves us in also accepting notions of efficient cause of such a kind as presuppose the reality of time*—which is my key contention through the rest of this chapter.

Yet the dead hand of Hume infects even those who reject positivism. Often they occupy themselves merely in insisting upon the possibility and need to give a more realist account of the necessity of the laws of nature—still, that is, they are preoccupied with generalization and with universal laws. They thus tend to neglect the question of what necessity attaches to individual instances and of what is involved in the relations which these laws imply must exist between the objects which the instances of the laws refer to.

The supposition that physics is only concerned with laws ignores the fact that its laws all have instances, whether these instances are found in astronomy or whether they are found in the supposed discussion or explanation of the phenomena of a microscopic kind inferred from what goes on, for instance, in Geiger counters, in Wilson cloud chambers, or in other more recent more sophisticated pieces of apparatus. Indeed, it is on that connection with particulars and particular phenomena that the gearing of any general theory or proposed system of laws to reality in a key respect depends.

Upon examination, it turns out that all discussion of individual instances of physical law has a narrative aspect analogous to the narrative elements largely constitute of history and of many parts of biology (e.g. palaeontology and, with it, accounts of actual cases of evolution, as well as descriptions of pieces of growth and behaviour in the life history of living things, obviously involve narrative). Whether we are describing astronomical developments, geological events, or events in laboratories, we have, as in biology and history, the situation wherein the cause precedes the effect in time, or the cause is the agent already established in existence, whose action brings about the effect, and wherein the reality of temporal ordering is presupposed in the possibility of causal ordering.

Thus, we find that the general account given of the necessity attaching to laws is parasitic upon, or implies, the existence of particulars with active powers which are able to generate within a given context effects of some specified kind, in general powers more than adequate to achieve the effect concerned, as the flood is commonly more than adequate to carry away the bridge. Then, at a certain level of abstraction we formulate laws on the basis of which, taking what is, for certain purposes, an overall view of the situation, we are able to see these effects as predictable, and even accurately predictable, according to those laws.[30]

(b) *The questions raised by the development of modern physics*

At this point two groups of questions arise, the first a rather peculiar-seeming group of questions generated by certain recent cosmological theories, and the second a group of questions arising in connection with any classical or quantum physics involving such notions as "fields" and the applicability of wave-models alongside particle-models, i.e. any moderately 'normal' physics whatsoever. Both groups of questions have to do with the difficulty of identifying anything one might call the 'stuff' of the Universe, and the consequent difficulty of any giving privileged place to substances, i.e. to certain particulars with active powers. Yet it is crucial to my general view that physics, as well as biology and

[30] Cf. M. Scriven, 'Explanations, Predications, and Laws'. It should be noted that a flood, and other natural phenomena such as storms and explosions, are not paradigm examples of 'first substances', although they would all seem upon reflection to involve the exercise of active power by something. A theory of identity, such as the notion of "first substance" requires, arises at a level of further reflection or analysis.

history, involves reference to particulars with active powers—even though it may be that in physics they constitute only one of several focuses or nodes in the total framework of explanation.

In the history of the development of modern physics, at the first stage, that represented by Newton, the picture may have appeared quite straightforward. Mass, momentum, and energy appeared to be three things, each of which remained constant in quantity, while only one of these, namely, mass, represented a measure of the quantity of the stuff of the Universe. At the same time Euclidean geometry and straightforward non-relativistic conceptions of time and simultaneity remained inviolate. Within this framework vast advances could be made, associating a finite velocity with light and with the propagation of fields and force, the development of theories of electromagnetism, statistical mechanics, and thermodynamics, and the utilization of this antique conception of mass to overthrow the phlogiston theory, allowing for the foundation of modern chemistry. However, with the advent of quantum theory together with the development of relativity theory, the joining of quanta of energy to massive particles of less velocity as parts of the stuff of the Universe seemed established, together with the interconvertibility of mass and energy. Certain of the apparently sacrosanct principles mentioned relating to space and time were abandoned. But still, the main structures of relationship between time and causality remained.[31] The alternative geometries utilized still represented variations on a structure still mainly Euclidean, and it remained clear that other things were conserved in the Universe besides its supposed stuff (this mass-energy), e.g. momentum, electric charge, etc. The principal extra candidates for belonging to the stuff of the Universe were not things which were conserved as such, e.g. momentum, but, amongst others, fields.

[31] This has been brought out clearly in the recent work of Penrose, Geroch and others (R. Penrose and E. H. Kronheimer, 'On the Structure of Causal Spaces'; R. Penrose, *Techniques of Differential Topology in Relativity*, and other publications; R. Geroch, 'Space–Time Structure from a Global Viewpoint'). Geroch (pp. 84–91) distinguishes from the point of view of the physicist two relations, "*p* being in the past of *q*" (an absolute or invariant pastness seems to be in view), and "*p* causally preceding *q*", and it appears that the naturally to be expected relations between these can be established as theorems within all common mathematical models of the Universe. (Geroch (pp. 91 ff.) and Penrose are also interested in stronger conditions corresponding to the prevalence of a global determinism, but perhaps unsurprisingly these have no such in-builtness to plausible relativistic cosmological models.)

At each stage in the development of physics, as so far described, it has seemed that the search for explanations involving fewer *ad hoc* hypotheses (e.g. the epicycles resorted to in order to preserve the Ptolemaic theory of the solar system, or the Fitzgerald–Lorenz contractions brought in to explain the absence of an *observed* variation in the velocity of light) might have radical consequences. These developments destroyed geocentrism, and involved the rejection of the conservation of mass and the sacrifice of absolute simultaneity, straight lines producible for the same length any number of times, and parallel straight lines, along with any use of the word "straight" in conforming to our human perceptual and imaginative *Gestalt*; in marrying wave-models with particle-models in the quantum theory, they removed microphysics and the basis of theoretical chemistry from the control of the imagination. But, nevertheless, they seemed, in the same act, to embody an advance in the direction of simplicity and elegance and to give closer access to the real as it is in itself. They had not thus far operated self-destructively in this regard because they had not undermined the possibility of envisaging the genuine exercise of active power by things exercising active power, within a framework in which the interrelations of time and causality were unaffected. In addition, all conceptions about space and light were such as to conform to the ancient or Euclidean picture in the limit so that the functioning of normal visual perception remained untouched.

It is against this background that we have to consider the two groups of questions mentioned. The first of these is that group of questions raised by the peculiar suggestion in some modern cosmologies, that it is mathematically convenient to treat the sum of mass-energy in the Universe as approximating to zero, so that somehow it is a space-time continuum with certain properties, rather than mass-energy, which persists through all the so-called or apparent singularities in the history of the Universe.[32] The fact that potential energy is measured in terms of differences rather than absolutely, so that the level associated with zero potential energy is either arbitrary or a matter of convenience within mathematical theory, seems to leave us free, as it were, to associate the time of the so-called "big bang" from which the present state of the Universe

[32] Alan MacRobert gives a popularised account of some such theories in 'Beyond the Big Bang'.

has arisen with a time of approximately zero mass-energy and potential energy followed in the "big bang" by the sudden appearance of vast mass-energy and vast potential energy assessed as a negative quantity; as the Universe expands, potential energy is converted into kinetic energy and potential energy, whatever it consists in, diminishes.

When one has a theory or type of theory at such an early stage of development, as yet with no identifiable observational or experimental bearing, it is not clear how much it is right to reshape one's presentation of the metaphysical or philosophical constraints on scientific theorizing to allow for the possibility of the validity of such a theory. After all, it is just *conceivable* that such theories will turn out, like some of their popularized presentations, to have no better status than the works of science fiction. However, there are, even now, points of a constructive kind which can be made clear.

If a theory proposes that at the 'time' of the 'big bang' the sum total of mass-energy was zero, and remains zero ever thereafter, it is plain that it must be wrong to conceive of the stuff of the Universe as consisting in this mass-energy whose quantity is zero. Just as at an earlier time there was no temptation to identify either momentum or energy (despite the fact that these were supposed to be conserved) with the stuff of the Universe, and at a later time it became clear that it should be, if either, mass-energy, rather than mass *simpliciter*, which was to be identified as this stuff, so now at a later time it would supposedly become clear (indeed if such a theory were to prevail) that even this identification was a mistake. But then the question would arise of what it was, if not mass or mass-energy, which persisted through the history of the Universe, whose nature was expressed in the unfolding of this history, and whose nature was explanatory of this unfolding.

For instance, let it be imagined that the singularity identified by the time of the 'big bang' is such that nothing supposedly previous to it determined when and whether it occurred or anything about the history of the Universe since its occurrence. Consider, then, the conjecture that it is not just that there is no continuity between the stuff of the Universe in the unfolding following any earlier singularity and the stuff of the Universe involved in this present unfolding in the midst of which we ourselves live, but that there was no previous time because no previous stuff. Now, this will not be like supposing that the world began in 4004 BC and that fossils were planted by God to expose human intellectual conceit. Rather, there will be nothing in the present state of the Universe in the least analogous to fossils, nothing waiting to be explained in terms of any such previous time. By contrast if, on the contrary, some features of such a previous time are explanatory of something in the present condition of the Universe, then this can only be through something within the space-time continuum having a nature such that things earlier are symptoms of things causative of things later.

In other words, there is no way in which an actually empty space-time continuum can support by itself any explanatory role: the hypothesis of an empty Time without something or other in it with causal powers is idle. The effect of different cosmological theories is, not to reinstate a quasi-Newtonian space-time existing independently of things in it, but to press upon us changes in what we conceive the stuff of the Universe, what is in space-time, to be.

Thus, even in seemingly the wildest theories, the links and interdependence between time and causality are not broken.

We come now to the second more down to earth and more certainly non-imaginary group of problems raised for us within modern physics. These are the problems raised by the need to situate *fields*, as things in regard to which wave-models are as or more appropriate than particle-models, within any metaphysical account which gives pre-eminence to substances as exercisers of active powers.

It is evident that *fields* have an explanatory role in physics which is not easily compatible with regarding them as particulars or with identifying the particulars therein concerned, although certainly we appear to have in the exercise of fields some exercise of active power and it is difficult to understand logically how there can be an exercise of active power without this having a logical subject.[33]

However, the initial explanation of the very notion of "field of force" seems to depend upon the notion of an object possessed of mass, electric charge, or magnetic polarity, either as the active cause of the field, or as the possible object of its action. It is not easy to see how the concept of field can get off the ground, i.e. be intelligible

[33] R. Harré and E. H. Madden do not seem to have registered the seriousness of this difficulty in their work *Causal Powers: A Theory of Natural Necessity*, right though they are to attempt to give proper recognition to the place of fields in physics (*vide* Mary Hesse, *Forces and Fields*, pp. 197–206, 222–4, 291 f.), so powerfully argued by Michael Faraday.

and be geared to reality in any way at all, without some such links being made either with particular agents setting up fields (i.e. masses, charges, poles, and such like) or with patients, i.e. things influenced, or types of thing which, if present, would be influenced by the field.

In any case, be the problem of fields what it may, it remains the case that any talk of space whatsoever involves some supposition of locatable subjects of active powers, some entities which are passive as well as active, *and in so far as they are passive and capable of being acted upon, are so in certain types of physical way that involve some measure of localization.* As a result it appears that at least physics, unless it abandons completely the concept of space, cannot be purged of subjects of active powers, whether these be particles (independently existing or not) of rather a small kind, (e.g.) quarks, or whatever. It will be noted in any case that the notion of the exercise of active power appears whether it have an identifiable subject, in the sense of a uniquely identifiable subject, or not—the problem with fields might be that one has an active power which might be ascribable, subject-wise, to a variably limited or extensive subject of discourse.[34]

(c) The status of temporal order: a supplementary source of argument

In addition, the exercise of active power itself appears to imply some notion of time such as allows that the subject of the active power pre-exists and persists through a period of, or moment of, exercise of the active power, and such as allows that the effect have some identifiability or persistence at a later time. At least it implies some quasi-temporal ordering.

Yet, if we concede that physics and not only biology and history involve subjects in the strong Aristotelian sense, i.e. subjects of passion and action (you will recall that Aristotle introduced the notion of subject of change), there will still remain the question of the significance of the assertion of temporal *direction.* Temporal order is not neutral, but has a 'sense', whereby in some key non-temporal respect there is a pointing in one direction rather than another. Temporal direction, we must say, has a role in respect of

[34] See n. 33 above.

explanation other than merely consisting in that explanation's proceeding in one direction rather than another. It must not be just that there is some asymmetry in nature in respect of temporal order but that this asymmetry must have some explanatory significance.

Now, within physics there has been a tendency to associate temporal order at the microscopic level with increase of entropy; that is, with increase of disorder of a certain kind. If we look at particular examples this is readily intelligible to the layman. I throw a cricket ball at a wall, all its molecules are ordered, albeit in a small respect, with a slight extra motion towards the wall, then, somewhat later, there is a slightly warmer piece of wall and somewhat warmer cricket ball on the ground with scattered motions of a more disorganized kind, warmer but in the disorganized way characteristic of an increased entropy.

But we must go beyond this. If we look at biology and history, it is evident that there are kinds of order or criteria of order of which physics takes no cognizance. That is, the notion of "order" transcends the particular criterion lower entropy, which is of interest to and measured by the physicist. It embraces kinds of order not noticed by physics as such *qua* kinds of order but not excluded by it, because they are picked out by concepts which do not belong to physics although physics does not exclude their applicability. The forms of non-physical explanation which thus cohere with the forms of explanation adopted within physics without belonging to physics involve temporal directionality in distinctive ways associated with and dependent on increases of these other kinds of order. The author sets to and writes a book; a speaker from being silent begins to speak; and there is some at least limited order in the sounds spoken or the marks written. In either case an order not noticed by, or not identifiable by, the physicist as such, and evidently, likewise, in the biographies which make up human history and the history of communities, there is a directionality, as in a child's growing up and as in the development of institutions, which involve time having a certain direction coincident with that of the increase of the relevant kind of order. Thus the acceptance of the increase of entropy within physics is quite compatible with an understanding within many limited contexts of an increase of order of other kinds, not cognizable within physics as such, but taken cognizance of within biology and history.

However, let there be no mistake, the acceptance of explanations even in physics, and not only in these other areas, presupposes some notion of time which is such that temporal order does not *consist in* the order within the type of explanation given but rather is *presupposed by* the particular explanations physics recognizes as appropriate and valid.

In this context, the most fundamental ways in which temporal order is respected within the structures of modern physics must never be forgotten.

In the first place, the conception of light as travelling with a finite velocity, a velocity never to be exceeded, has become ever more deeply embedded in the structures of modern physics, and ever more inseparable from the conception that no causal influence, exercise of force, or signal, can be transmitted faster than light *in vacuo*. As a result, throughout modern physics the principle that an effect never precedes its cause has become more, rather than less, absolute in status in a sense of being more deeply embedded in theoretical structures.[35]

In the second place the very conception of our Universe as a unity (whether or not it had precursors 'before' a supposed 'big bang') reinforces the conception of some ultimate cosmically all-embracing temporal order.

In the third place, it is only insofar as the system of physical law is conceived of as closed that there can arise any possibility of symmetry with respect to time in physics. It is true that the statement of classical and quantum mechanical laws is symmetrical with respect to time, but it is only insofar as what is indeterminate is conceived of as 'arbitary and lawless relative to all possible contexts' that the symmetry with respect to time is absolutized, in the same act as the philosophy of what David Bohm calls mechanism is reimposed.[36]

But now we need to consider a key further objection and difficulty.

[35] The way in which this is true was already plain in the work of Einstein, but the more recent systematic theoretical work of Penrose, Geroch, and others, assessing and analysing the constraints operative on modern cosmological theories in general, has articulated the notions of temporal and causal priority in a more precise form even within contemporary physics itself (see n. 31 above).

[36] This is argued by David Bohm in *Causality and Chance in Modern Physics*, pp. 36–9, 64–5, 160–4, drawing on the thought of earlier papers of his.

(d) *Objections based on the cosmological unity of the Universe*

What we now have in front of us is the need to consider a key objection of a more scientific-seeming kind, drawn from relativity theory and consideration of the unity of the Universe.

Since the seventeenth century the Aristotelian conception of the Universe as a unity has been confirmed by the success story of the development of physical science. But beyond this, it has also been extended as well as reinformed by the tendency to substitute deism for theism, so that God's freedom in relation to creation is supposedly limited to the perfect and complete *initial* rational ordering of a Universe within which *afterwards* the perfection of the order initially conferred makes any further exercise of freedom not only unnecessary but a kind of violation of order, an irrational meddling. (This deistic view was in a way a natural development of a predestinationist conception whereby the whole Universe is, as it were, the object of a single over-arching, all-embracing decree of Divine Providence, a decree fixed 'before the world began'.)

This metaphysical view of the Universe as a unity such that the past is perfectly inferable from the present or the future, and the future perfectly inferable from the past or the present, underlies the tendency to reduce explanation to the recognition of conformity to law, and to regard time as at the level of appearance only. However, it remains an entirely unproven and indeed quite wild metaphysical view, making human as well as divine freedom in regard to individual goings-on a matter of appearance only.

If one supposes that God upholds the Universe in existence by means of or through upholding things in their natures, natures all of which are such as to involve that these things stand in appropriate types of relation with other things, past and contemporary, then one can continue to regard the Universe as a unity without denying the reality of time and so without denying divine and human freedom in respect of the future. If one adopts this view of God's operation, then one's acceptance of the reality of time will indeed involve only such reality as the temporal character of things and the character of their relations involves. There will therefore be no need to suppose that there is any absolute *simultaneity* throughout the whole Universe, but only such absoluteness in relation of *being before* and *being after* as is presupposed by the causal relations implicit in the relations of things.[37] Hence acceptance of the special theory of relativity and of the observer-relativeness of certain judgements of temporal relations (*viz.*, it turns out, judgements of temporal relation between events where these are so related as to be unable to exercise causal influence over each other) makes no difficulty for the acceptance of the reality of time required according to my general argument. Such acceptance does nothing to qualify the openness of that which all observers must agree is future, i.e things which at that juncture any observer whatsoever would have to concur in regarding as future relative to that juncture.

As to the unity of the Universe, cosmological theories still bring to bear upon the explanation of what happens at a particular place and time only that which is past from the standpoint of all observers at that place, whatever their velocity. In relation to the future, these theories still have the character of theories only making predictions, not of theories which describe what is already actualized. Accordingly, although they treat the Universe as in a certain sense a unity, nonetheless if the future is included in this unity they do not treat it as a unity completely actualized in all its 'parts'. If we conceive of a much closer integration of substructure and background in physical law, the holism concerned need not involve—and, if time is irreversible, will not involve—conceiving what is future to a substructure as part of its background but only an integration between the substructure and whatever counts as its present and past context as background.[38]

Thus, reflection on the implications of relativity theory and of the view of the unity of the Universe assumed in modern cosmological theories shows that acceptance of these things is incompatible with any absolutizing of Time, i.e. any erection of simultaneity or of temporal frameworks as wholes to a status prior to or independent of causality in its setting or world of causal agents in action. But this does not drive one to conceptualism, positivism, or any other form of idealism. Rather, it reinforces a rejection of any view of

[37] That this is so is plain from all the writings in theoretical physics from Einstein onwards, but is made particularly precise by R. Geroch in 'Space–Time Structure from a Global Viewpoint'.

[38] David Bohm, *Causality and Chance in Modern Physics*, expounds the relevant holism on p. 139 and irreversibility on pp. 160 ff., within the context of thinking of the object of science as the 'totality of matter in the process of becoming' (pp. 168 ff.).

Time as a vessel, or river, or presupposed framework, antecedent to Nature, without in any way prejudicing the view that the natures of things are such that, in the unfolding of their existence and relations through time, particular temporal relations are the reasons for the absence of certain possibilities of causal relation and for the presence of others, rather than vice versa.

However, the position which I have taken up depends upon our establishing the 'reality' of temporal and causal relations jointly, and the firm securing of this waits upon our clarification of the need for, and the nature of, a realist account of causal action within any theory of explanation, the subject of my next chapter. One of my key contentions, both in Chapter III and in Chapter VII, will be that the search for explanation, the attempt to identify 'causes', of itself establishes a method (at least in principle) for distinguishing between 'things and their relationships as they appear to us' and 'things and their relationships as these are in themselves'.[39] This thesis in regard to 'causal action', crucial in philosophical history, is highlighted in Chapter VI. The upshot of my arguments in the present group of chapters (II–V) is to establish a realist view of causal relations—provided only that the epistemological queries (briefly indicated in Section 3 immediately below) are indeed successfully or adequately dealt with in Chapter VII.

The function of this chapter has been to make clear that, if a realist view of causal relations is genuinely established, then a realist view of temporal relations, realist in the strong sense indicated, will in the same act be established inseparably from it.

3. A FOREWARNING OF THE EPISTEMOLOGICAL QUERIES TO BE DEALT WITH IN CHAPTER VII

We have dealt with the, as it were, intra-scientific methodological questions about the relative status of time, causality, space, substance, and nature. But the philosophy of science cannot set aside the more general philosophical questions which require mention now, for consideration in detail later. Let it be noted that I have presumed the inadequacy of any positivistic view of natural

[39] Possibly the main value of Bernard Lonergan's work *Insight* lies in his understanding of this last point, shown in Part I of that work.

science. Now, there is the question as to whether (i) this is a matter merely of the insufficiencies of positivism methodologically, so that the scientist has at least to pretend to reject positivism or proceed as if he rejected positivism, if he is to make any discovery, or whether (ii) it is a matter of a metaphysics presupposed to his science itself. And I am using the term "positivism" to cover a multiplicity of different views of which the regularity view of natural law is the simplest and one might reasonably say the most simple-minded. It is arguable, but still waits to be proved, that this is not only the simplest but also the only possible form that positivism can take and that all other forms of positivism upon examination turn out to be just tarted up versions of the regularity view of natural law.

But these things all relate to the question of one's general view of explanation. Obviously explanation is something one seeks in seeking understanding. The key question is whether understanding does not include an understanding of why things are as they are, and whether the inquiry as to why things are as they are may not carry with it, not only the rejection of positivism, but even the further metaphysical consequences which I suggest. That is, as I intimated at the end of the previous section, the subject of my next chapter under the heading of a discussion of the nature of causation, a question which has to be set in the context of a wider discussion of the nature of explanation in general.

Secondly, even if the human mind sets a demand for explanation and understanding, whatever these involve, there is the question as to whether this demand can be satisfied or whether it should even be assumed to be legitimate. It remains to be shown that the result of a search for explanation and understanding, whether it be some affirmative conclusion in regard to reality or some agnosticism in this regard, would have the status it purports to have, i.e. the status of an answer to a genuine question—that is, a meaningful question, the ignorance of the answer to which would be real ignorance. The second question is the question whether there is any genuine sense in supposing the human mind, with its demands for explanation and understanding, to have the capacity to get beyond the realm of 'appearance' to some other realm called 'reality'. I.e., even if the provision of scientific explanation presupposes methodologically or metaphysically some answers to the questions of the nature of time and of whether God exists it is not clear whether this takes us beyond the levels of methodology or even the explanation of appearance, to reality itself.

This puts us into the region of general epistemological queries of the philosophical kind which I plan to come to more systematically in my seventh chapter, postponing until then the discussion which others might consider primary. Our discussion will then confirm that it is foolish for epistemology to proceed in a reckless way which is a priori, not only in the sense of being non-empirical (not appealing to experience), but also in the sense of exercising an almost deliberate ignorance of the details and variety in the types of argument deployed in the fields of purported human knowledge.

III

Second Investigation: Explanation and Causation

We have established that the question of the reality of time, and its character, turns upon our understanding of causality or of real causal action.

However, we have on our hands two distinct concepts, the second-order concept of "explanation", referring to things in the area of thought and discourse, *viz.* certain statements, beliefs, accounts, or theories, and a first-order concept, the concept of "cause", which applies directly to the objects referred to in these statements, accounts, etc. Both concepts have an extreme generality on whose peculiar character we shall have occasion to concentrate later.

It will in the end emerge in Chapter VII that it is the first order (in this case, causation) which shapes and determines of the second order. Yet because of the way modern philosophy and styles of thinking among non-philosophers have developed, this cannot be taken for granted, and in the order of exposition one has to advance through the byways of consideration of explanation (many aspects and varieties of which are little relevant) in order to reach clarity about causation. Since time has a key role in systems of explanation within physics, and since in any case, rightly or wrongly, physics has been often conceived of as presenting the paradigm for what explanations should aim at, it is natural that we should start our investigation by reviewing the breadth of character of what is counted as explanation within physics.

1. EXPLANATION IN PHYSICS

Our previous discussion exhibited that, within the area of physics, there was a demand for explanation, an explanation of the nature of the things to which we might attribute active powers, such as molecules and their components, or fields and suchlike, and this was key to our earlier argument. But there was also a search for understanding which extends to the nature of topics of discourse which we do not conceive of as possessing active powers, but which are presupposed in our discussion of the nature of things in general and the general system of their interrelations, i.e. of the general order of the Universe, and such features of it as space and time. And it became evident that, although the physicist might retain an overlap with the ordinary man in respect of the statements he makes about particular observable phenomena and about their explanations with the causal relations involved, his general demand for explanation reached far further. It involved the overthrow, if not of all the concepts utilized in commonly received explanations, then nonetheless a rejection of or agnosticism towards some very popular presumptions, preconceptions or imagined presuppositions of these explanations, rejection, for instance, of Aristotelian conceptions of space, Newtonian conceptions of space and time, and suchlike.

All this leaves us in the position of puzzlement or inquiry as to what it is to explain something, and, if to explain something is to satisfy the understanding in some way, and, if understanding is a form of knowledge in the most general sense of the word "knowledge", then what it is to satisfy a certain kind of demand for knowledge or a desire for a certain kind of knowledge or a variety of kinds of knowledge. We need to get clear what sort or sorts of things may legitimately be supposed, if valid or true, to constitute proper satisfactions of the desire for understanding or knowledge just mentioned.

What I postulated in our previous discussion was that it was a mistake to suppose that the exhibition of something as an example of some more general rule or law could by itself legitimately be conceded to be a proper response to the desire for understanding or could therefore be properly counted as an explanation. Some individual instance (or some set of individual instances) or some less general law can be seen as consequent upon, or logically an instantiation of or deduction from, some more general law (i.e. the exhibition of some more general regularity in the behaviour of things); but this fact presents itself as a possible explanation only because there is some reason for supposing, or greater reason for supposing, that the more general law is consequent upon or an expression of something in the nature of things. I.e. there is more reason for such supposition than there had been in the case of the less general law or the particular instance which presented itself to us initially as if it had a more accidental character.

The offering of wider generalizations constituting an explanation, without the background of such presupposition about the generalizations expressing something resulting from the nature of things, appears to be wild. In particular, to go, in the way Hume proposed, from general statements about observed instances of a certain kind to all instances of that kind itself appears *in isolation* to be utterly wild.

In any case, if we look at actual cases we find no actual examples of such generalizations from observed instances of a certain kind to all instances of a certain kind which are accorded an enlarged probability *without there being some background* rendering such extrapolation reasonable. To generalize from several cases of observed instances to some yet wider class of unobserved cases without there being a presumption of this being resultant upon the nature of the things concerned appears unreasonable even if it is only extrapolation to the future which is concerned. But when it is realized that all the generalizations being discussed deal also with unobserved cases of past happenings and even with things in the past which did not occur but which at an earlier stage had been possible, the supposition that all that is involved in explanation is discovering a wider law, a more extensive regularity, becomes even less plausible. The very contingency of the future, with which I am concerned in this book, gives some colour to the idea that the natures of things present and past leave open the situation of the future, but when we have examples of things in the past which may previously have seemed thus to lie open, but which fell out in accordance with the natures of the things previous to and contemporary with them, the idea that this falling out was a matter of mere conformity to law, mere regularity, with no basis in the nature of things, has even less plausibility.

If explanation does not consist in the exhibition of something as an example of a regularity, set within the context of more general regularities, the question arises again as to what explanation is!

We must all along realize that in dealing with the concepts of understanding, of explanation, of order, and of knowledge, just as in dealing with the concepts of life, value, and existence, we are dealing with the most general concepts used or expressed in human language, concepts which are not used according to any one definition applicable in all contexts of their use, but rather in such a way that each different context requires different specifications or definitions or criteria of application—in the loose sense appropriate, different 'contextual definitions'. That is, they are concepts which in a mediaeval terminology would have fallen under the heading of terms, neither univocal (used according to one definition), nor equivocal (with nothing real in common between their uses), but analogical—or, what is the same thing, as will appear below, in Ryle's terminology "polymorphous".

Out of this group of concepts of forms of 'knowledge', it is understanding which it is most natural to conceive of as correlative with explanation. Now, in connection with this concept of understanding, we must realize that there are notions of understanding people, of understanding literature and characters in literature, of understanding works of art, in which a certain kind of knowledge is sought. Understanding appears always to be a form of knowledge in the least limited use of the term "knowledge". But these are not cases of the desire or possession of an explanation in the sense of concern to the physicist, the biologist, or the historian. However, when it comes to the question, not of understanding the causes of particular phenomena or happenings and how these spring from or are the result of the natures of things in relation and interaction with each other, but of an understanding, for instance, of the nature of time, it need not be the case that the kind of understanding concerned is more analogous to the kind of understanding involved in knowing why a particular event happened than it is to the appreciative understanding of the shape of some character in literature or the appreciative understanding of some work of art or type of work of art.

Thus I am not concerned to say that the whole goal of the physicist lies in explaining particular events and goings-on that we observe in the world, so-called natural phenomena. Rather, it may be that the physicist is in some respects a contemplative whose goal is to articulate the appreciation of the structures of nature, as the goal of some students of music may be to articulate the way in which the effects realized in their music by Bach and Mozart are achieved, or to articulate the character of the music we there meet with. My limited concern is rather this, not to say that the whole goal of the physicist is in explaining phenomena but that this remains part of his task and is involved in the rooting of his more general contemplation in a secure connection with real existence. (What he is involved with is not like the study of works of human artists exercising their imagination but represents a coming to a direct grip with the challenge presented by understanding nature or reality itself whether this be the work of some divine artist or of no artist at all.)

My concern has been to insist that this understanding within physics of reality itself must include the understanding or appreciation of causal relations as an absolutely integral (even if only subsidiary) element, for the reason that this reality itself includes phenomena successively related, involving things with active powers exercising these active powers. And, accordingly, the causal relations requiring to be appreciated are primarily relations of efficient causality in the sense which one encounters in Aristotelian tradition, within which a substance is conceived of as an agent which, according to its nature, has certain powers befitting that nature and which, in exercising these powers, acts and brings about effects. This subsidiary element in the whole of what is involved in the contemplative understanding of reality is incidentally what brings the physicist most into unavoidable contact and communication with the man in the street.

We ought at this stage to distinguish three different types of argument which might be discerned in what we have just said, stating these three types of arguments in the first instance in a very abstract and general way, but bearing in mind that what is at issue is the priority of the objects of understanding and explanation over the abstractions through or by means of which we explain and understand them.

In the first place, it can rightly be urged that it is knowledge of middle-sized objects, accessible to sight and/or touch, in the macroscopic world which alone brings the physicist into connection with the world of objective experience. As a result statements about things at this level of reality embody the generically least questionable and least eliminable part of physics (presumably, as well as clocks, rulers, and parts of clocks or rulers, there are also some things rulers and clocks are capable of measuring, indeed designed to measure). Therefore physics, as presenting explanatory theories, has to be such as to bear upon the explanation of phenomena at this level; and, if these phenomena as given cannot be described except in ways involving certain substances or spatially distributed bodies as subjects of active powers, this integration between physics and the task of explaining phenomena will include the incorporation in physical theory of whatever explanatory element there may be in the concept of a substance with active power. This argument identifies a certain epistemological constraint upon physics and physical explanation.

In the second place, there is an argument based on the philosophy of language according to which events and 'states of affairs' are of their nature proposition-indicated, rather than indicated by subject-expressions proper.[1] As such they are, to the extent explained in Chapter V, 'creatures of our modes of thought', presupposing such things as substances as logical subjects in the propositions concerned.

And, by a parallel argument, objects, such as fields, even if they themselves exercise active power, nonetheless presuppose space and, with it, things localized in space in the way bodies are, and therefore presuppose substances. For what is characteristic of the way in which bodies are in space is that they are in space as objects available to be acted upon according to their bodily mode of localization, e.g. at their surfaces or over the volume within which their parts are distributed, and therefore as substances, not just *qua* agents but *qua* things acted upon.

A further argument from the philosophy of language, outlined in Chapter IV, makes it plain that Space and Time, and in a different way the Universe, in so far as they are objects of, or presupposed by, the explanations of physics, also presuppose substances.

Accordingly, it follows that physical explanations, as directed towards real objects as such and not towards intentional or abstract objects, must have substances as one of their focuses. (By contrast it might be suggested that the objects of artistic appreciation or mathematical reflection would remain even if they existed only in the imagination or without application.)

The gearing into reality here concerned is not, as in the earlier argument, a matter of a gearing into the area of the possible range

[1] This argument can be supported out of modern writings from Russell and Moore to Geach, but I leave it to Chapter V to make it more precise.

of human perceptions, sensibility, and experience but rather a gearing into the possible area of linguistically articulable coherent thought, however loosely, flexibly, or analogously the concepts utilized are used.

In the third place it can be argued that it is part of the nature of 'natural' or 'physical' explanation as such, as opposed to (e.g.) aesthetic explanation, to bear on the real as such. By this I mean: to bear upon the real, *qua* real, not merely (for instance) as an object of appreciation and also therefore not merely as a possible object of a system of description which might receive a possibly elegant mathematical axiomatization. And we may argue that it is essential to the notion of the "real" in any proper metaphysical sense that it be involved in, or be the field of exercise for, causal relations. Now it is the network of actual and possible causal relations that gives unity to Time, i.e. it limits or excludes the possibility of actualities in no determinate temporal order, not vice versa. And for causal relations mere regularity is not enough; real causal relation and therefore real causal action is, as I argue in Chapter V, essential to distinguishing positive existence from that which we appropriately regard as a mere privation or the absence of something positive. Accordingly, it is not coherent to suppose that physics ultimately has to do with a space-time continuum to which the presence of substance is accidental, but rather what makes physics into a science of the real, *qua* real, is that the province towards which its explanatory theories are directed is a province constituted by substances with active powers, causally as well as spatially and temporally interrelated.

One can make the point in a more common-sense way. Although "explanation" like "statement" is a second-order term, the objects of the kind of explanation and understanding with which we are concerned (like the 'objects' of physical knowledge and statement) are rather things, or their natures and behaviour, than laws as such. In stating laws, we always pick out or 'abstract' certain aspects and lay down norms or law-like regularities relating to these abstracted aspects. But the existence of law-like statements and models (enshrining our so-called 'abstractions') applying at one level of the description and analysis of phenomena with a certain range of application and validity in no way precludes there arising law-statements and models applicable at deeper levels—or arising from some newly realized further integration of the substructure and background—rendering the body of physical theory ever open.[2] The development of theory gives better appreciation and knowledge of the real; but one must not allow the openness and apparent lack of 'absoluteness' which belongs to the ensemble of our theories to hide the fact that it is orientation towards the 'real' which makes these theories parts of physical science at all, and that the 'real' is always already partly known, and in particular known as including causal agents, substances with active power.

Accordingly, in the light of our grasp of the datum that explanation is correlative with understanding and of the fact that the understanding with which we are concerned is understanding directed towards the real (it is not artistic work or the study of pure mathematics that we are concerned with), these three types of consideration converge in reinforcing our insistence that the physicist's contemplative understanding of reality must include as an integral and cardinal part of itself an appreciation of those causal actions and relations which are also objects of knowledge and concern to the ordinary man.

It may be asked whether these very abstract and general arguments I have been using are proper. I have urged that it is a precondition of the use of the notion of explanation in physics that notions of cause and active power be allowed a certain use in regard to the objects of physics. And I have argued that it is a precondition of the application of the notion of "real existence" in a metaphysical sense to anything at all that the notions of cause and active power be allowed application to any candidates for real existence. Now, it could be objected that I am employing second-order argument in order to support a first-order conclusion. To this objection two replies may be attempted: first, it is legitimate to utilize second-order reasoning in order to exhibit some proposed second-order theory as incoherent or in order to expose as nonsense some view that presupposes such an incoherent second-order theory; second, it remains the case that all second-order argument presupposes some first-order statements, e.g. statements of substan-

2 Cf. David Bohm's eloquent exposition of these ideas in *Causality and Chance in Modern Physics*, pp. 137–40, 164–70. It will be noted that he is at pains to point out how within a specific domain of consideration we can determine the errors so well (p. 167) as to attain a certain practical validity corresponding to the autonomy spoken of as arising at different levels (p. 140). But it remains that, for him, the object of attention in our science is not these abstractions but the 'totality of matter in the process of becoming' (pp. 168 ff.).

tiality and persistence through time or of the sharing of a common language-potentiality with others. Therefore, in the arguments of this section, I am not violating the requirement I identify in Chapter VII that second-order discourse has no priority or independence over first-order discourse. This is the requirement that we recognize the priority of the question of conditions of existence over the question of the conditions of human thinking about existence.

2. THE NATURE OF CAUSATION

(a) Types of semantic universal

Supposing that what we have said about explanation is granted, we are driven to a consideration of causation, and to the problems of defining either notion.

In relation to the notion of explanation, we must not expect any non-circular definition. Rather, we must *expect* that the notion of explanation, like the notions of truth, provability, statement, definition, and proof, and like such notions as knowledge, understanding, and thought, with which they are correlative, will lack any non-circular definition covering its multifarious applications. The indefinability of the notions of object, set, truth, proof and definition and the lack of formalizable or mechanically applicable criteria for their application have been demonstrated within the restricted field of philosophy of mathematics and proof theory, i.e. within 'logic' in an extended sense.[3] However, once demonstrated in these areas these things have implications for the whole of human knowledge since these subjects, logic and mathematics, are occupied with concepts universal in their application so that conclusions about the nature of logic and mathematics, and the concepts therein found, cannot be conclusions of restricted implication in regard to the kinds of knowledge they are relevant to.

It is clear that the concept of explanation is like the concept of statement in being a second-order concept, a concept which we use in describing or referring to our thought about the world, whether this be written or unwritten, spoken or unspoken. In this, it lies in

[3] See n. 4 below.

contrast with the concept of causation which, at least in the modern use of the word, has to do with a certain type or variety of types or span of types of relation between things known, so that, whereas explanation is, as it were, something within the human mind, causation, if there is any such thing, is something out there in the world, a relation obtaining between things as they are in themselves. It is not that things as they are in themselves have natures of a non-relational character, but rather that they have natures which are precisely such as to involve them in relations, and among these relations causal relations are perhaps the chief or most significant or notable.

Here, we meet with what is largely a gap in modern discussions. The idea that the most general second-order concepts such as "truth", "statement" and "definition", "provability" and "proof", cannot, without contradiction, be supposed definable in any formalisable language is a commonplace to those familiar with the semantic paradoxes and the work of Tarski and Gödel.[4] Likewise, the absence of any non-*ad hoc* way of avoiding the *logical* paradoxes in set theory[5] has removed any realistic expectation

[4] The semantic paradoxes were dealt with by Russell by the elaborate 'ramified' theory of types, involving a metatheory of orders, and by Zermelo by means of the introduction of the undefined notion of definiteness. The systematic difficulties obstructing the resolution of the semantic paradoxes, that is those turning on notions of truth, designation, or definition—involved when a system of speech refers to the relationship of its own expressions to actual or possible things in the world—suffice to show the obstacles to defining these notions. But Tarski produced a demonstration of what, against this background, might have seemed evident, *viz.* that no formalised language can contain its own metalanguage of such a kind as to allow it to consider in general the truth of its own statements. Gödel, more surprisingly, proved (at least if one presumes natural number theory to be consistent) that, even in the extraordinarily limited area of the theory of the natural numbers, no effectively applicable criteria for the application of the concepts of proof and provability could be provided by any consistent system of axiomatization whatsoever. A useful review of the results concerned is offered by W. and M. Kneale in *The Development of Logic*, pp. 652–72. The paradoxes are attractively reviewed by S. C. Kleene in *Introduction to Metamathematics*, and E. W. Beth in *Foundations of Mathematics* and analytically examined with the various proposed types of solution in A. A. Fraenkel and Y. Bar Hillel, *The Foundations of Set Theory*.

[5] The supposed importance of the distinction between the logical and semantic paradoxes was highlighted in F. P. Ramsey's demonstration (in the title essay of *Foundations of Mathematics and Other Logical Essays*) that the simple theory of types, without ramification or theory of orders, was adequate to secure the avoidance of the logical paradoxes, and his attraction to a notion of type akin to Frege's notion of level in order to avoid the logical paradoxes and espousal of Tarski's theory of metalanguages in order to avoid the semantic paradoxes. However, the axiomatic approach, associated with Zermelo and Von Neumann,

arguing, without inconsistency, for any non-circular account of the most general, formal, first-order concepts such as "object" (co-extensive with the second-order concept "subject of discourse") and "set". But, what needs to be recognized is that the most general, non-formal first-order concepts, such as "substance", "act" and "power", or (concepts distinguishing different ways of standing in time) "disposition", "state", and "episode", or (concepts of the activities we should think of as having various modes) "cause", "live", "know", "understand", and "think", are in no different situation from the most general second-order concepts, all are, to use Gilbert Ryle's expression, "polymorphous",[6] or in mediaeval terms, analogical in character.

Of all these terms it is characteristic that, although we can to some extent contextualise them, limit the contexts within which they occur (commonly doing this by means of other concepts no less polymorphous), and to that extent indicate their meaning or limit the ways in which they are capable of being understood, we cannot give them any non-circular definition or any explanation that purges them of this polymorphous character. Ryle observes that there is no single set of criteria covering the vast variety of types of what we call thinking. Likewise the word "know" is used in connection with propositions or facts, subjects or areas of knowledge, languages, skills, people (sometimes in connection with mere acquaintance, sometimes greater intimacy as with colleagues and friends, and, in older usage, of sexual relation), emotions, and the objects of perceptual knowledge. Various languages utilize

seems more open in its *ad hoc* character, granted that the simple theory of types involves the oddity of different types of number for counting different types of things (thus inventing a category distinction where none seems possible) but still fails to give an account of the categorial structures noticed by linguists, besides being in its own terms nonsensical (i.e. the propositions of the simple theory of types violate the theory itself). It is plausible that the most revealing logical paradox, that of Cantor, associated with Cantor's diagonal procedure, has the same root as the Richard paradox, and that it is mere superstition to suppose that the inability to enumerate real numbers was the result of there being too many of them, all in real independent existence, while the inability to enumerate definitions of real numbers stemmed from some queasiness in the concept of "definition" of a real number. However, the key observation to be made is that, if the simple theory of types has no privilege, then there is no established non-*ad hoc* way of avoiding the logical paradoxes, so that the most general, formal, first-order concepts such as "object" and "set" are in no better situation than the semantic concepts of "truth", "definition", etc.

[6] Gilbert Ryle, 'Thinking and Language'.

different verbs for some of these cases, but there is no uniformity in the way different languages divide up the ground and there is some similarity or analogy between cases which allows the different usages of different languages to be readily appreciated by a language-learner.

These cases of thinking and of knowing are not at all like the much abused example of the term "game", taken from Wittgenstein. It is plausible that there is a group of requirements, *A*, *B*, *C*, *D*, and *E*, such that a thing counts as a game if it exhibits any important sub-group of these, and this is the most popular explanation of Wittgenstein's notion of "family-resemblance" (even if foreign to his intention). If the term "game" can be dealt with in this way, then the analysis suggests that there will be other more fundamental terms, e.g. *A*, *B*, *C*, *D*, and *E*, in regard to which no such analysis is needed, probably contrary to Wittgenstein's own understanding of the situation. But in the cases of "thinking" and "knowledge", although most of the types of thinking and knowledge have some connection or other with actual or possible propositional thinking or knowledge, it is implausible to suggest that the unity or interconnection which gives rise to the spread of uses of the terms concerned is located in this heterogeneous set of connections with the propositional, rather than in some antecedent similarity or analogy of which these connections with the propositional are merely derivative symptoms.

This book is not the place to develop in detail, or defend in detail against sophisticated objectors, this theory of 'semantic universals'. That there are semantic universals is evident from a consideration of the most general terms, first-order and second-order, which I mentioned earlier. Of course, in any particular language a particular semantic universal may not be realized in any single word, or indeed by any word whatsoever: its existence for the people using that language may consist in certain aspects of their capacity to learn or appreciate other languages, and may turn out to have been expressed, if at all, by that people through some special intonation or stylized adaptation of language to context. I use the term "semantic universal" to include coverage of the case of the semantic significance of syntactical rules, where these have semantic significance, and not solely to indicate certain possible lexical items—it is to be noted that no lexical item has any semantic significance apart from at least some context of syntactical rules.

However, for our investigation, the terms of key importance are not those most general terms which formal logic and proof theory attend to, but first-order terms like "substance", "live", "cause", and "act". And in regard to these any analysis, such as that proposed for "game", is even more implausible than in the cases of "think" and "know".

(b) Contextual definition of the notion of causal agent

It is evident that the uses of the conjunctions "why", "because", "as", and "if" are very various. In order to get at the notion of "cause" as a first-order notion, distinct from the second-order notion of "explanation", we have to distinguish among these uses, but we must expect to find that we can only do this in informal ways, distinguishing certain general types of context in discourse from others, rather than by, as it were, plucking out from an apparent morass some single isolated sharply definable concept.[7]

In giving our contextual explanation of the meaning of the word "cause", in the sense with which we are concerned, we need first to make it plain that we are concerned with relations of so-called 'efficient' causality. That is, our contextual account must suffice to exclude the concentration of attention on answers to "why"–questions in terms of the nature or constitution of things, or in terms of their functional or structural setting within some larger framework, perhaps even a cosmic one, or in terms of a relation to something conceived of as their goal (as distinct from an efficient cause working towards that goal, i.e. I distinguish the goal, as so-called 'final' cause, from the agent who seeks that goal and that agent's dispositions of desire and belief). Next we need to recognize that, although the salient pre-condition picked out as cause in one of the ways just mentioned is typically precisely that which is stated or reported in the answer given to a "why"–question, nonetheless such events and states presuppose the subjects with active power whose doings, states, and alterations they constitute, and it is rather these subjects of active power that are, above all, causes. That is, we

[7] The interconnection or correspondence between different uses of the word "if" and different families of use of modal expressions such as "can", "cannot", and "cannot but" with their equivalents, and the need for appropriately modulated 'contextual definition' of each, was explained by David Braine first in 'Varieties of Necessity'.

need to make it plain in our contextual explanation or account that we are concerned, not with the imagined 'causes' of things in modern empiricism, such as events and experiences, but rather with the substances with active power enjoying what Locke and his predecessors thought of as real causal connection with their effects.

Events and experiences, in order to be conceived of as being in causal relation, have to be temporally discrete, and if time is continuous cannot be contiguous. Accordingly, they cannot be immediate to each other, since, as Aristotle showed, no multiplication of intermediates will restore immediacy.[8] Hence no 'real' causal action, as of one thing towards another, is possible so long as events and experiences are not conceived of within a context of agents or substances exercising active power. Upon reflection, it appears that causation or agency always involve (i) a prior persistent agent or exerciser of active power, (ii) an exercise of active power whereby the agent brings about or initiates the effect, (iii) some identifiable and normally persistent state or object, envisageable as resulting from the action of the cause. The existence of (i) does not normally logically imply (ii); and between states of the cause or agent, (i), and the existing or obtaining of (iii) commonly there is no logical connection. Accordingly, if causation is conceived of as a relation between events and/or states, then Hume's demonstration that there is no logical linkage between cause and effect is liable to hold.

These restrictions on which answers to "why"–questions, and which uses of the word "because", we need to attend to, and with what direction of interest (*viz.* towards the agent, and not just towards the event, or, if towards the event, only towards it as the action of some agent), still leave us with a concept of cause for which no adequate reason has been given for supposing it to be less wide than the concept of real existence—for example, no reason for supposing it to be restricted to the area of that which is known through the senses.

(c) *The variety of modes of causality*

Let us now fill out more fully our picture of the generality of the concept of cause or causal agent. If we consider the notion of causal

[8] *Posterior Analytics*, I, 20, cf. 22–3.

agent, which is no different from the notion of "exerciser of active power", since the achieving of an effect involves such exercise in order to bring about a change, then we find an enormous variety in the *modes* of exercise of active power, the *modes* of causation or agency.

Firstly, there is an exercise of agency according to the mode exemplified by knowing, desiring, and choosing beings. Then there are exercises of agency by living beings according to lesser degrees of knowledge as allow of the notion of desire but not of the notion of choice. Then there are exercises of agency by living beings which are not in accord with knowledge at all inasmuch as the term "knowledge" is not literally applicable to them, but which is still in accord with various kinds of teleology, as in plant and cell growth and regeneration. Then there is the most popular case, although this may seem to be the most degenerate, the case of agents which are inanimate and exercise active power in the manner of billiard balls.[9]

The received philosopher's picture of the supposedly pre-philosophical view of causation is that it consists of pushes and pulls whereby one tangible object exercises agency in regard to another. And in the wake of Cartesian dualism it has been fondly imagined that the person who rejects materialism is committed to supposing that, when a human being exercises agency, what is involved is a complex of such physical pushes and pulls preceded by a strategically placed non-physical 'push' whereby the mental acts upon the physical, the mental element in intentional action being a supposed 'volition'. In this way the conception of irreducibly different modes of agency has been lost sight of or repudiated in an attempt to reduce all agency to the material or mechanical model, or to mysterious mentalistic variants of this, introducing no new model.[10] It has been the task of philosophical psychology, as

[9] My concern is with the differences in what it is to be a causal agent in different cases, not with the criterion of identity of exercises of causal agency or 'causal acts'. The different modes of causal agency differ in the extent to which intentionality enters into the description of their exercises and this can affect ways of speaking of the identity of the action. A fully human (fully intentional) act will have many physical movements 'brute' to it and is not discrete from these movements, but the kinds of identity involved are not always strict, i.e. will not allow complete interchangeability of all relevant designating expressions *salve veritate*. The adage 'the agent does what happens' has to be interpreted with care.

[10] The insight to be found in Anscombe's discussion of knowledge without observation or Hampshire's discussion of intentional knowledge is presaged within

exemplified in Ryle and Wittgenstein,[11] and more recently Anscombe and Hampshire, among others, to exhibit the inadequacy of this way of viewing intentional action. Action is a particular mode of causal agency, to which intention is integral as a modal or adverbial aspect and in no way an external previous cause. It is as absurd to regard intention as merely an extrinsic cause of action as it would be to regard looking at a thing as merely an extrinsic cause of seeing it: all seeing is a 'seeing lookingly', and likewise all 'fully human' acting is an acting intentionally.[12]

When I speak of irreducibly different modes of 'causality' or 'causal agency', and speak of material or physical causation as a degenerate case, it is vital to seize fully the implications of what is being said. What is involved is at the same time a certain Aristotelian and holistic view of man and animals and a family of conceptions associated with Aristotle and Aquinas about the way in

modern times in Schopenhauer's account of our knowledge of the exercise of will, as described in David Hamlyn, *Schopenhauer*, pp. 82–3.

The conception that the 'I is not intimate with itself through and through' (Hamlyn, p. 84) in Schopenhauer is sound, but the idea that this limitation arises only from inner knowledge being subject to the form of time, as if causality was something that had only to do with the outer (along with space), is false. Unfortunately, for Schopenhauer causality belongs to the world of phenomena given in representations, a world which he supposes subject to the principle of sufficient reason. Hence, it remains that for him still agency is a matter, until intentions are observedly carried out, of a knowledge of mere intentions and 'therefore a matter of the intellect alone', a variety of observational self-consciousness. Hence it is fallible and rests in the intellect and observational self-consciousness alone, or it is so realized as to have its infallibility tied to whatever reliability there is in relation between states of the brain and subsequent action.

It has been said that there is nothing new under the Sun. Certainly in the association or identification of mind and brain there has been nothing new since the 17th century.

[11] The approaches to Ryle and Wittgenstein are in a way notably different. Ryle draws attention to the infinite regresses and absurdities to do with causation and identity in separating volitions from bodily movements and offers argument very reminiscent of Aristotle in his *Ethics* and *De Anima*. Wittgenstein occupies himself with particular examples in a way calculated to exhibit the absurdity of supposing that the language of intention, meaning, understanding, and thinking, any more than the language of pain, refers to some inner introspectible 'somewhat'. Both Ryle and Wittgenstein are engaged in setting forth criticisms which are above all directed against Hume—directed against the nonsense, for instance, of saying that 'the will' is 'nothing but the internal impression we feel and are conscious of when we knowingly give rise to any new motion of the body, or new perception of the mind' (Hume).

[12] This comparison was made by the author in his B.Phil. thesis at Oxford in 1965 entitled 'Explanation and Modality'.

which the same term may have a spread of use not covered by any single definition and yet be used 'properly' in all of them. It is not that there is some simple univocal common feature called 'causing' exhibited in purity in the material case of (e.g.) pushing, and that other cases exhibit additional non-causal features, but that the other cases exhibit other features which are still features in virtue of which the relation spoken of is to be called 'causal'. Nor is it that being a causal agent consists in a disjunction of possible alternative ways of being a causal agent, because, in the case of living bodily things, the causal features of agency are not separable. Somewhat as, in the Aristotelian picture, sensitive life did not exist separably from vegetative life, but was a different mode of life *qua* life, and a richer mode,[13] so it is in the case of causality. For instance, the intentional bringing about of an effect, which always involves a 'seeing to it' that the effect occur in situations at least slightly variable, includes a material element, but represents the provision of a causal explanation of the effect which is richer or more multi-aspected than a purely material explanation would purport to be. These characteristics of the concept of "causal agent" and similar concepts will prove crucial to our argument in Chapters VII and VIII.

Once it is realized that these are irreducibly different modes of agency, each of which is properly regarded as a case of 'agency' in the primary sense of that word, rather than in any secondary or transferred sense, there is no room for strange theories of agency such as have been proposed as a result of considering the mode exemplified in 'pushing' and 'pulling' as the only one. It is only this false starting-point which has allowed the development of theories in which all that is objectively knowable in causation consists in some conformity to law and everything else in the concept of cause arises from the insinuation of some perspective whereby the idea of necessity can be smuggled in, either the necessity of a consequence from premisses or the necessity of obedience to orders or to law.

By means of this presentation of some of the variety of modes of

[13] The translations as well as the commentaries show a discrepancy of interpretation. Hamlyn's commentary (p. 94 in the notes to his translation of Aristotle's *De Anima*) conforms to the idea that there is not one, univocally defined or fixed, proper genus to which different differentiae are added, but different forms of life not thus contained in a well-defined or identified genus. This probably accords with Aristotle's own understanding and intention even if it is not required by his text.

exercise of agency, we have given some content, of a non-formal kind, to our earlier remarks about the formal or transcendental generality of the concept of causal agent, the generality which left it open that it might be co-extensive with the 'notion' of real existence. We have also supplied what will prove to be the lynchpin of part of our discussion of the personal nature of the First Cause in Chapter VIII.

(d) *The problem of 'typical ambiguity' in causal principles*

It might appear that the lack of any demonstration that the ambit of the concept of cause was narrower than the ambit of the concept of "real existence" does not suffice to prove positively the co-extensiveness of the concepts. In one way this will be demonstrated in Chapter V through an examination of the 'concept' of "real existence". But, in another way, it has to be realized that for many purposes the question at issue is not that of the ambit or extension of the concepts, but of the range of application of principles whose statement hinges on the use of these concepts.

The fact that there are many modes of causal agency, so that no one set of criteria suffices to capture all of them, suggests that we should envisage that principles of causality and even the more restricted group of principles of agentive causality will be, in Russellian terminology,[14] typically ambiguous, or, in an older terminology, only of an analogical generality, not susceptible of any uniform kind of application. This was the root of the objection of Duns Scotus to the supposition that such concepts as "being" and "cause" were analogical, not univocal: he supposed that if these terms were not univocal, then there would be no principle using these terms which bestraddled the gulf between the finite and the infinite in such a way as to allow any argument from finite being to the existence or nature of God. And one of the most notable studies of the structure of Aqinas's thought in this century, *La participation dans la philosophie de S. Thomas d'Aquin*, by Louis-B. Geiger, by his argument, as if unconscious of what is involved, concedes this point. In Geiger's way of thinking the principles of causality

[14] Russell had the conception in *Principia Mathematica* that no generalization or law applied to things of different type without ambiguity, so that even the principles of logic and set-theory (e.g. the union of two disjoint sets each of two things contains four things) are 'typically ambiguous'.

will apply only analogously in different types of case and, as a result, it appears that the sense and validity of their application in each new type of case will need to be the object of a separate intuition or exercise of understanding: no adequate account is given of how the very possession of a certain concept (whatever is involved in the understanding or 'having' of that concept) might ground either a single principle (whose statement involved analogical concepts) or a spread of analogously related principles.

This problem is crippling to the possibility of any reaching back to God in metaphysics, or so-called natural theology, if this is conceived of as in the style of rationalists as proceeding from certain universal premisses such as the law of sufficient reason, understood as statable in terms of univocal concepts.

However, as we shall discover in Chapter VI, the structure of causal argument to God's existence is not at all that which the rationalist envisaged.

We do not in one sweep apprehend that every complex actuality needs a cause so as to have a univocal generalization covering both the case of the causation of the coming to be of temporal things by other temporal things and the case of the causation of the continuation of things whereby the future is given existence or of the coming of things into existence from nothing. Rather we apprehend the causation of temporal things by temporal things and grasp the need of a cause of the continuance of the temporal as a condition of that natural causation. The applications of a principle that the complex or composite needs a cause are analogous and all the concepts in terms of which such principles may be stated, concepts such as "positive existence", "causal agency", "actuality", and "potentiality", are analogical,[15] as Aristotle made plain.

[15] Aristotle treats 'positive existence' (i.e. existence within the categories) as a term whose uses are united by relation to a focal meaning and treats the notions of "actuality" and "potentiality" as used analogously. These are two different ways in which for Aquinas a term may be used 'analogically', since Aquinas's use of the notion of "analogy" covers both cases of use in relation to a focal meaning in which the term under consideration does not preserve its proper sense and cases of analogy in the sense of 'likeness' in which it does preserve its proper sense. In the text of my book when I use the term "analogical" I have not bothered with these elaborations, but I have had the case of analogous uses in which the proper sense of the term under consideration is preserved principally in mind. In my view, although there is a relationship as to a focal meaning between the uses of "real existence" of substances and of positively existing non-substances ('*hexeis*' or 'havings' as opposed to privations), the use of "exist" of different kinds of substance presents a case of analogous use rather than of focal meaning, so that in every case of special interest it is analogy in a restricted sense or 'likeness' which is concerned.

We know of the actual application of these concepts in respect of cases of secondary causality first, and from this know a posteriori of the coherence of the statements applying these concepts in these cases: nothing here waits upon any further reflection such might lead towards theological or cosmological conclusions. Then at a second stage we recognize that the conditions of this coherence include the operation of God's primary causality in causing the reliable continuance of the natures operative in secondary causality. What unites the different applications of the word "cause" is not a single univocally understood general proposition, but the understanding of a single analogically understood concept, an understanding realized in a variety of different types of application. Such is the shape of metaphysics explained in Chapter VI and vindicated against objection in Chapter VII.

IV

Third Investigation: Existence in Logical Theory

Our preoccupation in this book is programmatically to set forth a demonstration of the existence of one living God. This will be one God, knowable via metaphysics in respect of existence and life and not capable of being set in opposition to, or contrast with, the God of Abraham, Isaac, and Jacob. Accordingly, it is vital for our reflections to consider what is involved in the ascription of existence in a statement to some grammatical subject. We are well accustomed to ascribing existence, in a yet to be made clear metaphysical or strong sense, to the subjects of being acted upon and of change, susceptible of successive existence, possessed of boundaries, denizens of time and space. But it is vital to establish that, in the same strong sense, we can also ascribe existence to an agent who or which, although immediately active everywhere in time and space and therefore in one way properly said to be in time and space, is in no way localized spatially or temporally and in no way capable of being acted upon. For this is how God is supposed to be.

However, all our considerations of existence are liable to be vexed by the diversity of interest as between the logician and the metaphysician. Let us first consider existence in those respects in which it is of concern to the logician as such. We need to do this in order to clear the ground and get into a position to show that there does indeed exist such a metaphysical or strong sense of the word "exist" as I have referred to, and what this sense is.

Our task in this chapter is not to put forward some particular step in an argument to God's existence, but to remove some of the systematic obstacles to interpreting such an argument correctly, and to make clear the links between statements of existence in the 'strong' sense and causal agency. We need, in effect, to rewrite logical theory if metaphysics and epistemology are to be freed from the presumptions set by the logical theory exemplified in Russell and Quine, and freed from the unnecessary task of unpicking the extraordinary constructions engendered by them.

INTRODUCTORY CAVEAT

In considering the history of logic, we need to distinguish between logic in its strictly formal, symbolic, or mathematical side, and, on the other hand, what is variously called the philosophy of logic, logical theory, or the philosophy of language. By "logical theory" or the "philosophy of logic" I mean, for instance, the subject in which one discusses what is meant by meaning, denotation, and connotation, or one explains such distinctions as those between intensionality and extensionality, subject and predicate, general and singular statements, or one treats of truth and proof, identity and the nature of sets and numbers.

There may be a certain ambivalence in our judgements in regard to some notable figures, supremely in the case of Aristotle. Sometimes we find discussions in the philosophy of logic that anticipate in their quality, and even in their detail, the discoveries of the philosopher Frege, the discoveries which opened out the modern development of formal logic: the philosophical works of Aristotle, e.g. the *Categories* and the *Metaphysics*, chief among them, and stray remarks in Plato and in Aquinas provide examples of this.[1] On the other hand, we find works of formal logic of great stature which are yet innocent of any such philosophical sophistication. The formal logic of Aristotle, based on the notion of term rather than the notion of predicate (which figures more significantly in the philosophical works), is the supreme example of this. On the one hand, this work constituted the foundation of the subject of formal logic itself. On the other hand, because of its basic ideas, it was the source of the quagmire into which mediaeval logic[2] later plunged, so that the ingenuity and acuity of such men as Peter of Spain, Ockham, and Buridan had no lasting fruit.

Accordingly, when, in modern times, we meet mathematical

[1] The notion of predicate represents the fundamental idea which in Frege allowed the unification of the logic of general terms with both propositional logic and the logic of relations. Except that temporal indications might be omitted, the same notion holds a key place in Aristotle and Aquinas, prefigured in the *rhemata* which Plato contrasted with nouns (*onomata*) in the *Sophist*.

[2] The impossibility of encompassing relations or multiple quantification in a logic of terms and the confusions about identity resulting from thinking of general terms as having denotation and general statements as expressing kinds of identity were to represent two of the false difficulties experienced in this tradition—false difficulties illustrated in Geach's discussion 'Nominalism' in *Logic Matters* and in W. and M. Kneale, *The Development of Logic*.

logicians of the stature of Russell[3] and Quine, we must not assume that their quality as mathematical logicians will carry over into their discussions in the philosophy of logic or logical theory. Against this background we find it is unsurprising that there have grown up certain dogmas or commonplaces which do not stand up to much examination. The favourite among these is the dogma that the only existence is that which is expressed in formal logic by the existential quantifier.[4]

This dogma is open to objection on three grounds. Firstly, in associating existence with the existential quantifier, it ignores the presupposition of existence which arises whenever any quantifier whatsoever is used, *viz.* the presupposition that generalization or quantification is over a non-empty domain: it therefore overlooks the fact that the quantifiers cannot consistently be conceived to have one domain, but rather are separately used or applied in connection with each of a multiplicity of very different domains, universes of discourse, or subject-matters. Secondly, this dogma ignores the tensing, and sometimes the associated modality, which is found in the normal or paradigm cases of generalization or quantification. Thirdly, it ignores the interrelations of general and singular statements, and the primitiveness of the existential presuppositions associated with statements which are 'singular' in

[3] In the case of Russell, indeed one might say that the philosophical discrimination exhibited in his work in logic or logical theory underwent a steady decline from the high point at which it stood in 1903 in the work *Principles of Mathematics*. Already in 1905 in his article 'On Denoting', and in *Principia Mathematica*, Russell had begun to ignore or set aside some of the distinctions marked in ordinary language painstakingly observed in *Principles of Mathematics*, thereby losing even at that early date any potential for dealing with the problems of intentionality and ignoring the real possibility of considering finite non-empty collections as in a way consisting of their members rather than as functions or concepts, and, in the theory of complete symbols and logically proper names, giving a false place to ostension. The idea that a concept could be identified with the class which was its extension (i.e. the things which 'fell under' it or which it 'applied to') and given an extensional criterion of identity, all the while being no less a 'logical fiction' than other classes, opened out a short way and a cheap one of dealing with all abstract objects: e.g. a time was a class of events, the events happening at that time, as if to 'happen at' was a way of 'being a member of'.

[4] This dogma had been a commonplace for many years to judge from the attitudes of many philosophers, but has received notably direct expression from W. van O. Quine when he says, 'Existence is what existential quantification expresses' ('Existence and Quantification', p. 97), and has received a recent and ably argued defence by C. J. F. Williams in his *What is Existence?*—whose espousal of Quine is expressed on p. 152 and p. 294, although he purports to discern vacillation in Quine himself on p. 216.

the logicians' sense—existential presuppositions which cannot be analysed as if they were general propositions.

Let us consider each of these in turn.

1. THE MULTIPLICITY OF DOMAINS OF QUANTIFICATION

The first of our observations must be this. It is not legitimate or coherent to take as one's domain of generalization or quantification 'all things' but always necessary to take some more limited concept as marking such domain. And from this it follows inexorably that a distinction will have to be made, in respect of statements within such an area of discourse, between existential statements using the existential quantifier, expressing the instantiation of some property within that domain, and existential statements expressive of the non-emptiness of the domain.

Let us first explain the point more informally.

If I say "There are tame tigers", it is plain that I have made a statement which has the same truth conditions as the statement "Some tigers are tame".[5] In the latter statement I appear no longer to be asserting the existence of tame tigers but to be, in some (at this stage as yet unexplained) sense, *presupposing* the existence of tigers. If I insist nonetheless that there is no other kind of existence or sense of the word "existence" beyond that kind or sense which is expressed by the use of the existential quantifier, and that existence is not tied to any predicate or class term such as "tiger", I may alter "Some tigers are tame" to either "Some animals are both tame and tigers" or, if you prefer to formulate it this way, "There are animals which are both tame and tigers". I began by seeming to associate existence with tame tigers; I then seemed to associate it with tigers; I now seem to associate it with animals. The advocates of the dogma that the only existence there is is that which is expressed by the existential quantifier will wish to drive me yet further, *viz.* to say "There are things which are animals and are tame and are tigers", and will tell me that the concept of "thing" is a non-concept (intensionally empty) or a formal[6] concept. However, at this point,

[5] The "tame tigers" example was the one taken in G. E. Moore's classic discussion in 'Is Existence a Predicate?' (from a Symposium with W. Kneale).

[6] Wittgenstein, *Tractatus* 4.126 to 4.1274, tells us that formal concepts are not concepts proper and that every variable is the sign for a formal concept. He appears

the suggested dogma has led its protagonists into absurdity, as if existential statements said nothing about the world directly but said only that, whether in some abstract or non-abstract domain, there were objects in regard to which certain modes of speech were appropriate (subject-expressions and quantification), and which proceeded to assert some predicates of these things, as if existential statements were a peculiar *mélange* of the metalinguistic or second-order and the first-order.

Let us now move towards a more formal or precise explanation of the point at issue.

(a) General existential statements: two functions and two types of meaning

Some people may, of course, prefer different English idioms. Instead of saying, "Some things are both tame and tigers" or "There are things which are both tame and tigers", they may prefer to say, "Of some thing, it holds that it is tame and it is a tiger" which they may reformulate, "Of some thing, x, x is tame and x is a tiger", possibly preferring this presentation to "There is a thing, x, such that x is tame and x is a tiger", which still seems to associate existence and even tensed existence with things (or, in variants of earlier versions, with animals). But no point of substance is at issue in these further changes of formulation. And against all these variants, the objections I am making hold equally.

Against me the counter-objection may be made that, whatever kind of existence is concerned in such general statements, whether of tigers, or of animals, or of some wider class, this can be and is still properly expressible using the existential quantifier. That is, it is alleged that one can always introduce a predicate term that applies to every member of the domain of discourse with which one is dealing, say the predicate "F" or "$F \ldots$", and assert "Something is F", or in modern style "Of something, x, Fx", i.e. formally "$(\exists x)(Fx)$", to convey that the domain is not empty, that there are F's, even though "F" is a concept of this peculiar kind.

I say 'of this peculiar kind' because for the required *meaning* to be conveyed it is not enough that "F" be a term which just

to have in mind what would later be spoken of as metalinguistic concept: e.g. an 'object' will be that which is, in the appropriate sense, picked out by a subject-expression or quantified over.

accidentally happens to apply to all and only the members of the domain concerned, but rather it is required precisely that "*F*" express membership of the domain concerned. In other words,"*F*" is more like what Wittgenstein called a formal concept, than a proper concept, although in fact there are more distinctions than one to be made where Wittgenstein in the *Tractatus* made only one under this heading, "formal/proper".

"*F*" will be a term which, in effect, for the purpose and meaning concerned, it would be more appropriate to attach as a subscript within a quantifier than to allow to appear within the scope governed by the quantifier. If one expresses "Some tigers are tame" by saying "Of some animal, *x*, *x* is a tiger and *x* is tame", or "Of some animal, *x*, *x* is a tame tiger", and makes some such mode of expression the basis of one's understanding the symbolism involved in the use of the existential quantifier in modern predicate logic, then by this act one has debarred oneself from treating "Animals exist" as having the same type of analysis as "Tame tigers exist"—"There are animals" as "There are tame tigers"—or at least from treating them as having the same type of analysis within the same context of application of formal logic, i.e. within the same 'formalised language'.

There exists an illusion that because people can learn to apply the formalism of modern symbolic logic within a context of habituation to sloppy or absurd presentations of its meaning and readings of its formulae, it follows that these presentations and readings must be O.K. after all. In this way, it is thought that notions such as "ambiguous denoting" and readings such as "Of some *x*, *Fx*" or "There exists an *x* such that *Fx*" along with corresponding types of reading in the case of the universal quantifier are O.K., as it is said, 'because the language-game is established', whereas in fact they are parasitic upon explanations in other terms, likewise drawing upon natural language, but not essentially dependent upon a misuse of it. I mean, for instance, the misuse whereby "*x*" appears first in a context requiring a general term, and second, in apposition with it, as an index-marker or dummy singular term or 'name' in an argument place, in the mathematical sense of the word "argument". "Of some *x*, *Fx*" is neither flesh nor fowl, nor even a respectable hybrid; surreptitiously, in the "some *x*", either "*x*" is being thought as if it were a general term and the "*x*" in "*Fx*" is being left hanging loose, or a general concept (covering the domain of generalization) is being thought but not expressed, so the thought is "some thing/animal, e.g. *x*, . . .".

It is not that there is nothing in common between "There are substances", "There are animals", "There are tigers" and "There are tame tigers". But it is a mistake to suppose that what these have in common is 'what is expressed by the existential quantifier'. By means of the existential quantifier, presupposing the non-emptiness of some particular domain of logical subjects of predication ('things'), we assert that at least one thing in this domain (this is compatible with many or all) has a certain property. More accurately, we do not say "At least one thing in the domain of the things which are *F* has the property of *G*-ness" but "Of at least one *F*, e.g. *x*, *Gx*", e.g. "Of at least one animal, *x*, *x* is herbivorous" rather than "At least one thing in the domain of animals has the property of herbivorousness".

Accordingly, some general statements of existence are most appropriately represented as asserting the instantiation of some property, from some point of view accidental, within a domain whose non-emptiness is presupposed. These have their sense most appropriately represented by means of the use of the existential quantifier (*prenex*, i.e. having the whole statement within its scope), the non-emptiness of the domain being presupposed by the statement, rather than its direct intent. The verb in these cases is not contained in the quantifier, but rather signifies instantiation than existence as such.

But, conversely, other general statements of existence, given the context which is partly indicative of their sense or intent, are rather concerned with the non-emptiness of a certain domain than with the instantiation of properties by logical subjects ('things') within that domain. This is a non-emptiness presupposed as much by singular as by general statements: e.g. if a man says "Celestial bodies exist", he is saying something as much presupposed in singular statements about the Sun and other named celestial bodies as in any general statements to the effect that some or all celestial bodies have this or that property.

The person who makes this last-mentioned statement might, of course, be occupied with the logical or philosophical point that some bodies are celestial, rather than all bodies being earthly. But more commonly he will not be occupied with the possibility that there are no celestial bodies, as if to inquire whether celestial bodies have no better status than the Sky (conceived of as a material dome), phlogiston, centaurs, and dragons. Rather we will be stating the subject-matter routinely presupposed in some piece or area of discourse, e.g. astronomy. By contrast, reverting to the case of supposed tame tigers, although it might be that some wild theoretician was concerned to discuss, as if it were a clearly established group, the situation of domestic tigers, and as a preliminary routinely to assert their existence as a group, it is more likely that context will exhibit that the person who asserts the existence of tame tigers is not intending to register an established domain of inquiry but to mark the perhaps quirkish instantiation of a certain property, *viz.* tameness, within an established group, *viz.* tigers.

Many existential statements, according to context, will invite different interpretations in the different contexts: sometimes what is involved will be a empirical or metaphysical interest in kinds of being already agreed, or presupposed in the discussion concerned, to exist, noting this presupposition; at other times, a logical or philosophical inquiry may be under way within which the existence of the same kinds of being is not presupposed, but, in the spirit of academia or the pursuit of truth or at least of logically approvable classification, questions are being raised as to whether this or that genus of a rather general kind exists.

(b) *Universal statements*

The manoeuvre of representing "Some tigers are tame" in the end as meaning "Of some thing, *x*, *x* is a tiger and *x* is tame" or "There is a thing, *x*, such that *x* is a tiger and *x* is tame", so that in every general statement of the kind technically called 'existential' or 'particular',[7] the generalization is over the domain of all 'things', recurs with universal statements.

On the face of it, "Tigers belong to the cat family", "Every tiger is a cat", and "All tigers are cats" imply or presuppose the existence of tigers. However, if we construe them as meaning "Every animal, if it is a tiger, is a cat", we are at the same time making a more ambitious or grandiose claim about all animals and involving

[7] In Aristotle 'according to a part' is used as a technical expression for statements of the form "Some *A* is *F*", i.e. statements which imply that some part or other of a class *A* has the property *F*.

ourselves in a smaller commitment in respect of existence: i.e. we now appear only to commit ourselves to the existence of some animal or other, and no longer to the existence of tigers as such. However, if we are led by Frege (before he knew of the existence of the logical paradoxes) and Russell (who attempted to remove the theory of types dealing with the logical paradoxes into the area of the linguistically second-order, what would be later called meta-theory[8]), we will construe the original statements as meaning "Anything, if it is a tiger, is a cat" and will again have landed ourselves in the absurdities involved in attempting to generalize about all things or subjects of discourse at once. (If 'logical type' is a metaconcept, not belonging to first-order theory or the object-language, we can presume the existence of a corresponding coextensive first-order concept applying to the same objects, what Kant would have called a schema.[9]) The question as to whether or not, in order to make our symbolic representations correspond to the ordinary meaning in speech and writing of such statements as "Every tiger is a cat", there is a need to supplement "Anything, if it is a tiger, is a cat" with the statement "There are tigers" is entirely subsidiary, a point of secondary interest arising from the consideration of what might appear accidents of natural language. The main point remains that of the absurdity of generalizing about 'all things', an absurdity which exhibits the rationale underlying what might otherwise appear merely accidental in natural language.

We can say this with certainty because it is an implication of the discovery of the logical paradoxes (of which Russell's is the best known among philosophers, and Cantor's perhaps the most fecund in the possible significance of the way it arises and in the directions in which it points) that any attempt to quantify over all things or

[8] Russell's view seems to have been that statements of differences of type between entities were statements about the terms for or about the propositions (singular or general) in regard to the entities in question, rather than about the entities themselves. His theory of orders was separately developed to deal with the semantic paradoxes. Both involved language about language within a single language or theory, but both have the consequence in that no part of logic can be applied to all logical subjects or all propositions. However, by means of a theory representing all logical subjects except certain basic ones ('individuals') either as referring to fictions of various different types (all ranked as classes) or else as intended to refer to or generalize about linguistic entities, there was left behind a residual sense in which logic applied at once to everything.

[9] Kant, *Critique of Pure Reason*, p. 181.

non-imaginary, non-fictitious, non-ideal, subjects of discourse as such will generate paradox. And, besides the so-called logical paradoxes arising if any attempt is made to assess the number of these things or to consider the relation of these things to sets or collections of them, we have the so-called semantic paradoxes if attempt is made to discuss the meaning or truth of statements concerning them. Accordingly, the attempt to generalize about all things or all subjects of discourse is quickly self-defeating.

Therefore one must insist that all statements within a subject-matter, whether general or singular, and, if general, whether universal or existential, presuppose the existence, not just of some subject of discourse or other, but of the very subject-matter which the statement concerned has to do with.

(c) *The nature of logical laws*

The system of logical laws should be thought of as being, as it were, a hat which fits isomorphically on to each of a multiplicity of different subject-matters, i.e. applies isomorphically to each of them. In this generality, they should be thought of as including the laws of set theory (including there the laws of natural arithmetic) and the laws of the theory of relations and order, which like the laws of propositional and predicate logic apply isomorphically to each of those subject-matters to which they do apply, granted that whether certain of these laws do have application or not to a subject-matter may be affected by the cardinality or number of the objects comprised in that subject-matter. Each subject-matter carries with it its own family of types of statement and its own family of interrelated, different sorts of object, which, in a many-sorted theory, will not lie all of them within the range of significance of the same predicates.

There is a role for a general theory of sorts or logical types, which need not be given the basis Russell suggested (i.e. the vicious circle principle need not be one's criterion for determining the range of significance of predicates). And there is also a role for statements such as "An individual is not a set", "A set is a set", within a general logic, but such statements have to be handled with care and do not themselves conform to all the laws which belong to this general logic, i.e. to the laws of logic as an integrated group, geared to application to each of many subject-matters, but only to some more limited sub-group of them.[10]

2. THE IMPORTANCE OF TENSE

There is, in the dogma that the only existence is that which is expressed by the existential quantifier, a self-defeating blind eye turned to the problems of tense and of the distinction between the extensional and the non-extensional. Logicians tend to treat quantifiers as if for practical purposes[11] they were tenseless, to be interpreted, for instance, "there is, was, or will be", in the case of the existential quantifier. But, if they understand the resulting quantified propositions extensionally, as if they were equivalent to (i.e. the same in truth-value as) some infinite or finite conjunction, in the case of universal propositions, or equivalent to some infinite or finite disjunction, in the case of existential propositions, then they have the problem that the status of some of the subjects of the supposed disjoined or conjoined propositions are future individuals

[10] There are many precedents for this. For instance, in Von Neumann set theory, there are laws which apply to those classes which he calls sets, but not to all classes as such (i.e. not to those classes which are not sets and which he calls proper classes). Again one can have a logic which embraces all the common laws concerning "and", "or", and some sort of "if . . ., — —", so far as these laws do not involve negation, what we might be tempted to call an 'implicative logic', whether or not this is what Griss intended in his negation-free variant of intuitionist logic. Or again, one may conceive of a logic which adds to this 'implicative logic' some but not all the common laws concerning negation, e.g. the disjunctive syllogism ("p or q; Not-q; Therefore, p") and *modus tollens* ("Not-q; If p, q; Therefore, not-p"), perhaps one form of *reductio ad absurdum* ("p implies *Absurdity*; Therefore, not-p"), but not the law of the Excluded Middle and not the common-sensically parallel form of *reductio* ("Not-p implies *Absurdity*; Therefore, p") - whether or not this logic is conceived of along the lines of the doctrines of the 'intuitionists' Brouwer, Heyting, and their school.

[11] My phrase 'for practical purposes' is intended to give a half-hearted nod towards the existence of efforts on the part of logicians to perform face-saving operations, not intended to infect the presentation of logic in teaching manuals, in order to preserve some semblance of truth or plausibility. Supposedly, the quantifiers are either omnitemporal so that they mean (e.g.) "there is, was, or will be some . . . (perhaps, 'thing'), x, such that Fx", or else they have no internal tensing, so as to be expressed better (e.g.) by "Of some. . .(perhaps, 'thing'), x, Fx" so that there is no tensing except within the sense of the "F . . .". Russell and Quine have theories of tense compatible with the above, and the normal presumption in teaching is that the problem may be ignored. My generalization (I used the word "tend") does not of course apply to A. N. Prior or Lukasiewicz, or to a multiplicity of logicians in pre-modern times.

whose status is at present only that of possible existents. Hence, unless they assume that logic (on the face of it a branch of human knowledge and inquiry) can only be sensibly conducted from a Boethian or Platonic 'God's eye' point of view as if time were unreal, and as if the distinction between the possible and the actual were a matter only of causal or epistemic relation to the state of the inquirer, they are involved in offering a strangely out of the way view of the quantifiers.

In all this, they are offering a view which conforms neither to the needs of thought about ordinary past and present individuals, nor to the needs of logic as applied to mathematics. (Bringing time into the exposition of intuitionistic logic represents a popularization, disguising rather than revealing the status which the intuitionist philosophers of mathematics accord to real numbers and to the non-denumerable infinite.[12]) Nor does it represent any requirement on logic as applied to laws in physical and biological science where omnitemporality is never asserted without an implication of necessity arising from 'nature', and associated with application to future possible non-actual as well as future possible actual instances.

Accordingly, if there is to be a logic which has application, not only to areas, such as that of natural numbers, to which time is irrelevant, but also to areas to which time is relevant, then this general logic must allow for uses of quantifiers and of the word "exist" which are internally tensed. And it cannot be right to regard non-ideal or real existence as an accident as Avicenna did, let alone as an accident in the category of time,[13] a suggestion recurring in Geach but well rebutted by Dominic Banez in the sixteenth century.[14] (Geach's attempt to explain real or metaphysical existence as opposed to the existence presupposed whenever a proper name is used and any 'actuality' or existence directly associated with the quantifiers, by reference to tense, must be a

[12] The intuitionist school of mathematicians require that what exists, at least in the case of mathematical objects, should be definable or identifiable constructively, and think of such definitions in mathematics as typically formed generatively or progressively. This requirement is presented for the purposes of the imagination in temporal terms. Cf. A. Heyting, *Intuitionism: An Introduction*.

[13] Étienne Gilson, *Being and Some Philosophers*, pp. 77–80, cf. 88.

[14] Dominic Banez, *Scholastica commentaria in primum partem angelici doctoris*, 1584: translated in *The Primacy of Existence in Thomas Aquinas*, pp. 31 f. and pp. 40 f.

mistake. In this connection, one may in any case note that blindness can come and go so that things whose existence consists essentially in the absence of something positive, e.g. also holes, may have temporal existence without having the real existence or the existence as other than a mere subject of discourse which Geach and others are in pursuit of.[15])

The statements of arithmetic are supposed to present a clear case of a use of the copula, in both singular and general statements, to which time is irrelevant. However, they do not constitute any primary or paradigm case, either of singular or of general statements, since their very meaning is such as to be parasitic upon or to presuppose actual or possible application in the assessment and comparison of sets, normally (albeit not necessarily[16]) sets of

[15] Geach wants to explain that sort of existence ('present actuality' in *Three Philosophers*, pp. 90–1) which he regards as quite different from the existence expressed by "there is" (supposedly expressed by the existential quantifier) as the persistence of some form in an individual (cf. pp. 58–60, in the context of pp. 53–60 of 'Form and Existence', 1954/55, as reprinted in *God and Soul*). He presumes the same distinction in 'What Actually exists' (cf. pp. 65–6, as reprinted in the same collection). Unsurprisingly, granted this background of thought, he wishes elsewhere to explain God's eternity in terms of a special kind of immutability and everlasting duration, and has special problems with God's foreknowledge. However, to make the predication of temporal activity explanatory of actuality is a fundamental mistake—as if actuality, any more than what is expressed by "there is . . .", was a predicate.

Moreover, the explanation given leaves him prey to queries as to which things count as 'individuals' for which (or whom) persistence (or, rather, the persistence in them of the same 'form') would constitute actuality. In informal discussions, he represented speaking of holes, blindness, or darkness as individuals (particular instances are what are under consideration) as 'not a proper way of speaking': but it is coherent, although paraphrasable and perhaps dependent on this for its intelligibility, and the *root* of what debars these things from being considered actualities does not lie in an absence of temporal predication. Augustine, in explaining how evil is a privation in *On Free-will* and elsewhere, and Aquinas in his treatment of evil as a privation, both alike deny that privations or absences are any sort of being or nature (and I understand from G. E. M. Anscombe that the same point is strongly made by Anselm in discussing *nihil*, *malum* and, as an example, blindness in *The Fall of the Devil*: but none would explain nature or positive being primarily by reference to temporality; in every case, a connection with the requirements for the possession of active power would be more significant.

[16] Frege emphasized that we can count things of any sort, angels as well as men, numbers as well as anything in anyway concrete. Evidently we count groups and sets whether of concrete or of abstract things, types as well as tokens and sometimes it is not clear whether we are occupied with tokens proper, (e.g.) we may count the words in an article or the accents in a line of blank verse, without strictly counting types, but not individuating even editions, let alone copies of the work in question, and so the so-called 'tokens' are not particulars in any metaphysical sense. Meanwhile many things commonly called concrete are not combinations of form

concrete things. True, the person who grasps how the natural number system goes on and on might be in bed seeing how high he or she can count up to that night, without being occupied in counting anything. But the number system goes on in this way only and precisely because at each stage it is understood that given that a class of n things of a certain sort is possible, there could be a class with just one more, i.e. a class with $n+1$ things of the sort concerned (accidental obstacles arising in the context of this sort apart); so the sense of counting is still relative to possible application.

In any case, the significance of general statements in arithmetic, so far from presenting the clarity of situation appropriate to a paradigm case, is, and not surprisingly, a matter of doubt. The view that numbers have an existence 'out there', and in their infinite number exist straightforwardly in extension, is anciently and modernly countered by the view that the numbers have only a potential infinity and should not be thought of as objects, existing in their sets, independently of application, applicability, or thought. This query can be traced in Aristotle and has received vigorous restatement in the thought of modern mathematicians, both the intuitionists and their precursors.

Yet what is decisive is this: the singular statements of natural arithmetic, such as "two plus two", need to be understood schematically as general laws, e.g. "two things and two things are four things". (More precisely stated in ways guarding individuality, disjointness, and a concern with logic, not with temporal processes such as putting things physically together: i.e. stated, for example, in such a way as "the union of two disjoint pairs is a quartet".) As general laws they have a generality which gives them application to abstract as well as concrete subject-matters, imaginary as well as

and matter in the way bodily things are (i.e. not concrete in the Aristotelian sense), or not clearly so, e.g. noises (since, although matter, *viz.* air, as well as pattern, is involved, it does not seem that the relationship of the air to the noise is the same as that of bricks and bodily parts to houses or animals). However, nothing in my argument turns upon the word "normally", but only upon the gearing of statements about numbers to the possible application in the same sphere of numbers.

It belongs to a quite separate argument, not here relevant, to establish that all the areas of possible application for the number system are in some way dependent, in respect of their existence, nature, and identity, on some relation to substances, unless it be that the objects of sciences which are intrinsically topic-neutral (objects such as numbers and truth-functions) have an 'existence' involving, but not depending upon or consisting in, the possibility of relationship to substances.

real worlds, and future possible as well as future actual worlds, although never any two of these at the same time. Accordingly, it seems incorrect to understand their meaning in purely extensional terms.

Extensional statements are primary, only because extensional singular statements are primary, not because universal statements of an extensional kind have priority over universal statements with an omnitemporal or modal interpretation. It is these singular statements to which I now wish to turn my attention, along with their presuppositions of existence.

3. THE PRIMACY IN LOGIC OF SINGULAR STATEMENTS

The dogma that the only existence is that expressed by the existential quantifier is myopic in a further respect.

All general or quantified statements, whether 'universal' or 'existential', i.e. whether they be of the sort "Every *A* is *F*", or the sort "Some *A* is *F*", have intrinsic to their sense a complexity greater than that of singular statements of the sort "*a* is *F*". It may be that the only way of telling whether or not a statement of the form "*a* is *F*" is to be interpreted extensionally or non-extensionally is to examine whether or not it is implied by some universal statement of the sort "Every *A* is *F*" or to examine whether it implies a particular statement of the sort "Some *A* is *F*". But it remains that it is impossible for anyone to understand these two types of general statements except as, respectively, implying, or implied by, the appropriate sorts of singular statement of the sort "*a* is *F*", understood under the presupposition that the object denoted by "*a*" belongs to the class "*A*". And this is not a matter simply of the conditions of learning but of internal and contemporaneous conditions intrinsic to understanding as such.

One may add, any class of general statements presupposes a certain range of possible singular statements with their singular presuppositions so that to analyse these last in terms of the same class of general statements involves vicious circularity.

Therefore, the philosopher who says that 'to be' is to be the value of a bound variable,[17] or who represents existence as a presupposi-

[17] The meaning and logical syntax of this saying is obscure, but on the face of it, since, given a statement "(*x*) (if *x* is a man, *x* is mortal)", it would be natural to take

tion of singular statements, besides not having escaped the problems of tense already mentioned, has undermined the possibility of truth in the view that the only existence of concern to the logician is that expressed by the quantifiers. On the contrary, every use of the quantifiers presupposes, not only the non-emptiness of some limited or special domain of discourse, but also, insofar as it presupposes that this domain of discourse is not empty, the applicability of the term 'existence' to the members of this domain and the primacy of this applicability over whatever use or uses of the word "exist" are associated with quantifiers—i.e. it presupposes what one may naturally call 'singular existential statements'.

By the expression "singular existential statement", I mean a statement in which existence is grammatically predicated of what is logically, not just grammatically, a singular term, e.g. "Socrates existed", which is the existential presupposition of both "Socrates was bald" and "Socrates was not bald". Obviously, I am committed to the rejection of the doctrine that all existential statements are general statements (general statements of the subspecies traditionally called 'particular' to contrast them with those that were labelled 'universal'), and uninterested in the grammatical distinction between "Some man is bald" as grammatically singular and "Some men are bald" as plural. As general statements presuppose the non-emptiness of some domain of generalization (the interesting or important kind of 'general existential statement'), so singular subject-predicate statements presuppose singular existential statements.

In fact, general existential statements of the kind merely expressed by, rather than presupposed by, statements governed by the existential quantifier are of all existential statements the least interesting, important rather for the confusions which failure to analyse them properly generates than for their own sake—for it is only when quantification or generalization is properly analysed and understood that the roles of domains of generalization and of 'objects' as the values of bound variables or subject of singular statements can be fully appreciated. But the instantiation of a property in a domain or by an object should not be allowed to cast into the shadow the existence of the domain or of the object which it presupposes. In brief, the achievement of modern predicate logic,

individuals, e.g. Socrates, Plato, etc, as amongst values of the bound variable concerned, the saying would seem to associate 'being' with these individuals.

thanks to Frege, in giving a clearly differentiating analysis of propositions expressing the instantiation of properties (singular propositions and quantified or general propositions), was indispensable to any later possibility of understanding the more underlying kinds of existential statement: but to suppose, even with some feeling of triumph, that 'the existential sense of "be" is definable in terms of instantiation'[18] is to exercise the special capacity for missing the main point exhibited by some of those dazzled by formal logic—at the same time making the refuting of them a task which threatens almost endless logic-chopping (of the sort experienced by those who reject the ontological argument).

It is not often realized that the doctrine that to be is to be the value of a bound variable gives primacy to singular existential statements, and is quite opposite to the doctrine that the only existence is that expressed by the quantifiers. I note that the possibility of empty domains of discourse in the case of universal statements, besides being of questionable coherence, makes no difference to the need to distinguish between (*a*) the non-emptiness of a domain, (*b*) the possession of a property by something in a domain, and (*c*) the existence of a designated member of a domain.

SUMMARY OF THE POSITION

Thus, besides the existential quantifiers, we have also uneliminably general statements of the existence of classes or domains of discourse, and, as well as these, presupposed by them (since they of their nature imply the truth of some singular statements or other affirming the existence of some members of the domains of discourse concerned), singular statements of existence, some of the most fundamental of which are uneliminably tensed. Moreover, this tensing of the existence in the singular statement does not turn the singular statement into a subject-predicate statement in the category of time. Rather the tensing is part of the presupposition, as much as the existence is part of the presupposition, of the use of the term that figures as *grammatical* subject of the singular existential statement and which appears as the *logical* subject of various other properly subject-predicate statements in the logician's sense of the

[18] C. J. W. Williams, *What is Existence?*, p. 294.

word predicate as utilized by Aristotle, Aquinas, and Frege.[19]

The notion of presupposition might appear to have been smuggled into the discussions of logicians and philosophers on the basis of a merely linguistic excuse, highlighted in Strawson's criticism in 'On Referring' in *Mind*, 1950, of Russell's supposedly seminal article 'On Denoting' in *Mind*, 1905. But, however unsatisfactory Strawson's arguments may have seemed to some (including Russell), two things remain perfectly plain as preliminaries, and, once recognized, leave the propriety of some such notion in logical theory indisputable.

Firstly, a presupposition is as much a truth condition of a proposition as is an implication, inasmuch as the non-truth of either carries with it the non-truth of the proposition itself. Secondly, there is evidently a distinction between what a proposition or statement expresses (says or asserts) and what it presumes as agreed between speaker and hearer and which it therefore does not present as the point at issue, quite apart from other distinctions between what the asserter of the proposition in the proposition itself is, by the 'illocutionary force' (J. L. Austin[20]) or the 'meaning' of the 'statement' made ('sentence' in Gilbert Ryle[21]), centrally or principally saying, and implications which are either subsidiary or ones which arise from relations with other propositions.

But these preliminaries make it quite evident that there is plenty of room, in logic, not just in psychology or linguistics, for a distinction between "presupposition" and "implication". The trouble in early modern discussion, as initiated by Strawson, was that, not only was this not made plain, but also the attempt was made to smuggle negation into logic on the back of the notion of presupposition, instead of vice versa.

The notion of presupposition here involved is the notion of a

[19] Cf. J. Thomson, 'Is Existence a Predicate?'. I also note P. T. Geach's exposition of Aquinas's notion that words like "some" and "every" indicate how a predicate applies to a subject (*Reference and Generality*, pp. 201 f. and 209). This dovetails with Geach's decision, following W. E. Johnson, to call such words 'applicatives' (Geach, *Reference and Generality*, p. 73 f., Johnson, *Logic*, vol. ii, p. 97).

[20] J. L. Austin, *How to Do Things with Words*. The illocutionary force as distinct from the perlocutionary is part of the public sense of the statement in question.

[21] The way in which Gilbert Ryle uses the word "sentence" to cover the things other philosophers refer to as 'statements' or (as some use the word) 'propositions' can be seen in early papers such as 'Categories' (1938) as well as late papers such as 'Use, Usage and Meaning' (1961)—reprinted as 12 and 31 in *Collected Papers*, vol. ii.

statement whose truth is a truth-condition both of the statement presupposing it and of the negation of that statement (and along with the negation, statements equivalent to the negation, so-called contradictories as distinct from mere contraries). The notion of negation is not a merely linguistic one and cannot be explained without circularity in terms of the notion of falsehood since any definition of falsehood has to include the notion of *not* being true. And, if the notion of negation belongs to logic, and not just to grammar, then so does the notion of presupposition.

Accordingly we are in a position to dismiss as having been refuted those philosophers who tell us that the only existence to be explained or dealt with by the logical theorist is that expressed by the existential quantifier. But we have on our hands the task of explaining the sense or indicating the sense of singular existential statements.

Upon reflection, if we review the way in which singular existential statements arise within each different subject-matter or domain of discourse, we must conclude that the role of this species of presupposition is in each case to make a claim that the thing to which existence is ascribed is non-mythical, non-ideal (contrast ideal points and lines in projective geometry), non-imaginary, non-fictitious, etc. That is, it is in brief to mark a claim that the thing concerned is a genuine subject or topic of discourse, (i) about which questions are to be raised with affirmative or negative answers, and (ii) predications about which will be implied by universal generalizations about the domain of discourse concerned and will imply existential generalizations about the domain, and (iii) which are the proper subjects of determinately true or determinately false identity statements (by contrast with merely possible or hypothetical 'individuals'), and therefore possible objects of counting and being collected together into sets. Thus, by contrast, for instance, even if one concedes that in *some* sense "Pegasus was a horse", "No horse is winged" will not imply "Pegasus was not winged" and "Pegasus was not winged" will not imply "Some horse was winged". This matter, of being a genuine or 'actual' subject of discourse, falling within some well-established domain, is exemplified as much by abstract as by so-called concrete objects, as much by numbers, colours, regions of space, and moments and periods of time as by the changing and causally acting things which we prefer to consider real or concrete.

There is room for a theory of intentional objects, i.e. objects towards which thought, imagination or desire are directed. Within such a theory there may be a place for such statements as "Pegasus was a winged horse", "Cerberus was a three-headed dog", and such like, and, along with these, certain seemingly paradoxical uses of the phrases "there is" and "there are", such as "There are objects of belief, such as Apollo, the Styx, and Pegasus, which do not exist" and "There are kinds of thing, such as centaurs, gryphons, and dragons, which do not and never have existed". However, this theory of intentional objects, in Brentano's sense of the word "intentional" (quite different from Aquinas's meaning for the word), ought not to be the basic framework within which the logic of that which exists is built up as a substructure, as perhaps happened in the theory of Meinong: nor should the basis preferred be a theory which allows empty singular names a status on a par with non-empty ones, an approach associated with Lesnieski. Rather it ought to be developed as an excrescence parasitic upon the logic of that which exists backed up by theories of human speech, intention, and imagination.

Thus, "Cerberus was a three-headed dog" may be explained as meaning "In the Greek system of belief, there was believed to be a three-headed dog called Cerberus". Other such glosses may be needed to deal with Mr. Pickwick, points at infinity in projective geometry, and so on. It may be that all such cases can be dealt with by a theory according to which adequate explicitness will always reveal the expression for the intentional object as occurring within an embedded sentence. Accordingly, it will be developed within a theory of non-truth-functional 'operators' on 'propositions' (sometimes called non-extensional embeddings of propositions)—although there is a further question as to the nature and interpretation of any such supposed 'quantification' over propositions, and as to the nature of 'propositions'. Or it may be that we sometimes attend directly to a term according to its sense (also in common modern custom called by logicians its 'intension').

In either case our theory of intentional objects will involve a second-order and perhaps metalinguistic element in the meaning of particular statements about particular intentional objects. It may be that the objects of imagination and desire, 'ideas' in Frege's classification,[22] cannot be dealt with in this way, but have a being

[22] Frege, "The Thought".

which simply consists in being objects of imagination or desire. But all such speculation is secondary within logical theory, and I mention it here only to indicate its secondary status—and also to indicate its entirely different character from the main line of metaphysical discussion to which I am about to turn my attention.

Aquinas distinguishes, on the one hand, an *esse*, existence or being, which is associated with any use of the copula whatsoever, differentiated according to the different uses of the copula, and, on the other hand, an *esse* which is also describable as an *actus essendi*, a so-called 'act' of being.[23] The first is that which is the concern of the logician as such, bearing in mind that the logician distinguishes predications in a strict sense in which existence is presupposed, and which has sometimes been supposed to be associated with the applicability to the subject in question of at least some affirmative predicates, in some privileged sense of the word "affirmative", from 'predications' in some looser sense according to which even intentional objects are subjects of predication. The second, or *actus essendi*, is not called an 'act' in virtue of being something done by some agent but in virtue of its being the opposite of potentiality, signifying either the actualization of a potentiality, or the absence of potentiality. It is this second sense or use of the words *esse* or existence which remains obscure and which we now need to pursue.

[23] *S. Th.*, Ia, Q. 3, Art. 4, resp. and ad 2.

V

Fourth Investigation: Existence in Metaphysics

PART I. THE EXISTENCE OF SUBSTANCES

1. THE CENTRAL CHALLENGE

It is plain that the materialist is not *as such* a nominalist. He is not occupied in any game of declaring the numbers, colours, spaces, and times to be fictitious, mythical, imaginary, or 'ideal' in any technical sense in use in any speciality outside philosophy. Rather he is occupied in denying to these and to any other non-material things some notable characteristic which he will at one time call existence, reality, or real existence, but at other times actuality, at other times substantiality (appearing as *ousia* in Greek and as *esse* in some Latin writers), and at other times concreteness. And sometimes in his argument he will imitate Aristotle, who in the *Metaphysics* tells us how universals and numbers are not causes or principles: they in no way do or explain anything themselves, let alone suffer anything; they do not enter the composition of anything as elements; and so are not to be ranked with *ousiai* or beings in the proper or primary sense.[1] The materialist has been anxious to go further and add that likewise minds and things mental lack this 'concreteness' (no longer in the original Aristotelian sense of a *concretum* of form and matter) and neither do nor suffer anything. In disputing this position, the mentalist has wished to insist that this property of actuality or 'concreteness' belongs only to minds and things mental.

The straightforward and natural assessment to make in this situation is that we are involved, in these disputes between materialist, mentalist, and others, with some second quite different

[1] Relevant passages in Aristotle include *Metaphysics*, Alpha, 5–9, especially 9, Zeta, 1, 7–9, 13–16, Lambda, 5, and Nu, 6; and *Categories*, 5. His comments on chance deserve comparison (*Eudemian Ethics*, VII, 14).

use or set of uses of words such as "exist" and "real" from those uses with which the logician *qua* logician is concerned. Some philosophers have imagined it to be necessary to carry through a programme of reductionism, exhibiting all the statements referring to the kind of abstract entity under discussion as dispensable and eliminable because reducible to some equivalent set of statements only referring to concrete objects, before they could justify giving the abstract entities a different metaphysical status in reality.[2] But it seems a mistake to suppose that a denial of reality or of actuality, or of the real existence spoken of by the empiricists when, following Aristotle,[3] they said 'only particulars have real existence' or 'the only real existence is particular existence', depends upon there being such an equivalence, or on the carrying through of such a reductionist programme. One can hold a theory such as that of Bishop Berkeley whereby the material objects and numbers enter into our discourse only as by-products of our modes of speech, without asserting any such extensional equivalence. And the content or meaning of such terms as "abstract" and "real" seems different from the content or meaning of such terms as "reducible" and "irreducible".

If the materialist feels committed to nominalism he seems to have offered a quite unnecessary hostage to fortune. There is no need to stigmatize classes, material objects, persons, numbers, colours, and so forth as 'logical constructions', 'logical fictions', or 'fancies of the mind' in order to state the metaphysical views respectively concerned. If I deny that a class has any metaphysical reality over and above the reality of its members, I do not seem to *mean* that talk of classes is dispensable, but to be using the expression 'metaphysical reality' in some sense not belonging either to logic or to epistemology.

Of course, the word "metaphysical" is a kind of cypher, signifying some preoccupation transcending that of *physis* or the natural and something supremely general and fundamental without

[2] W. van O. Quine often makes clear his view that if talk about one kind of entity can be eliminated in favour of talk about another (e.g. talk about classes in favour of talk about individuals) then the former kind represents a mere convenience of speech and no real enlargement of ontology, e.g. 'Two Dogmas of Empiricism'. It is because talk of classes cannot be thus eliminated that they represent a real problem for Quine to any nominalist ontology. In this way reducibility in the strongest sense is made the test of any application of Ockham's razor in ontology.

[3] *Metaphysics*, Lambda, 5.

being second-order or metatheoretical in the manner of logic or epistemology. The denial of the existence of metaphysics is the denial of the existence of any first-order inquiry wider than that of 'physics' or the natural. However, the statement that the only real objects or only real existents are natural (counting the phenomena of *psychē* or animate life as natural) is itself, prima facie, a first-order statement not belonging to the science of the natural, but (it is idiomatic to say) 'meta-physical'.

The most popular candidate in post-rationalist philosophy for serving as a criterion of metaphysical reality has been temporality: experiences and events, along with the supposed subjects and objects (both inner and outer) of experience, and all the objects of physical science seem to satisfy this criterion. Materialists have wished to add the criterion of spatiality, but it seems implausible to regard this as part of the meaning of the term "real": the statement that all real objects are spatial is more obviously not an identity than the statement that all real objects are temporal. However, the status of Time is a matter of philosophical dispute, and the reality of Time seems plausibly (as we have seen) to be regarded as a derivative corollary of the reality of things with positive existence, enjoying temporal relations (and, in this sense, 'in time').

The notion of "positive existence" is an obscure one, but some of its applications are intuitively obvious. To take a favourite example of the late Gilbert Ryle, if I have a barrel of beer and a bung, in a certain sense of the word "thing", I have two things, whereas, if I lose the bung, although I now have a barrel of beer and a hole, in the relevant sense I now only have one thing, since a hole is not positively 'a thing' but has an existence which consists only in the absence of a positive thing in a certain context. Likewise, where collections of concrete things are concerned, if I have a sheep and a goat and I count myself as thereby having two things, I do not then have three things, a sheep, a goat, and a pair of animals.

The most plausible candidates for being proper criteria for 'real' ('thingly') or 'positive' existence are those of being an agent or cause of change and/or of being the subject of change. These criteria may have been suggested by Plato, when, in the *Sophist*, he says that to be is to act or to suffer (though Ryle may have been right in suggesting that Plato intended these in a very generalized sense[4]).

[4] Ryle, in his lectures on the *Sophist*, in the early 1960s suggested that Plato meant: being the subject of an active or passive verb, in a sense opposing these to

However, the presence of these criteria, especially that of being a cause or agent, is very prominent in Aristotle and Aquinas[5] as well as later among the empiricists. The subjects of active power and events are both said to cause things, and unmistakably count as particulars rather than as abstract or general entities; and, if it were the case that causal relation consisted in the conformity of appearance to law, then experiences merely as such might count as causes. Clearly imaginary and fictitious daggers kill nobody, and it is therefore plausible to tie real existence to causation. Whether this view is right and what kind of causation is the relevant one are matters for later discussion. But at least the main lines of the problem of the meaning of the word "existence" have now been set.

However, first, we must draw attention to a complication which will have to be catered for in any view.

2. EXTENSIONS OF THE NOTION OF POSITIVE EXISTENCE

At first sight it may seem that the notion of what has positive existence is straightforward. It is not regions of space and time but rather the things which occupy these regions, three-dimensional bodies, fields, systems of radiation, events, and suchlike; not holes or vacua, but substance or substances, even if only the ether. Already we have on our hands the problem of distinguishing between an event and the absence of an event, as to which in a particular case counts as positive. We shall return to this later. But at an earlier and more obvious level we have the problem that certain things such as knowledge and sightedness seem evidently positive and to count as assets or things 'had' or 'possessed', whereas ignorance and blindness appear to have the nature of privations, i.e. the absence of something positive in contexts where its presence would have been possible, fitting, appropriate, or even due.

This case is very different from the case of the existence of

predications of rest, identity, or otherness—the term "verb", *rhema*, being understood in an earlier or less fixed sense than that later canonized by grammarians.

[5] *S. Th.*, Ia, Q. 48, Art. 1, obj. 4 and ad 4. Aquinas also makes act or operation the criterion of subsisting: thus, the Persons of the Trinity are distinct *subsistentiae* because they have notionally distinct acts, and the soul of man subsists because it has an operation which is not the operation of any bodily organ.

colours. We can conceive colours as having an existence simply in a logician's sense, i.e. which consists in their being subjects of discourse albeit of an abstract kind. True, the discourse within which they figure as logical subjects has certain presuppositions which are contingent and which depend upon actuality, e.g. which relate to light radiation and its properties, the perceptual capacities of human beings, and the system of classification of colours within the tribe or culture concerned. "Positive existence", in the sense in which we ascribe it to knowledge, only arises in connection with colours so far as we are concerned with a particular physical property, e.g. emitting light of a certain range of wavelengths, or as a particular experience within a certain tribe or culture: neither is strictly speaking a colour since the second is rather the experience of seeing a certain colour, not the colour itself, and, as to the first, we have the experience of seeing a certain discriminable colour normally in connection with a wide variety of mixtures of disjoint ranges of wavelength and not solely with a single, simple range, as seen in 'normal' conditions of lighting and contrast.

By contrast, when we think of knowledge, we are commonly occupied, not with distinguishing one of a range of abstract entities called intellectual traits, but with the knowledge of an individual as a particular enriching the Universe, and it is embarrassing to have such ill-defined or variably-edged particulars regarded as primary subjects of real existence. It is, therefore, attractive to take Aristotle's view that these so-called accidents have existence only in a derivative secondary sense, *viz.* in the sense that it is in virtue of them that substances are positively qualified in a certain way: in the *Categories*, it is unclear whether their existence is prior to their qualifying of substances or vice versa, but, in the *Metaphysics*, Aristotle is clear that their qualifying of substances is what is primary, and that to say that a certain accident exists is not to say anything over and above that substances are positively qualified in a certain way.[6] Aquinas expressed the Aristotelian view in these terms, that the sense in which so-called accidents such as knowledge have positive existence is this: by having them, the substance which has them exists more or less, or exists perfectly in this or that respect or in varying degrees deficiently.[7] One may

6 Aristotle, *Metaphysics*, Zeta 1, *et al.*
7 Aquinas, *Comm. Met.*, XII, lect. 1, 2419–20; and *Quodlibet.* IX, Art. 3.

notice here the near equivalence in English between the notions of existing more fully and living more fully. There is an idiom in which neither existence nor life are all or nothing things but in which, in regard to one and the same subject, existence and life have the capacity for a variation in degree, so that the person may come to fuller existence or fuller life (this is quite different from any doctrine about degrees of reality possessed by different types of subject of discourse and therefore by implication by different subjects of discourse, numerically distinct).

In this development of the Aristotelian view, we shall say in regard to any substance that it has, firstly, a primary mode of existence associated with its identity and with its kind (perhaps there is a connection between the possession of identity and the criteria of identity dictated by belonging to a certain kind). In regard to this fundamental mode of existence we can say that it is all or nothing and has no degrees: it is exemplified in the statement that "Socrates exists", which is a presupposition of all other statements about Socrates (i.e. is even a presupposition of the subject-predicate statements "Socrates is a human being" and "Socrates is Socrates"). And then, in addition to this, we shall say that Socrates has, secondly, what, in Book Delta of the *Metaphysics*, Aristotle describes as accidental existence according to non-substance categories;[8] and, by contrast with the other case, this will be in many cases susceptible of degrees, whether these degrees be quantitatively measurable as when we are talking about size, or whether they have order only, without mathematical measure, as when we talk about knowledge and courage. One of the reasons given by Geach and others for supposing that there must be some kind of existence or actuality not dealt with in contemporary logical theory is the phenomenon of intensive magnitude, e.g. in the case of luminosity, which is not to be explained in terms of any mixture of opposites, but in terms of degrees in some positive quality.[9] Upon examination, it turns out that such intensive magnitude always belongs to something that can be regarded as the activity of something with active power, the realization or exercise of some nature. Accordingly, it is not

[8] Aristotle, *Metaphysics*, Delta, 7; Thomas Aquinas, *S. Th.*, Ia, Q. 5, Art. 1, ad 1 and 3, and *Comm. Met.*, XII, 2419–20.

[9] P. T. Geach in *Three Philosophers*, pp. 92–4 and *God and the Soul*, pp. 61–2.

difficult to conceive this intensity or degree of existence in a quality as consisting in the degree to which the existence of something is realized, not its substantial existence but this existence according to non-substance categories.

In this exposition of an Aristotelian view, what is required in order to explain the kind of positive existence we indicated forces and invites us to ascribe real existence only to substances. And whatever view we take of the identity of the metaphysically real, whether Aristotelian substances, empiricist ideas or experiences, events, or anything else, it will be open to us to adopt a view having the structure implied above. This structure is as follows: firstly, a set of senses or uses of the verb "to exist" which are correlative with each subject-matter and with the application of the laws of logic to it, i.e. a set of uses of the verb "to exist" belonging to logical theory, in which different uses are distinguished according to the types of predicate sensibly applicable to the type of subject concerned; and, secondly, a quite different use of the verb "to exist", applicable to that which is metaphysically real and expressing this metaphysical reality in the non-predicative style appropriate to it; and, thirdly, a derivative set of uses of the verb "to exist", connoting being in various relations to the metaphysically real in virtue of which the metaphysically real gains its proper fullness or is enriched or enlarged.

This approach achieves for us a certain liberation. There is still a question as to which qualities and which relations positively enrich the existence of their subjects and enrich the Universe which contains them, but we are now in pursuit of one kind or a small range of kinds of subjects of real existence, in the primary sense, and not a plethora of kinds allowing for each and every kind of so-called accident.

We are now free to address ourselves to two main questions:

(i) Which things have real existence in this metaphysical sense? is it Aristotelian substances or Plato's ideas, or the experiences the empiricist speaks of, or events, or something else? and in virtue of what are they said to possess it? We are not at present asking ourselves what this so-called 'real existence' is, but only which kinds of thing are subjects of it. It is natural to suppose that this 'real existence' is the existence as an act or actuality which Aquinas distinguishes from existence as a mere fact in the sense merely of something asserted by means of the verb "to be" in the composition of a proposition.[10] We might then contrast 'real' or 'positive' existence with existence as a mere fact (something stated in a true affirmative existential proposition), as is suggested in Section 10 below. But our general argument does not depend upon this, and at present our question is only as to which things have real existence.

(ii) What is the relation between this metaphysical use of the word "exist" and the logician's use? if the doctrine that existence is not a predicate is to be upheld, how are we to explain how existence in the metaphysical sense fails to be a predicate? and does the metaphysician's use represent a distinct 'sense' as if there was complete equivocation? and, in that case, is temporality internal to the metaphysician's use?

3. THE LESSONS OF THE PHILOSOPHY OF LANGUAGE

Man is a talking animal. It may be, since the notions of knowledge and understanding have an extreme generality, that God and the angels have a mode of knowledge which is not such as to require the presence in the Divine or angelic mind of any verbalizable or language-related entity whose structure realizes or expresses the object of propositional knowledge. I.e. if God knows that P, there does not need to be, in God, any form of words expressive of P, and his knowledge does not consist in any disposition in relation to the meaning of such a form of words, so far as this meaning is a linguistic entity or construction. However, in man's case, the situation is quite opposite. Accordingly, the philosophy of language can exhibit certain metaphysical views as incoherent and absurd.

The view of F. P. Ramsey[11] that the same proposition is expressed indifferently by the statements "Socrates is wise", "Socrates has wisdom" and "Wisdom Socratizes", presupposes a view of propositions which prescinds from their relation to speech, writing and the structures of these statements. No adequate reason has been given for supposing that we have any kind of acquaintance

[10] Aquinas, *S. Th.*, Ia, Q. 3, Art. 4, ad 2.
[11] F. P. Ramsey, 'Universals'.

with or knowledge of any such entities, although we may know of the property of 'having the same truth-value as the statement "Socrates is wise" ', and may conceive of the open set of all possible statements with this property, including, for instance, this statement's conjunctions with any true mathematical statement.

In order to understand such a statement as "Socrates has wisdom", we have to know the meaning of the term "wisdom", supposedly denoting a 'universal' or signifying a 'general abstract idea'. And, as this term appears to be a singular term, presumably, if we had to know its meaning independently of knowing the meaning of the predicate "is wise", we would have to know its denotation, i.e. which entity is referred to by it. In addition we would need to know what relation is signified by " . . . has — —". It is not the relationship of ownership, as between a man and his house, nor any other such supposedly empirically known relationship. If we are debarred from identifying this relation in terms of the notion of predication, as if such expressions as "has wisdom" were to be understood derivatively from such expressions as "is wise", we seem to be debarred from any route whereby we might obtain a knowledge of this relation. Thus a Platonic erection of 'universals', conceived along these lines, into entities primary in reality and in knowledge is not coherent, and the reasons for its rejection are not based in epistemology, as in Strawson's critique[12] of Ramsey, but in logical theory or the philosophy of language.

In sum, we have no access to the understanding of statements about universals, conceived along these lines, except via a prior understanding of statements within which the corresponding predicates are used. All apparent mention of such universals is parasitic upon such prior understanding of these predicates, used and not mentioned.

Plato's own espousal of universals, or 'Ideas', would seem to be based on rather different considerations from those of Ramsey. But the same general conclusion as to the dependence of our understanding of statements about universals and identification of them upon a prior understanding of statements in which the relevant predicates are used still obtains.

It appears natural to us to make remarks in which universals are seemingly overtly mentioned and are even objects of some kind of fairly direct ostension. E.g. we speak of the letter "s" as one of the twenty-six letters of the alphabet; we use words to refer to themselves, e.g. saying that 'The

[12] P. F. Strawson's critique of Ramsey's treatment of universals is to be found in *Individuals*, ch. 5.

word "cat" has three letters' and 'The name "Socrates" has eight letters'; we distinguish 'water' from 'oil' and speak of 'fire' as an 'agent' or a 'stuff'; and we speak of such species of animal as 'Man' or 'the horse', noting whether the one spoken of is quadruped, warm-blooded, hooved, graceful, or otherwise, and speak of 'red' and 'blue' as different colours: and, in all these cases, we take as the grammatical subject noun of our statement, not a by-formation (in Aristotle's *Categories*, a"paronym") of a predicate or any nominalisation (such as horsehood, redness, wisdom, or courage), but the term itself, whether it be called a "predicate-term" or anything else.

However, in regard to every one of these cases, the cases which inspired or misled Plato in his speaking (e.g.) of 'the just itself' or 'the equals themselves' as well as some of those of the kinds just mentioned, it is clear that the situation is precisely as stated in the text above. I.e. all access to understanding of such statements about universals and any apparent mention of universals, as if they were objects in some sense, is parasitic upon a prior understanding (whatever this involves, granted that what is implied or signified is not univocally, or according to one definition, the same in every case of the application of the same term) of what the statement in which the relevant terms are predicated signify in regard to their subjects. We begin with a recognition that this and that are horses, this and that are "s"s, this and that are red, these and those are equal, and so forth, and, contrariwise, that this or that is not a horse, and so forth: and it is this, and not some emancipation from the shadows, which is a logical pre-condition of understanding those other statements.

As "wisdom" would thus appear to be what we might call a nominalization of the predicate "is wise" (being wise" is a more primitive nominalization of the same predicate), so the expressions "the collision between Smith's car and the wall" and "the sight that Smith had of the wall" would appear to be nominalizations of the complete sayings or statements "Smith's car collided with the wall" and "Smith saw the wall". That is, so-called events and experiences appear to be picked out in our speech and writing by nominalizations of complete sentences, analogously to the way universals are picked out by nominalizations of predicates. And, just as we have no access to understanding of statements about universals except via prior understanding of statements in which the relevant predicates are used and apparent mention of universals is likewise parasitic on this, so we have no understanding of statements about experiences or about events except via the prior capacity to understand the statements in which they are reported, not referred to or denoted. Parallel remarks to these we have made about events and experiences apply to states, qualities conceived as particulars, and so-called accidents of every other kind.

Accordingly, those forms of empiricism, and what we may call scientism, which make experiences of events primary in reality or knowledge, are ruled out of court by the same types of consideration as rule Platonic 'Ideas' or 'Universals' out of court. It is not only that the identity of these objects could be known only indirectly via prior understanding of discourse not referring to them, as if they might be fundamental in reality although not primary in human knowledge; it is also that the natures of these objects can be understood only in terms of the same background, so that, speaking metaphorically, we can regard these objects as creatures of our linguistic mode of thought and knowledge. The approach I am proposing is not new; Aristotle explains how universals would do nothing either for metaphysics or epistemology; I merely point out that parallel considerations apply to the much vaunted 'events' and 'experiences' which we have sometimes been imagined to have presented to us primitively—from this position, it is then imagined, to find ourselves free to 'conceptualize' or 'order' in alternative ways.

The other favourites of the older empiricists included 'ideas', but if these are neither universals, nor experiences or events of perceiving, but the objects of experience, perceiving, or sense, then it will be true of them that, as Berkeley said, 'their *esse* is *percipi*', and, as in Berkeley's account, they will be subordinate in reality, both to the subjects of the operations of perceiving, and to these operations, besides being, in Berkeley's terms, 'beings' in name only,[13] and in our terms, intentional objects of some kind.

Time and Space, and times and spaces, will fall victim to the same kind of argument as universals, experiences, and events.

Aristotle seems to have originated three theories in respect of the nature of space and spaces. Firstly, he proposed in *Physics* a theory whereby a space or volume is the innermost boundary of a containing vessel, and presupposes the vessel in the manner in which a surface presupposes, or has no existence except dependent on, the body of which it is the surface.[14] Secondly, in the *Categories*, we can locate a theory whereby spaces are the immediate or proximate subjects of predications of spatial magnitude, so that, although metaphysically it is a body which is primarily subject of colour or magnitude, nonetheless, as colour is ascribed to it in respect of its surface, so spatial magnitude is

[13] Berkeley, *Principles*, para. 89.

[14] Aristotle, *Physics*, IV. 4–5.

ascribed to it in respect of its place.[15] And, thirdly, we can trace both in the *Categories* and in the *Metaphysics* a theory whereby place is a category or class of predicate and spaces are somehow predicates.[16]

Aristotle also presents a parallel set of theories of Time, as if not recognizing its peculiar status within statements. But he also in the *Physics* separately developed a theory of time as the measure of change,[17] thus making time intimately dependent on the ascription of change to a subject (i.e. on a predication in the category of 'passion'), every change being conceived of as a movement from a state at one time to a state at another time. Yet this account, while making time co-ordinate with change, rather than prior to it, nonetheless gives undue prominence to measure as opposed to what is more fundamental, namely order, and leaves it obscure how the existence of time is to be reducible to some system of predications. Leibniz, making the existence of spaces and times, and of Space and Time, consist in the existence of a system of spatial and causal relations, might be conceived of as having given more satisfactory development to the Aristotelian endeavour of explaining how space and time are not extra or prior entities besides substances and their relations, but metaphysically dependent on the system of substances.

For our own part, so long as we recognize that to enjoy spatial relations is integral to the nature of any bodily substance, an account which makes the existence of space consist in nothing more than the existence of actual and possible spatial relations seems adequate. A parallel theory, in respect of time, needs, as we have seen, to be supplemented by a realization that temporal conjunctions and tense-indications are more analogous to non-truth-functional operators such as "if", "as", and "because", and to modal operators, than to any non-sentential relation or predicate, together with a realization of their special relation to causation.[18] It might be argued that, in appealing to the philosophy of language in order to refute certain metaphysical theories, I am violating the principle which, in Chapter VII, will appear crucial to the whole

[15] Aristotle, *Categories*, 6.

[16] Aristotle, *Categories*, 4, and *Metaphysics*, Delta, 7, and Zeta, 1.

[17] Aristotle *Physics*, IV, 11–2.

[18] This point was explained in Chapter II, Section 1(*c*), with its reference to P. T. Geach, *Logic Matters*, pp. 311–6.

enterprise of this book, the principle, namely, that the resolution of first-order questions should never be made to turn on the answers to second-order questions in any way which gives the second-order discussion any independence of, or priority to, first-order considerations. However, in the first place, the philosophy of language does not proceed independently of first-order knowledge of human beings, and their modes of thought and knowledge, which are precisely modes of thought and knowledge implying cognitive involvement with the physical world, human beings being by nature animals. In the second place, if a metaphysical statement, as the content of a linguistic act, turns out to be incoherent or nonsensical, it would appear to belong to the philosophy of language to notice this; some theories, e.g. the theory discussed by Aristotle[19] in *Metaphysics*, Gamma, that contradictions can be true, do not get off the ground.

4. THE PRIMACY OF SUBSTANCE AS THAT WHICH EXERCISES ACTIVE POWER

Considerations of the philosophy of language rule out, from any position of metaphysical primacy, entities which are, in the metaphorical sense indicated, creatures of our modes of speech and thought or which are merely objects of mental acts such as the supposed 'ideas' of Bishop Berkeley. Whatever has metaphysical primacy, i.e. exists in the metaphysically privileged sense, is either unknown to us (so that the concept is for us purely formal) or else denotes some class or classes of entity which the philosophy of language would allow us to consider primary without this involving any incoherence.

The structures of speech and writing set before us a structure with many aspects not described by Aristotle, or not described accurately by him. However, the groundwork of later discussion was laid down by Aristotle, and his treatment will suffice to illustrate and enable us to pick out features which will recur in any more developed account. We have a distinction between two types

[19] Aristotle provides other nice examples of this, e.g. the theories discussed in *Physics*, I, that what is is one and motionless (as opposed to natural things being, all or some of them, in motion) or that change requires only two principles, rather than at least three.

of predicate, one which we may call episodic or 'live', situated in the categories of action and passion, and another which we may call static, situated within the categories of substance (second substance), quality, quantity, and place. The category of relation overlaps most of the other categories and includes predicates of both these kinds. The category of time indicates something which differentiates episodic from static predicates. Action and passion presuppose a primary substance which is the agent or subject of change, and this is the subject of all the other kinds of predicate. Typically, predicates signifying second substance or quality indicate a nature or disposition, a power, liability, or tendency realized sometimes or always in a certain kind of action or passion. Accordingly, if the philosophy of language throws up any first candidate for consideration as being metaphysically primary, it is these subjects of action and/or passion, the possessors of these natures among them, which if subjects of passion will be things which are in place, these things which are logically the primary subject of tensed predication—the first substances of Aristotle's *Categories*.

Yet, at this point some person will raise the question as to whether we have gained access, not just to true descriptive statements about objects of experience, but to true descriptive statements about realities as they and their relations are in themselves, not dependently on, although within the reach of, human experience and thought.

At this point, one kind of criterion alone enables us to make the necessary discrimination: namely, the kind of criterion put into our grasp by our capacity to distinguish genuine from makeshift patterns of explanation, and our access to knowledge of real causal relations, involving not just conformity to law, but exercise of active power. That this is so is shown by arguments from the philosophy of language, associated with points in the philosophy of science which have been adverted to in Chapters II and III as well as in this section, to be confirmed by the epistemological and methodological considerations exhibited in Chapter VII.

In brief, only the real in so far as it is real can *cause* anything in its own right, i.e. by the exercise of its nature in an appropriate context. Inter-subjectivity does not secure reality in the relevant sense: there can be objectivity in the sense of reliable or guaranteed inter-subjectivity in relation to visions, without the content of the vision being thereby really and substantially present.

It is this criterion which allows us to deny the real existence of skins on the surfaces of liquids, although these do play a role in a makeshift explanation of surface-tension phenomena, and which conversely allows us to envisage fields and electromagnetic radiation as possible candidates for real existence in the metaphysical sense, despite the absence of ways of referring to them less abstract than that in virtue of which such-and-such a proposition or system of propositions is true. But if scientific explanation requires the admission of fundamental entities distinct from Aristotelian first substances, although intimately related to them, these fundamental entities will not be events, powers, or any other creature of language, but these 'principles' or 'things in virtue of which . . .', such as fields.

This criterion also invites us to distinguish between mere heaps or aggregates, including what Aristotle called accidental unities,[20] the explanation of whose nature and behaviour is, in principle, reducible to the natures, interrelations, and behaviour of their parts, and, on the other hand, natural unities whose nature and behaviour is not thus reducible. It may be that the fundamental particles whose parts the physicist calls quarks are natural unities in this sense, the quarks not having independent existence,[21] and, if Aristotle was right, living bodily things, plants, and animals, including men, are also natural unities. It may be that systematic considerations make it easier to judge that the higher animals are natural unities and therefore primary substances, not just in a sense belonging to logical theory or the philosophy of language, than to make this judgement in the case of fundamental particles, when these are not relatively isolated. (Molecules and ions may be only somewhat better candidates than fundamental particles—I mention this merely to indicate an example of the systematic problems, here the problem of the complete reduction of chemistry to physics, which are liable to arise.)

Our discussion in this book has to be at many points programmatic. At this stage, let us merely postulate as reasonable, without giving full proof, that some Aristotelian substances will remain among the substances which, exercising active power, are primary,

[20] Aristotle, *Metaphysics*, Zeta, 16–7, cf. 7–11.

[21] J.C. Polkinghorn, *The Particle Play*, ch. VII.

not just in explanation *ad hominem* but in real determination or causation of what goes on in Nature.

It will be noticed that I have given precedence to action, causation and agency over 'passion' or being the subject of change. Many actions are internal to an agent and do not consist in bringing about a change of state in something else. As to changes, in so far as they are not conceived of as being brought about by an agent, but, for example, consist in the continuance, as it were (e.g.) in virtue of inertia, of some change, for instance change of place, they can be considered logically as actions. Passion, in so far as it is not conceived of as involving causal 'agency' in any way, even the agency of nature or inertia, does not by itself clearly involve substantiality or real existence: e.g. changes in intentional objects as presented to the visual field in dream or in hallucination do not imply substantiality in their so-called 'subjects'. Therefore, it seems right to give pre-eminence to what Locke called real causation and to agency.

To avoid misunderstanding, we must quickly deal with two obvious objections.

(i) An objection arising from the structure of matter

Firstly, I must not be understood by what I have said to have made what has metaphysical reality or exists in the primary metaphysical sense into a recherché or esoteric matter or an issue accessible only through the mystifying discussions of philosophers.

If the parts of an accidental unity have this metaphysical reality, then so does the accidental whole constituted of these parts, in the same and not a secondary sense, albeit only in virtue of the reality of the parts. If bricks are real, so is my house. True, we have conceded that it may be a matter of some difficulty to decide where the ultimate locus or focus for purposes of explanation lies, e.g. whether at the level of fundamental particles, or of atoms, ions, or molecules, or of macroscopic organic wholes such as plants, animals, and human beings, or even the Universe as a whole. And, true, one has to bear in mind the validity of Aristotle's statement that a substance, as a complete realization (*entelecheia*) or actualization (*energeia*), cannot be composed of substances existing in complete realization or actualization themselves[22] (i.e. the

[22] Aristotle, *Metaphysics*, Zeta, 13, 1039a 2 ff.

ultimate focus of causal explanation cannot be in this respect ambivalent). But it nonetheless remains the case that such problems do not bring into doubt the reality or substantiality of the material parts or the accidental unities which the ultimate substances may be made up of or go together to make up. All alike have the reality we speak of, if any of them do, even if it is only in virtue of some that the others, smaller or larger, have this standing. Therefore the Sun and the Universe, animals and plants, molecules and ions, and fundamental particles, all have metaphysical reality, if any of them do.

(ii) An objection arising from consideration of the Christian doctrine of the Trinity

Secondly, I have spoken indifferently of the order of explanation and the order of causal explanation as determinative of what we should consider to exist or to be real in the metaphysically significant sense. In this I have, for the purposes of exposition, temporarily omitted mention of many important complications.

However, in Chapter III we recognized that the term "explanation", unlike the term "causation", was a second-order one. As such, the term "explanation" applies not only to so-called 'factual' or 'causal' explanation but also to epistemic connection between possible objects of thought or knowledge and also to so-called "final causation", in which the cause is not a psychological state such as intention or desire, but its object (one species of what we called an intentional object). Desire and knowledge presuppose the substance desiring of knowing of which desire and knowledge are predicates or attributes. But the object of desire, *qua* known, and the object of knowledge are prior to and, *in a non-factual, non-causal sense*, 'causes' of the states of desire and knowledge: e.g. fact is the cause of truth in a proposition, rather than vice versa. In the case of sensory knowledge the object known operates in a certain context causally upon the knower, and the state of knowledge is caused in the ordinary causal or factual sense, but this need not hold in general. However, this will not create a difficulty for us when the objects known are facts or true thought-contents and when the objects of desire are no more than intentional objects.

We can therefore, without worry, allow our attention for the most part to be limited to factual or efficient causal explanation, henceforth often referred to in this book simply as 'causal explanation': this will preserve us from attributing metaphysical reality to thought-contents or intentional objects as such, and will occasion difficulty only when we have objects of knowledge or objects of desire which are metaphysically real according to the criterion of being factually or causally explanatory and which are presupposed by acts of knowledge or desire on the part of something else metaphysically real, without there being any factual or efficient causal determination involved.

However, I am not aware of any cases in which this (what might appear purely academic) problem actually arises, except in discussions of the Christian doctrine of the Trinity. There the personal acts of knowing and loving in which God's life is represented as consisting presuppose the Persons as objects of love and knowledge without there being any order of efficient causal priority between Them: so even in God there is some distinction, at least according to Christian doctrine, pertinent to some kind of explanation, between subject (*hypostasis*, *subsistentia* or *suppositum*[23]) and actuality or positive fact, such as may be signified by a true statement and not by a subject-expression, but no distinction pertinent to efficient causal explanation. This distinction, between subject and actuality or else between subject and positive fact, is a key one, but at this stage it has been relevant only to mark that the Christian doctrine of the Trinity has not been excluded. (In Sections 9 and 10 below I shall make clear that this key distinction, between subject and actuality or positive fact, figures within metaphysics, but it will be the existence of causal agency which enforces this recognition by the metaphysician, not discourse about the Trinity.

[23] From the time of the Cappadocian Fathers (Basil, Gregory Nazianzen, and Gregory of Nyssa) in the late fourth century onwards, the word *hypostasis* was uniformly used by theologians to denote the Divine Persons, the Father, Jesus as His Son, and the Holy Spirit, as the ultimately most concrete subjects of predication in the Godhead.

Aquinas in representing this tradition in the Church uses the word *subsistentia* as the most literal available Latin rendering of the Greek *hypostasis*, the word *substantia* having other uses long established, and he identifies what it means with Aristotle's 'what is neither said of nor present in a subject' in *Categories*, 5 (*S. Th.*, Ia, Q. 29, Art. 2).

To be a *suppositum* for Aquinas seems to be something signified by a singular term (ibid.): if *supponere* means 'to go proxy for' and is equivalent to 'to stand for' so that *suppositio* can come to mean "way of referring to" (P. T. Geach, *Reference and Generality*, p. 84), then to be a *suppositum* is to be one of the objects stood for by the terms with one of various kinds of *suppositio*.

It is only in Section 10 that I explain why it seems preferable to speak of 'actuality' rather than merely of 'positive fact'.)

The considerations I have just mentioned show that the over-simple view of God's simplicity whereby, in God, there is no distinction between subject, nature, and existence, as well as no distinction between these and the activity of life of God, savours more of Platonism than of Aquinas himself, who in his account of the Trinity is careful to distinguish subject or *suppositum* from act or actuality, whether *esse* or *essentia*. As we shall discover that there is no sense or coherence in the conception of God as a transcendent One of whom no tensed statements are true or significant, it is evidently wise to be circumspect in the way in which we explain God's simplicity and timelessness and in the way in which we draw the logical divide between God and creatures.

To summarize our position, then, we can say this: to have real existence is to have the active power to 'efficiently cause' something, i.e. to act causally. It is not that active power is what real existence means, is, or consists in, but that the possession of it is sufficient criterion of real existence. In the case of a personal being such as God, what we speak of will be presumed to have reality in itself in virtue of immanent or internal acts of self-understanding and loving without being necessitated to act causally: but still the mark for us of its reality will be its capacity to act causally.

5. EXISTENCE IN THE METAPHYSICAL SENSE IS NOT A PREDICATE

When the logician or the logical theorist makes the correct observation that existence is not a predicate, what he says marks the fact that existence, whether in the sense of the non-emptiness of the domain over which generalization or quantification is made, or in the sense expressed by the existential quantifier, or in the sense of what is presupposed in the use of a singular term as such, is not expressible as a predicate (represented by a predicate-letter) in modern predicate logic. Nor is it denoted by a term in Aristotelian logic. Nor does it have a set or class as its extension.

From the point of view of the logician, the existence of a thing is not a predicate but stands always as a presupposition in the logician's sense of subject-predicate propositions about the thing, pre-condition equally of the truth of the proposition and of the truth of its negation. In this sense, the existence of a thing is a presupposition of all ascriptions of a nature as well as of all attributions of properties to the thing. In this way a presupposition of existence arises in the case of so-called abstract objects such as numbers and colours as much as in the case of so-called concrete objects.

However, when it has been said by philosophers in the past that existence is not a predicate, they have often had in mind other considerations of which, in the cases concerned, these logical features might appear merely symptoms. For instance, it has been regarded as absurd to view existence as a property, even a definitional property, of a thing for other reasons than that existence is not a predicate in the logician's sense. This has been argued in some such way, in effect, as the following.

On the one hand, if the existence of a thing is not its nature or part of its nature, then, whether mentioned in any merely nominal definition or not, it will be external to any real definition of the thing, i.e. in any definition expressive of the real nature of the thing. In this case the existence of the thing would be either explained by the thing's nature (and in this sense 'caused by' or 'consequent on' its 'essential principles') or else not explained by its nature (and in this sense 'accidental to' it); in the latter case its nature would have to be such by its essential principles as to be capable of such an accident;[24] therefore, either way, the nature of the thing would be prior to its having existence as a attribute, i.e. to its existence. But this is impossible because there is no 'nature of the thing' as a non-abstract entity, nor any 'thing', prior to a thing's existence, so the nature, as an abstract entity, cannot be prior.

On the other hand, suppose that the existence of a thing is its nature or part of its nature. Then either there is no real definition of the thing because the thing does not have a nature expressible in language (this is part of its incomprehensibility); in this case the existence concerned (we would have to call it an 'actuality' or invent some other term) must be distinct from what is asserted in

[24] It will be noted that my argument is parallel but not identical to the argument in Aquinas, *S. Th.*, Ia, Q. 3, Art. 4. Where something in regard to a thing is not explained by the essential principles of the thing, Aquinas argues that, if it is not identical to the thing's essence or nature, then it must have an external cause. I do not dispute this but the weaker argument in my text suffices.

asserting existence (the 'fact' or 'true proposition *qua* true')[25] and, since it is thus in no way something asserted, for this reason cannot be a property. Or else the assertion of the existence of the thing would be part of its real definition; in the latter case the existence of the thing would be an element in its nature which would involve its being explanatorily prior to this nature and thereby explanatorily prior to itself, since there is no existence of a thing without the existence of a thing with its nature.

Thus the hypothesis that existence is a property has been supposed to involve vicious circularity, whichever way it is taken.

The circularities concerned have been supposed to be vicious, not merely because of the structure of speech or thought and the customary ways of representing these among logicians, but because the canons of explanation have been violated. Existence is what gives actuality to properties, and so is explanatorily prior to or explanatorily presupposed by them, and cannot flow from or be included amongst them without self-causation being involved.

Therefore two questions arise.

Firstly, what is it for existence not to be a predicate or property in this more profound and general sense, a sense not belonging to the philosophy of formal logic but to the theory of explanation? This we will discuss in Part II below.

Secondly, is existence in the metaphysical sense after all a predicate in the more pedestrian logician's sense? To consider this is a vital preliminary to considering whether the word "exists" is used completely equivocally as between the logician and the metaphysician. For, if assertions of existence in the metaphysical sense are of subject-predicate form, they have a quite different logical structure from assertions of existence in the logician's sense and there will indeed be complete equivocation.

To this second question, once posed, the answer is obvious. For something the expression for which connotes being a subject of active power (a 'first substance') not to exist in the metaphysical sense is for it not to be a subject of active power or 'first substance',and this can only be by its being non-existent in the logician's sense.

[25] This distinction between existence as an act or actuality and existence as a fact or something asserted in the composition of a proposition is made in Aquinas, *S. Th.*, Ia, Q. 3, Art. 4, ad 2.

6. HOW MANY SENSES DOES THE WORD "EXIST" HAVE?

If I am right that the word "exist" has distinctive features in metaphysics of which the logician and logical theorist takes no cognizance, the question arises as to whether the word "exist" as used in metaphysics has a meaning entirely separate from the meaning it has in logic and logical theory.

We have spoken of a spread of uses of the word "exist" recognized in logical theory, one use for each distinct logical type, i.e. for each most general sort of thing so far as the most general sorts can be distinguished according to the kinds of predicate of which it is significant or sensible to ask whether or not they apply to things of such a sort. In each of these uses the word "exist" will be applied affirmatively according to different criteria and so, in a certain technical logicians' way of thinking, will be being used in a different 'sense' (or, more accurately, in each case, in three interrelated senses corresponding to the non-emptiness of the domains of quantification, to the existential quantifier, and to what is presupposed in the use of a singular term).

However, the different uses thus distinguished by the logical theorist in his consideration of the application of logic will be formally or structurally the same in each case, i.e. will have something of the unity which is possessed by the uses of the word "true" in connection with each of vastly different genera of statement, and it would be natural in both cases to speak of a unity of sense because of the unity of function in relation to knowledge, thought, and speech.

In addition, we spoke of a spread of uses of the word "exist", as between that which is metaphysically real in the primary sense and those positive qualifications or accidents which enrich this existence: the latter would indeed be metaphysically real in some secondary or derivative sense in somewhat the way explained by Aristotle—i.e. their reality would consist in one of a certain range of relations to what was primary, i.e. to substances. Our question only concerns the relationship of these senses of "exists", along with equivalent expressions such as "is real" or "is actual" and suchlike, to the sense or use of the word "exists" allowed by (indeed insisted on by) logical theory in respect of whatever things do exist in the primary metaphysical sense.

If, by the word "exist", the logician or logical theorist intended a second-order term, meaning the same as 'is a non-mythical, non-fictitious, non-imaginary, non-ideal . . . subject of discourse', then we could say straightforwardly that the logician's meaning of the word was quite different from the metaphysician's. However, on the contrary, the logician understands the word "exist" in respect of each subject-matter in regard to which it has application as a first-order expression in regard to that subject-matter. That is, in whatever is the relevant way, the existence of something is that in virtue of which it is a subject of discourse of the kind indicated. In the case of entities such as numbers and universals (e.g. colours), what is relevant is only something appertaining to the structure of our thought and discourse, what we may call sometimes the *ordo dicendi* and sometimes the *ordo cognoscendi*. But where considerations about explanation indicate that we are speaking about subjects of discourse which are 'ultimately real', the existence asserted routinely as an element in the application of logic or 'thinking straight' is *ipso facto* the metaphysical existence or actuality we have been speaking of.

When we say of the number three that it lacks the property of being divisible by two, or when we say that two plus two cannot help equalling four, we speak metaphorically; and it is plain to us that the essential meaning of the expressions "lack" and "cannot help" to which we are committed in respect of these statements is contained in the formal element shared by all metaphorical and non-metaphorical uses of the expressions concerned, *viz.* the element of negation implicit in the word "lack" and the element of necessity expressed in "cannot help". Yet in other cases the material content of such privative and modal expressions would seem to be core to their meaning.

When it comes to the spread of uses of the word "exist" we are presented with a parallel problem.

On the one hand, as talking animals, all the structures of speech that we use are shaped by the structures that we find it appropriate to use or which force themselves upon us in connection with concrete subject-matters. When we say "There is . . ." or "This . . ." we speak as if a '*demonstratio*' was possible or was actually going on, and the very metaphor we use, *viz. demonstratio*, reveals the shaping of our thought about the abstract to our thought about the concrete.

However, on the other hand, we do not thereby imply that there is anything concrete about the abstract, either by our mode of speech or by the metaphor we use to describe it: we imply nothing with the least tinge of such falsity. Accordingly, in the light of this second consideration, *it might seem* that the most formal part of the meaning of the word "exist" was that which was represented by its spread of uses identified by the logical theorist distinguishing the areas of application of logic.

Yet, it seems that this suggestion, that the most fundamental part of the meaning of "exist" when used of concrete entities, characteristically agents in time, is this most formal element, is highly dubious. It is more natural to suppose that the fundamental element lies in the expression of what I have called metaphysical reality, associated closely with agency.[26] The same instinct is revealed in the common-sense prejudices against supposing holes, spaces, and times, rather than their occupants, to be real, and in general exhibited in the common-sense distinctions between what things count as positive and what things count as privations.

Accordingly, the right approach would seem to be this. We shall say that there is, firstly, a fundamental use, or family of uses, of the word "exist" in regard to the metaphysically real, so that the statements in which the word "exists" is thus used carry an implication of being metaphysically real: this may count as one primary sense. Then we shall say that there is, further, on the one hand, a spread of senses of the word "exist" expressing various determinate relations to this use in regard to the metaphysically real or substance, and, on the other hand, a tightly governed set of uses of the word "exist" arising by abstraction (what we may label 'formal' abstraction), described and put into application by the logical theorist.

In effect, then, we should first distinguish one fundamental sense, connoting metaphysical reality or substantiality, and a spread of derivative uses arising from it, still carrying some metaphysical connotation arising from the fact that they express some determinate relation to substance, and, in the most important cases, something we tend to call positive existence (all of these exhibit the same formal features which are noted by the logician and logical

[26] The story was often told by the late Gilbert Ryle of how he and a companion, discussing the nature of existence, determined to ask a waiter what he meant when he said of a thing that it existed, and after a pause he replied, 'I mean it's alive'.

theorist). At this point, we should secondly distinguish, as a distinct sense (or group of three senses), those uses of the word "exist" in the application of logic in which these formal features remain, but all connotation of substantiality or determinate relation to substance is absent from what is expressed. That is, this seems the most appropriate way to use the word "sense" in this case.

The rules governing when we should speak of a logically distinguishable use of an expression (one recalls Wittgenstein's dictum in the *Tracatus* at 5.451 that when a term is introduced it should be introduced at once in all its combinations), and when we should speak of a distinct *sense*, are somewhat arbitrary. They appear to be determined by the varied concerns of logicians, learning-theorists, and others, in using the word "sense".

7. TEMPORALITY IS INTERNAL TO ATTRIBUTION OF EXISTENCE IN THE PRIMARY SENSE

So long as it was obscure whether existence, as something of which the metaphysician takes cognizance, was a predicate or not, it was obscure whether or not the use of the word "exist" by the metaphysician had any of the logical features of "existence" in logical theory, and therefore obscure whether we had here an example of complete homonymy, a merely equivocal use of the same term. Until we were clear that the use of the word "exist", which is primary in metaphysics, can coherently be considered a distinct sense, indeed the fundamental sense, of the word "exist" (rather than merely a distinct use of the sort recognized by the logical theorist as such, with extra connotations, e.g. of substantiality, merely arising from its context of application), we were not in a position to discuss whether temporality was internal to it.

We are now in a position to state that temporality, i.e. that which is indicated or expressed by tense or temporal qualification, is internal to existence whenever it is asserted in the primary sense. Human beings, in their linguistic modes of thought and knowledge, use tensed statements even in respect of existence, even in regard to God. The reality indicated by such statements may not be (indeed, in the case of God, is not, as we shall show) temporal, but our statements and thoughts indicative of it are essentially tensed, unlike those of number theory. That which has active power and to which we attribute real existence need not itself be temporal, but it must have at least some temporal effects for us to be able meaningfully to attribute real existence to it.

An object is typically in space in virtue of its having boundaries which are where it is susceptible of being acted upon in certain ways, e.g. touched. The primary way of being located in space turns out to be associated with the capacity to be acted upon, as well as to act. However, as we have seen, time enters into our thought and discourse in a very different way. Indeed, we find it natural to think of it as primarily a property of events and states, rather than of bodies. More precisely, it is rather a modal component of, or qualification on, the verb, copula, or predicate of a proposition, or on the proposition as a whole. Accordingly, existence, as presupposed by statements attributing states, actions, and passions, belongs to substances according to the tensing appropriate to attributions of action and passion and attributions of the nature exercised in the action or exhibited in the passion.

Since, if God upholds things in existence, and all other things are impotent in this regard, He, in His own nature, must be immediate to the upholding in question, so that for instance, if the upholding is present, i.e. the upholding of things now, His existence must be present, i.e. it must be correct to say "He exists now".

The theory has been proposed that God is present everywhere and at every time in the sense that He acts everywhere and every time, and He is present in that His action is present. Thus far, the theory is unobjectionable. However, the theory then proceeds to explain that God's action is here and now, or in space and in time, in virtue of His created effects being here and now or in space and time, so that for God to exist here and now or in space and time turns out to be nothing more than His effects being here and now or in space and time, which on the face of it makes God present only in a metaphorical, transferred, or derivative sense: it is His effects that are literally present and He might be as alien to space and time as the numbers or other abstract objects, or at any distance from or no determinate distance from his effects.

What has happened in this theory is that the unity of an action as such has been lost sight of or conceived of as compatible with there being intermediates between some primary will-act and its effects—in a view compatible with supposing that, in God, the primary act expressive of will in regard to creation is an eternal act, internal to

God's own mind, rather than the performance willed itself. If the communicating of existence as such, as in creating or upholding in existence out of nothing, as I argue in this book, excludes intermediates, then this dissolution of an action into supposed parts (some mental and some material) is excluded beforehand in the case of God's action towards creation in respect of existence. The temptation to such an analysis would be removed by a proper account of human action—I shall refer to the dualist mistake, of breaking an action up into mental and material parts, in more detail in my next section. The point is here that the time of God's acting is not 'in eternity' but here 'in time', at the time of what happens. In the present connection, it is sufficient to note that for "God exists now" to be non-metaphorical and therefore to be incompatible with the statement "God does not exist now but it is only His effects that exist now, while He remains in His eternity", it is vital to maintain the dependence of the tensing of existential statements about substances on the tensing of predications of their actions—and vital that the time of God's actions is the time of its 'effects'. In this way we resist the disintegrative account of an action which reduces it to a mental act of will and its effects.

Thus, the instinct of those philosophers who associated existence with being temporal was not wholly misguided. Their mistake was to think of things which act in time as being in some sense thereby implicitly *contained* within a temporal framework. No such inference can be made. If one describes what is contained in a temporal framework as 'in time', the mistake made by such philosophers as the empiricists, and the early Russell and Moore, has been to identify having a genuinely tensed existence with being 'in time'. This mistake is in fact much older. Many of the convolutions of scholastic discussions in this area, including those of Aquinas, involving peculiar types of time or non-time for the angels, intermediate between time and eternity, and inviting the suggestion that for angels succession involves 'time in the imagination', arise from the same confusion. The opinion I would offer as preferable is that, besides the temporality which enters into all human statements, discourse and thought as such in regard to the real, one can speak of temporality in a different way as *intrinsic* to the nature of a thing in order to signify that the very life or existence of the thing is realized in a succession of acts or activities, and commonly a succession of states—and, in living creatures, a developing succession of states arising from these activities. In this second way, everything which is supposed to have real existence, angels as well as natural bodily things, except God, will be temporal.

It does not seem right to take as ultimately separate questions the question as to whether a being or substance is the subject of contrary accidents and the question whether it is internally temporal in nature. If contrary things hold of the same subject in virtue of or involving something internal to that subject, then this can only arise by the contrary things being true as we say 'at different times', however this is to be understood. It is a mistake to confine the notion of time and of that which is temporal in nature to the natural Universe, understood as the sphere within which time has a continuity and measure because of the existence of continuous spatial motions, some of which are regular. The notion of the "nature" of a thing is on the one hand the notion of an object of understanding, i.e. what we understand in understanding *what* the thing is, and on the other hand is the notion of a root or principle of the activities in which a thing's life or existence is centrally and indispensably realized. If these activities central and indispensable to a nature are temporal, then the nature is compresent with and exercised in the activities in such a way as to be temporal in the same way. If the central indispensable realization of God's life and existence is not successive, then despite the tensedness of our statements about God, it will be false to describe the reality or nature they indicate as in itself temporal. But in the case of all other beings to whom or to which we ascribe action or passion, it seems that this is indispensable to the realization of their life or existence, and so accordingly they should be described as temporal.

It may be objected that I am putting forward a view according to which God, if He exists at all, is eternal, including no process within His essential activity and life. This, it is urged, would commit me to an objectionable view of God as simple, undeveloping and impassible, unaffected by desires or love, and so (thus the objection runs) not coherently described as living at all. However, the view I have put forward does not exclude there being logically distinguishable aspects of God's activity and life, or even certain kinds of priority between them, but only the absence of any distinction belonging to factual or causal (by which I have said I mean efficient causal explanation) explanation. It is because of this last that there must be no passage from unrealized to realized potentiality, and therefore no real distinction between potentiality and actuality, and there must be no causal relations between the respects distinguished, and no independence of the existence of one respect from another. We may better offer a furnace than a geometrical point as an image of the actuality which God's life and existence consists in, albeit the context is one in which no image is in the least proportionate: as it is said, God is unimaginable as well as incomprehensible. Love does not essentially consist in imaginable feelings, but in dispositions and activities of heart and will, wherein something is sought or delighted in: Aristotle demonstrated that pleasure, satisfaction, or delight was not a process or emotion involving stages, but a state or act. The doctrine that God is simple is essentially negative in content: there is no diversity in God involving causal order or independence, in the sense of the word causal which I have delimited; the doctrine says nothing about types of diversity not involving consequences in the sphere of causal explanation.

The question whether there is one temporal order or more than one is determined by the interrelation of the temporal and the causal and depends upon the question of the degree of integration or unity of the 'world'. It may be that considerations of God's providence and wisdom imply a close unity in the world, but I do not argue that the conditions of human knowledge imply such close unity and integration unconditionally. Likewise I have implied that all effects of which we can know are temporal, and that all causes of which existence could meaningfully be asserted must have some temporal effects, but I have not argued for the plausible view that all effects must be temporal.

8. EXISTENCE AS THE OBJECT OF THE ACTION OF A CAUSAL AGENT

(a) Existence and creation

We can now open out the perspective whereby all existing is the action of God.

According to Anscombe's analysis of action in her book *Intention*, so far as what happens is intentional, the agent does what happens (i.e. the specification of what happens according to intention is the same as the specification of what is done intentionally).[27] In order to give a coherent exposition to this analysis, one must distinguish between, on the one hand, what is done formally speaking, e.g. the potter makes the pot, and his making the pot is what is done, the specification of what is done including reference to the agent, and, on the other hand, what is done materially speaking. In the case of the potter, what is done materially is that his hands move in various ways germane to the moulding of the clay, the placing of it in a kiln, and the later removing of it from a kiln; this is, as it were, materially what is done by him. The distinction could be expressed in terms of a description of the action agent-wise, and the description of the action object-wise.

Aristotle says that, when action and passion are correlative to one another, the action is the change, and this is in the patient, not in the agent as such. He here has in mind common-sense examples such as the cutting down of trees or the moving of objects from place to place in which the changes to which the action is directed are in the objects or matter which the action has to do with, rather than in the agent *qua* agent, although, if the agent needs to use hands or other parts of his body in order to achieve the change, reference to these parts of the body and their movements may be integral to the material or object-orientated description of what is done.

As an aside, one should remark that, upon deeper examination, it may turn out to be one of the grosser defects in the dualist account of human action that it substitutes a division of the action into two parts, a mental part and a material part (albeit that the mental part is described in a way intentionally 'referring' to the material part), for this distinction between global, agent-centred descriptions of an action and material, object-orientated descriptions of its content. The unravelling of the absurdities of dualism belongs to philosophical anthropology (so-called philosophical psychology), and our argument does not depend upon the refutation of dualism; but, if dualism is incoherent in just this way, then its refutation would serve, by example, to exhibit the coherence of the pattern of

[27] G. E. M. Anscombe, *Intention*: detailed references and fuller explanation of her view are given in Chapter I, n. 16, cf. Chapter III, nn. 10 and 11.

description which I am contending that philosophical cosmology forces upon us.

In the case of the coming into existence of temporal things and their continuance in existence, there is no logical incoherence in regarding them as things done by God. They are happenings or goings-on and they are, it may be held, object-wise or materially speaking, what is done when God creates something or maintains it in existence. 'Creating' and 'maintaining in existence' are formally speaking what God does, i.e. what He does described agent-wise. He is held not to use any bodily parts or tools in order to create or in order to maintain the substance of the world in existence through time, so that, if we compare Him to an artist or writer, we speak metaphorically, since He does not utilize a finger or brush. But, speaking in this way, it is as if the existence of things is drawn by the movement of His finger or brush. Thus, for each thing its present existence is constituted or done by the present action of God; its past existence is the expression or realization of God's previous action; and its future existence, if it has any future existence, will be the expression or realization of God's future action; but, since the future in no way yet exists and God's future action in no way yet exists, this action is free and the future (so far as God's action is concerned) is radically open.

All this is not to say that God creates, as it were, instant by instant, as if periods were composed of instants (a view which Aristotle refuted in the course of resolving Zeno's paradoxes[28]). Rather God creates things as having certain natures, and it belongs to these natures that the growth, activities and other expressions of their existence are not instantaneous, so that the possession of a nature presupposes some continuance, subject to the chances of 'this-worldly' existence, but presupposes this only within the context in which at any juncture its future does not exist at all but is radically open in the way we have outlined, dependent upon the action of God whereby things continue.

It belongs to the way of thinking referred to above whereby the agent does what happens that the time of the doing is the time of the happening so that the time of God's acting will not be in eternity but in time, the time of the happening (the view we anticipated in our last section). One may add that in the same

[28] Cf. Chapter II, subsection 1(*b*) and its appended note.

general non-dualistic way of thinking the logically primary expression of will is in the action, not in the intention, so that the primary time of the willing of the action will be the time of the happening: it is only then that the matter is 'settled' and 'fixed'.

If the matter is viewed in this way, God is not the grammatical subject of existential statements about creatures but nonetheless remains the first and only unconditional agent of their existing. In the case of creatures, the grammatical subject of the statement of existence, although linguistically by a metaphor we may speak of it as exercising or as having existence, is not in any way an agent of the existing concerned. For the creature to be an agent of its own existence would involve its existence being a predicate of itself, and would involve vicious circularity, since it needs to exist in order to act and its action is an expression of its nature, nature and action being both alike in presupposing existence. Thus, in the formal description of God's action, i.e. its description agent-wise, He is the first and only unconditional agent or doer of all the existing, coming into existence, and continuing in existence, that we find in the world of temporally existing things, i.e. of things the realization of whose existence is successive, or more precisely the agent of all this so far as the existence concerned is real and positive. But, materially speaking or as described object-wise, creatures are the grammatical subjects of all this, and to say that their existence is an actuality is not to mark that this existence is their act as agents but only to mark that it is not a thing or substance but signified by a proposition (of which the creaturely substances are grammatically subjects) and of the character outlined in Part II of this chapter.

The positive existing of creaturely substances is the existing of things that cannot exist instantaneously but only through time, a time through which they exercise their characteristic activities and passions according to their natures. It cannot then be an existing which God maintains without their activities in exercise of their natures continuing, and their activities are in general theirs only, not His, i.e. not activities of which He is the subject. But His upholding does not causally depend on their activity but only vice versa, as is explained at the beginning of Chapter VI. And all the positive existing presupposed to their activities is His activity.

In this subsection my purpose has been to exhibit the conceptual features and coherence of the conception of creation of temporal things and the maintaining of temporal things in existence as the work of a personal First Cause.

(b) The logical features of God's existence

The question next naturally arises: what of God's own existence? If His existence is an actuality signified by propositions within which He figures as grammatical subject, is He describable as agent of His own existence?

To this question, our answer has to be carefully framed. God cannot be described as the agent of His own existence without the canons of logic and of explanation being violated. However, His existing, *qua* an actuality (existing is an actuality, not a substance), is identical and in no way distinct in the order of factual or causal explanation from His actions of loving and understanding; and these are things of which we can say that He does them. That is, His existing, *qua* existing, is not something God does but as an actuality, although not as an object of thought, is identical with His understanding and loving which are things that He does. That is, we cannot say that His existing is something He does, but we can say that it consists in something He does.

Hence all the positive existing presupposed to the activity of creatures, whether the positive existing of God or of creatures, either is or consists in an activity of God. Or rather this will be the consequence if our argument to God's existence in Chapter VI is valid.

PART II. THE 'RELATION' OF SUBSTANCES TO THEIR EXISTENCE: THE NOTION OF ACTUALITY

9. THE PRESUPPOSITIONS OF CAUSALITY AND CONTINGENCY: THE NOTIONS OF COMPOSITENESS AND INCOMPOSITENESS INTRODUCED

(a) *Preliminary remarks*

I come now to the key problem: the problem as to what is to be said about the relation between the first substances which are the typical primary subjects of existence and the fact or 'actuality' of their existing. Considering this problem will allow us to introduce the key terms "compositeness" and "incompositeness".

The overall structure of the argument of this book is that temporal things are impotent to give continuance of existence into the future to the temporal, and that the underlying reason for this is such that anything not intrinsically underivative (for internal reasons incapable of being caused) would be equally impotent. In particular, I shall argue that what is thus impotent is so because it is 'composite', so that anything composite would be equally impotent and any cause of temporal continuance must be 'incomposite'. It is important to recognize that an argument of this general structure may stand even though my particular explanation of compositeness may be faulted—a general programme is presented in this book whose separate parts need separate consideration.

I shall explain the notion of "compositeness" in very abstract or formal terms, in effect redrawing the picture presented in the early parts of Wittgenstein's *Tractatus*. We shall find that the matter stands somewhat in the way Anscombe stated it in *An Introduction to Wittgenstein's Tractatus*:

> Here it is worth remarking that the truth of the *Tractatus* theory would be death to natural theology; not because of any jejune positivism or any 'verificationism', but simply because of the picture theory of the 'significant proposition'. For it is essential to this that the picturing proposition has two poles, and in each sense it represents what may perfectly well be true. Which of them is true is just what *happens* to be the case. But in natural theology this is an impermissible notion; its propositions are not supposed to be the ones that happen to be true out of pairs of possibilities; nor are they supposed to be logical or mathematical propositions either. (p.78)

More accurately, what is death to natural theology is the logical structure presented in the *Tractatus*,[29] however this be based, rather than precisely the picture theory as such as presented in the *Tractatus*[30] on which Wittgenstein based it.[31] And the reason why the logical structure is death to any account of the intrinsic necessity in God's existence is that it renders impossible any account of ordinary intrinsic possibility, in particular the intrinsic possibility of non-existence represented in what is contingent and in what is caused, as we shall show in this section and Section 10 (the intrinsically possible is that in the case of which there is no intrinsic reason for its being impossible). In Chapter IV we showed that logically singular existential propositions are not to be reduced to general propositions (e.g. existentially quantified propositions) and so are as much to be described as 'elementary' with respect to the structures available in propositional and predicate logic as are any other propositions; and we are exploring the presuppositions of saying of these that some of them are contingent, so that in this way there is such a thing as contingent existence of ultimate (albeit not simple) objects. For Wittgenstein, there was no contingency in the existence of the ultimate objects (supposedly simple) referred to in our propositions and subject-predicate propositions about them were contingent but could only fail of truth by the predicate's not holding of its subject, not by that subject's not existing.

Such abstract presentations are difficult to understand except as illustrated through temporal examples such as I provide in Section 10. This makes it tempting to argue that the composite has to be temporal but I do not do this, for reasons I give in Section 11.

(b) The distinction between objects and 'facts' or 'actualities'

What, then, of the relation between a first substance and its existence?

The existing of a first substance ('its existence') cannot be conceived of as prior to the first substance whose existing it is. Nor can the first substance be conceived of as prior to its existing, since without its existing it is nothing. The question is as to what, if anything, may sensibly be said about the 'relation' between them.

[29] e.g. in *Tractatus* 1 to 2.063 and 4.1 to 5.135.

[30] e.g. in *Tractatus* 2.1 to 3.01 and 4.01 to 4.022.

[31] Roughly speaking, the logical structure is presented in chs. 1–3 of Anscombe's *Introduction* whereas the picture theory is presented in ch. 4.

The existing is not something predicated of some quasi-abstract subject (a kind or concept, an 'essence' or 'possible'),[32] but nor is it something predicated of the first substance itself, for existence is not any kind of predicate logically or metaphysically of anything. Hence to speak of this existing or existence as an attribute is empty, explaining nothing. From the standpoint of logical theory and the philosophy of language the substance and its existing are spoken of in different ways and the expressions signifying or denoting them are not interchangeable. But does this formal distinction or distinction in the modes of significance of our idioms correspond to any distinction in reality? and, if so, what?

We speak of one statement as, in the logician's sense, explained above, 'presupposing' another statement. But what is of interest to the general philosopher or metaphysician is something different, namely, one fact or reality picked out in one species or spread of linguistic expressions, presupposing another fact or reality, picked out by another species or spread of linguistic expressions, which is (*a*) distinct, not necessarily by either of the two being independent of the other, but by this difference amongst or between the modes of expression appropriate to them, and (*b*) related in some way of which cognizance belongs to the theory of explanation. I explained in Section 4 above that it is considerations concerning explanation that alone can give us access to the real as such, so that our statements are not merely true, but speak of objects real in themselves (i.e. refer to *supposita*), and state facts real in themselves (i.e. signify 'actualities' or 'positive facts'). A mere difference in the spread of linguistic expressions, appropriate to two thereby logically distinguishable subjects of discourse, does not of itself carry with it any presumption of there being some relation pertinent to efficient causal explanation between the two subjects distinguished.

It will be noticed that I am raising the question of the status of the distinction between two 'kinds' of thing whose character is indisputably positive (i.e. not to be explained in terms of the absence of something else): on the one hand, 'real objects' (the first substances of Aristotle's *Categories* or *supposita* of Aquinas or

[32] On the mistake of seeing existential statements, not as statements in regard to objects, but as statements ascribing properties to concepts or to essences whether conceived of as abstract and general or as *possibilia*, see W. Bednarowski, 'The Riddle of Existence', and É. Gilson, *Being and Some Philosophers*, pp. 74–135.

suchlike) and, on the other hand, 'actualities' or 'positive facts' as things signifiable amongst human language-users only by complete propositions (statements, judgements, or thoughts) but still essentially positive rather than privative in nature. This is the distinction I made in Section 4 in respect of the Persons of the Trinity, and is a distinction which arises in discussions of perception when we distinguish between the subject of a perception and the perceiving done by it,[33] and between the perceiving of an object and the object itself,[34] two distinctions ignored by Hume.[35]

It is useful to take up the term "*suppositum*" as a technical term to mean what has often been meant by "subject" or "first substance" in order to avoid the psychological or epistemological, Cartesian or Kantian, associations of the term "subject", and to hold back from the varied developments of Aristotle's notion of "first substance" in the *Categories*. I shall reserve the term "*suppositum*" to mean 'subject' in the sense dictated by logic and the philosophy of language made clear in Sections 3 and 4 of Part I.

In my argument in this section I deliberately leave it open whether or not the terms "actuality" and "positive fact" are synonyms, even though this may make my presentation seem more abstract and non-constructive. I do this to avoid its appearing that my argument at this stage depends upon the notion of "actuality", and the notion of "existence" as an 'act' or 'actuality', commonly supposed to be very obscure notions. True, it is very natural to suppose that we do form some conception of 'actualities' as objects of attention and knowledge in the world—e.g. episodes and situations—of a kind more global than 'facts', and to which facts stand, as it were, as aspects. These episodes and situations or 'actualities' might well seem to be better candidates for having reality or positiveness ascribed to them than any kind of 'fact', and it would indeed seem that it must be in virtue of some special relation to them that some 'facts' are to be described

[33] Cf. Sydney Shoemaker, *Self-Knowledge and Self-Identity*, pp. 63–80, on the distinction between the subject and his perception or perceptions. Berkeley, much earlier, likewise respected the Aristotelian distinction between a subject and its operations and passions, including its perceptions.

[34] The importance of the distinction between the perceiving and the object of perception is made plain by M. Merleau-Ponty in the 'Introduction' to *Phenomenology of Perception*. The same distinction appears in Berkeley in the opposition made between spirit and its operations (including perceiving and imagining) as things of which we have "notions", on the one hand, and ideas as the objects of these operations, on the other.

[35] In the *Treatise*, bk. I, pt. IV, sect. V, Hume argues that because the objects of our perceptions have contrary properties, the perceptions themselves must have contrary properties and therefore even the fancied subject of perceptual operations. In this his argument seems to depend on first confounding objects of perceiving with perceiving and then perceiving with the subject which perceives.

as 'positive' in some absolute sense. In terms of such a notion of 'actuality' it may seem possible to give a less arid or abstract, more constructive-seeming, account of the relationship between 'substance' and 'existence' and of the notion of "compositeness"; and this may be found in Section 10. But my argument does not at this stage depend upon this examination.

Some philosophers may query this use of the word "signifiable". But to use the words "refer" or "indicate" would be misleading, since these 'things' for which I have offered the labels "actuality" or "positive fact" are precisely things requiring a propositional form of reference or 'indication' as primary, i.e. a form wherein their nature and identity were 'expressed' rather than merely 'denoted'. Yet the problem which arises with supposed singular terms or proper names from the fact that some are empty (e.g. there never has been any Cerberus) arises in a less easily side-stepped form with propositional expressions which are false, and because of this 'indicate' no actuality and no positive fact (i.e. 'indicate' nothing) while still obviously in one sense 'signifying' something.[36] However, the answer is straightforward: the word "signify" has to be disambiguated, as between signifying (in the sense of expressing) a

[36] There has been throughout philosophical history a rivalry between two opposite tendencies in considering 'the significance' of the linguistic entities variously called statements, propositions, sayings (*sententiae*), and utterances, or of noun-clauses such as those expressed in English in the form "that *p*".

On the one hand, there has been a recurrent tendency to regard them as signifying facts or actualities of some kind, and to have their meaning or sense determined by which fact or actuality they signify or refer to. Thus knowledge that *p* might seem to be knowledge of a fact or actuality (even though it cannot be analogous to acquaintance since knowing the fact that *p* is the same as knowing that that *p* is a fact). However, we can say or believe falsely that *p*, and this cannot be a saying or believing of a fact. This was plain to Plato in the *Sophist* and the *Theaetetus*, to Anselm in his *De Veritate*, and to Wittgenstein in the *Tractatus*, and intermittently plain to G. E. Moore in his *Some Main Problems of Philosophy*.

On the other hand, these linguistic entities have been held to "signify" more abstract seeming things described as 'senses', meanings, 'propositions' (in the more abstract sense), or thoughts.

The complete nonsense which results from confusing these two uses of the word "signify" can conveniently be highlighted, as a result of the significance of tensing and as a result of the existence of contingency, because it generates such paradoxes as that marked by G. E. M. Anscombe by the label "the fallacy of being guided by the truth in identifying the sense of a proposition". In the case of a statement such as "The Germans will reach Paris", it might seem that up until May 1940 the statement (whenever uttered) 'signified' the occurrence referred to which took place in that month, and, if it had not taken place then but later, would have 'signified' the occurrence whenever later it did occur, but that, if at the time of utterance the occurrence had already taken place or if it was never later to take place, then the statement would 'signify' nothing.

sense and, through this means, signifying (in the sense of indicating or referring to) an actuality or positive fact.

Of course, there will be more than one distinction to be made. There are 'empty' 'names' for planets, dogs, and horses, *inter alia*, which have been wrongly supposed to exist, but also 'empty' names or designating expressions for supposed holes, gaps, and suchlike, likewise wrongly supposed to exist. The situation with propositional expressions is more complex: but again there are false statements of something expressedly positive in character, purporting to refer to, but not actually referring to, an actuality, and there are true statements expressive of privation, the first group failing to refer to actualities or positive facts for a reason analogous to the reason empty names fail to refer, and the second group failing to refer to actualities or positive facts for a reason analogous to the reason expressions denoting holes fail to refer to anything positive.

Linguistically, there is an impassable gulf between, on the one hand, expressions for objects, i.e. what the logicians call subject-expressions, and, on the other hand, expressions of the sort which the logicians call propositional. The latter are the expressions which, in English, are signified by noun clauses such as those represented schematically by forms like "that *p*",[37] i.e. expressions signifying what we may call thought-contents. However, there is a question as to whether this division in linguistically formulated speech or thought corresponds to any division independent of the modes of knowledge of language-using animals such as human beings, at least in the case where what is signified is a fact, i.e. *p* is true. I.e. does the distinction or distinctions between 'subjects' or '*supposita*' and 'actualities' or between 'subjects' or '*supposita*' and 'positive facts' involve a distinction internal to reality as it is in itself?

(c) Presuppositions of causality and contingency

If Wittgenstein, in the *Tractatus*, had been right that the ultimate statements about reality have only one way of being false, namely by their negations being true, and not two distinct ways, *viz.* one, by their negations being true, and another, by their subjects not existing (so that both they and their negations fail of truth), then the statement that existence is not a predicate would have no

[37] Linguistically, the noun clauses of the form "that . . ." are analogous to noun clauses opened by the interrogative adverbs "what" and "which" (not the interrogative adjectives or pronouns thus identified).

significance in metaphysics. If Wittgenstein had been right, there would have remained a distinction in our modes of speech between subject and actuality or positive fact, first substance and its existing, but no distinction in reality as it is in itself.

Such then is the problem: what argument is there that bears upon the question whether or not there is some distinction in reality corresponding to this distinction in modes of speech, and what kind or kinds of distinction in reality are to be made? There is an argument to be drawn from the nature of persons as subjects of personal relations and also an argument to be drawn from the nature of perception to the effect that there is some distinction in reality, but it remains obscure what distinction. If there is thus a distinction relevant to personal relations or to perception, is it one which is relevant to efficient causal explanation? Where bodily things (even living ones, such as persons) are concerned, personal relations and the relation involved in perception inescapably involve efficient causal relations. There might be supposed to be personal relations and cognitive relations in some way analogous to perception which did not involve efficient causal relation, e.g. the relations between the Persons of the Trinity, and God knowing 'by sight' of the world. But if these relations do not involve efficient causation, they would seem to be of a different nature from relations which do involve this, and which according to our argument are the relations of primary concern to metaphysics.

Therefore, if, as against the view mooted in the *Tractatus*, those statements which considerations about efficient causal explanation compel us to consider to be statements about the real are yet of subject-predicate form, i.e. have two ways of being false in the way mentioned, then the proposition that existence is not a predicate, but presupposed to any predicate, will have a key place in metaphysics.

It needs to be noted before we proceed that, for there to be an order relevant to explanation between two things, it is required that they be distinct as logical subjects, but not that the existence of either, in the logician's sense of the word "exist", be independent of the existence of the other. It makes no contradiction to say of a nature that it presupposes the thing whose nature it is, or that it presupposes the existence of this thing, or even to regard the thing as presupposing its existence, although presumably none of the three 'exist' independently of each other and plainly there can be no relation of cause to effect, even though there may be relations relevant to efficient causation between any of them.

We have referred to the impassable gulf in our modes of speech between objects and actualities or between objects and positive facts. If Wittgenstein was wrong, then this impassable gulf expresses a distinction intrinsic to the real as it is in itself. Now we cannot speak of substances or 'things' as such as contingent or otherwise, but only of their existence or of some activity or passivity on their part as being so. Accordingly, if there is any real contingency in respect of existence as such then this must be the contingency of the existence of some substance or 'thing', and *for there to be such contingency there must be a distinction (evidently not the distinction between a cause and its effect) between the subject which is said to exist and its existing*; and it will have been primarily or initially considerations of explanation, in particular explanation of the future, and indeed we may add of factual or efficient causal explanation, rather than any other sort of explanation, that have driven us to this conclusion. Likewise, it is vital that causation is never of a thing but always a causing that a thing exists or that such and such is the case,[38] so that in the same way *for a thing to be caused in any respect requires a distinction between the subject and its existing, whether the latter is directly caused or presupposed in a change which is caused.*

In either case the distinction between a thing and its existence is implied as an internal condition of the real possibility of being caused, whether caused in such a way as to be necessary or caused in such a way as to be contingent. What is caused, whether necessary or contingent, has an external or prior ground of possibility and, if necessary, an external or prior cause of necessity. There is nothing in the internal structure of the thing that exists and its existence to make either necessary. And for there to be room for

[38] Knowledge that so-and-so is not a relation (e.g. acquaintance) with an object with the result that if we speak of such knowledge as a knowledge of facts or actualities it is so in the sense that excludes facts or actualities from being objects (cf. nn. 35 and 36 above): it is the analogous logical point that we are making in regard to what is contingent or caused—except whereas we can speak of knowledge of facts or actualities as a knowledge of truths we cannot say that what is really contingent or caused is primarily a truth. (Some ordinary language argument to the same effect appears in Z. Vendler, 'Effects, Results and Consequences'.) Obviously, what is caused has to be distinguished from a product and is rather the existence or coming to exist of the product, but the key point is that what is caused is a *motus* or actualizing in a sense in which a product cannot be.

or no internal impossibility in there being a cause or for being contingent there must be the same internal structure as is required if there is actually a cause or real contingency.[39]

Therefore, firstly, since where there is a cause or real contingency the distinction between the thing that exists and its existing is a condition of having a cause or being thus contingent, the distinction is positively pertinent to the efficient causal explanation of the thing or its existence and must be regarded as a distinction in reality, i.e. a 'real' distinction; and, since the same internal structure is concerned, there must be the same 'real' distinction if there is to be even room for a cause or this contingency. And, secondly, wherever we have a reality which is not intrinsically underivative, i.e. not for internal reasons incapable of being caused, this real distinction must be present.

We can gloss this argument but not substantially add to it.

For an effect to be an effect it must be contingent upon other things whose actuality is not derivative from, dependent on, or contingent upon it. This causal background constitutes a field which under one aspect does not include but sets a basis of possibility for the effect but which under another aspect determines (efficiently causes) that the effect is actual. For there to be an efficient causal relation the effect has to depend upon its background for its possibility, not only for its actuality, and has to have its possibility prior to its actuality.

Even if the causes are necessitated in their actions so that the effect is necessary, this necessity is not internal to the effect which, considered as it is in itself, is of such a nature as to have a real possibility, not just a logical possibility, of having not existed. Any supposed necessity must be external and merely *de facto*, not intrinsic, if the effect is indeed an effect. Still there is an external ground of possibility and the possibility of the effect is prior to its actuality.

So far as the ontological structure of a thing or its existence or what is internal to it is concerned, there is no difference between that which is caused in such a way as to be necessary and that which is caused in such a way as to be contingent: in both cases there has to be a distinction between a thing and its existence if there is to be even room for an external ground of possibility.

In the same way, in what does not have this ontological structure whereby the thing's existence is really distinct from the thing, its focus, there is also no distinction which would allow one to state counterfactual

[39] This seems a valid application of Aristotle's principle that, if A is possible, nothing impossible must follow if A is supposed to be real (*Metaphysics*, Theta, 4, 1047b 14 ff.). If there is room for a cause or real contingency, the same internal conditions of possibility must be satisfied as if there is actually a cause or such contingency.

conditionals about its existence, e.g. so as to say that, if some cause had not caused the existence of the thing, the thing would not have existed. For there to be room for counterfactual conditionals there has to be the same distinction between a thing and its existence as would be implied by the contingency of the existence of the thing or by its having a cause of existence and external ground of possibility.

So much for the gloss.

For the existence of a thing to be necessary, of course, likewise involves the distinction of two logical subjects, the thing and its existence, expressions for which are not freely inter-substitutable. But this distinction is not as such positively pertinent to the efficient causal explanation of the thing. Rather, in the case of anything intrinsically underivative where there is no room for a cause or for an external or prior ground of possibility, the function of such an assertion of necessity is instead to represent any such explanation as excluded.

At this stage I deliberately state the matter in an exceedingly abstract or formal way because I do not believe my argument depends upon those further explanations which I offer in the next section, which serve to make it more accessible to the imagination or more constructive. But this next section may also help the reader to grasp the formal structure involved.

The inquiry before us has, therefore, been this: does the form of our statements give us access to the real in its nature and relations as these things are in themselves? And we reduced this inquiry to the questions as to whether ultimate statements about reality, those statements which considerations about efficient causal explanation lead us to consider as ultimate, include statements of subject-predicate form such as can be false in two ways. And we showed that the existence of real contingency, as connected with the reality of time, forces upon us the answer: yes.

Recognition of the reality of time thus forces upon us the recognition of real (time involving) contingency and this demonstrates that the distinction within our modes of speech between a temporal thing which exists and its existing corresponds to a distinction in reality. By "distinction in reality" I do not necessarily mean a distinction between two 'things', if "thing" is any concept belonging to a first-order science, even metaphysics; rather I mean a distinction between two logical subjects (subjects of which predication is made) positively pertinent to their efficient causal explanation and, because of this, something of which first-order sciences of the real *qua* first-order sciences of the real have to take cognizance.

(d) Introduction of the term "composite"

We are now in a position to introduce the term "composite", leaving open at this stage whether or not the 'composition' is between 'things' in any first-order sense. Temporal things, and anything whose existence is caused or contingent, are (we may say) 'composite', defining compositeness in these rather abstract terms: a thing (or, we shall equivalently say, a thing's existence) is composite if and only if the distinction between the thing and its existence is positively pertinent to the efficient causal explanation of the thing's existence. As Section 11 will make plain, being composite is equivalent to having a 'nature' which is a quasi-abstract object and to not being intrinsically underivative. And since all and only temporal things exemplify the distinction between 'being of a nature' and 'exercising a nature', or in Aristotelian terms 'first act' and 'second act', anything for which this other distinction is real will be composite. The point or strategic significance of this seemingly abstract concept of compositeness in philosophy will become clearer as we proceed.

The distinction of a thing into different parts only implies the compositeness of the thing in the sense we have explained if it implies the possibility of having an efficient cause or being contingent. There needs to be a reason for supposing the distinction between different parts of a thing, e.g. the parts of a natural or essential unity, to involve that the thing could have an efficient cause or have real contingency. Thus, the Universe would seem to have parts, but it has been alleged that it neither has nor is sensibly conceived of as having had a cause, there being supposedly no background to it which might leave it open to be appropriately supposed either that the background necessitated the Universe's existence or that the background left it contingent whether the Universe existed or not. Against this position some argument is needed, e.g. such argument as that provided in the text that the Universe's supposed temporal parts are so related that the earlier ones leave the later ones absolutely contingent with the implication that the Universe itself as a trans-temporal whole including the future is not a reality in the required sense. It does not have necessary existence considered as a single whole, since it has contingent parts whose contingency is not merely relative to it as an integrated whole, as it is not integrated in the intimate way required, a way which would require an interdependence or symmetry incompatible with the reality of time. (In any case our earlier argument would suggest that the Universe itself as a trans-temporal whole including the future cannot be a reality because it does not have effects.)

One may view the matter in this light. The Universe has been supposed to be *de facto* underivative. But the Universe is not a unitary thing but something some of whose parts at any stage lie in the future so as not to be actual in the present and so is only a candidate for being caused in certain respects, e.g. its inception or continuance. But, so far as its inception or continuance are concerned, there is nothing in its nature or the structure of its existence to preclude a cause of existence: it does not matter that, the Universe not being unitary, what makes it up lacks any absolute simultaneity, since whatever makes it up can have its existence caused by the cause concerned, by this cause's acting at each and every locus or juncture within it or, where inception is concerned, at some initial juncture or field of junctures. Thus, if the Universe were underivative, it would only have been *de facto* underivative, not intrinsically so.

The question of the existence of God thus turns crucially on the question as to whether or not there is a First Cause which is not composite in this sense: a being which is eternal, not temporal, and within which no distinctions relevant to efficient causal explanation in the way indicated are to be made. (If there are distinctions between Persons in the Godhead, these distinctions do not involve efficient causal relations between the Persons or differences in efficient causal relation to things in the world, although they may involve different modes of speech in regard to these causal relations and differences in personal relation to created persons.)

10. THE NOTION OF "ACTUALITY" AND THE POSSIBILITY OF A MORE CONSTRUCTIVE ACCOUNT OF "COMPOSITENESS"

Up until now I have couched my explanations in terms which are non-committal in regard to the question whether or not we can form any viable conception of "actuality" or "act" distinct from the notion of "positive fact".

But in Sections 1 and 2 we showed the need to distinguish between things with 'positive' existence and other things to which existence is ascribed, as if statements of 'positive' existence stated 'positive facts' in a sense in which other existential statements did not. Yet a 'positive fact' would seem *qua* fact to be merely, like any other fact, something statable in a true proposition. Therefore its positiveness would seem to arise from some special relation to something to which reality is to be ascribed in some stronger sense than the sense in which it is to be ascribed to facts in general. In the case of the primary statements of 'positive' existence, *viz.* statements of the existence of substances, it has been suggested that this reality is an 'act' of existing, the word "act" being used in an exceedingly formal or undetermined sense.

(a) *Informal approach to the notions of "situation" and "actuality"*

The relevant notion of "act" or (the term which I shall prefer) "actuality" appears to me best approached by considering our ordinary ways of speaking of situations, of acts, actions, and activities in the more limited senses of these words, and of events, occurrences, or episodes. For in these ordinary ways of speaking we appear to distinguish such things as situations, actions, and episodes from the 'facts' which constitute aspects of them. We shall end up by speaking of all these things, that is situations, actions, episodes, and suchlike, as 'actualities', thereby setting them in contrast with 'facts'. As will become clearer in (*c*) below, all have some kind of 'positiveness' which is not explained by lumping them together as 'facts' (i.e. what makes true propositions true): this has led some philosophers to attribute 'being actual' not only to so-called 'continuants' (of which substances would be a paradigm example) but also to 'occurrents'[40]—these latter would correspond to our 'actions' or 'activities'.

Let us begin by considering how in common speech we speak of situations. We speak, for instance, of the situation one finds oneself in at a particular time or the situation in a particular country at a

[40] These terms, "continuants" and "occurrents", from W. E. Johnson, are used by P. T. Geach in 'What actually exists' in *God and the Soul*, pp. 70–2.

particular time or the situation within which or against the background of which a certain event takes place, and we conceive of these as inherently global, involving as elements or aspects an unlimited or dense multiplicity of facts. Inasmuch as true statements reporting (or stating) facts state aspects of situations or of elements of situations, it is natural to think of facts as inherently aspectual, the aspects or facts picked out by the statements being distinguished according to the senses of the statements concerned, prescinding from indications of time and place. Statements ascribing natures to substances or to other things or asserting the existence of this or that thing or type of thing within a context will be statements of aspects of situations. The existence of a substance as a fact will likewise be an aspect of a situation.

Despite the fact that situations typically do not have discrete identity, we still ascribe reality to them inasmuch as we treat them as subjects logically of attention, inquiry, and attribution in first-order discussion. We are more inclined to ascribe reality to situations and suchlike (actions and episodes) than to facts because it is situations and suchlike rather than facts that are objects of first-order attention and inquiry. Facts are rather what is stated as a result of such inquiry than what is attended to in such inquiry. And the global and open character of situations would seem to be the result of their being determinative of aspects in the way reality is determinative of the truth of statements.

That we do indeed regard situations and suchlike as real in a strong sense can be seen from their role in efficient causal explanation. And this role of situations and suchlike in efficient causal explanation also secures that their relationship to facts as aspects should be more intimate than and different from the relationship of substances to facts. For in any true statement, "*p*, because *q*", "*q*" states a fact in a context in which it is implied that the fact that *q* is a key aspect of a situation or whole which results in its arising that *p*. This structure within causal explanation secures and explains how it is that situations (and in general the wholes of which facts are aspects) are not things—precisely because of their being thus proximate to the statement/proposition-mediated things which we call 'facts', and because of thus having facts as aspects in a more intimate and different way and sense from the way and sense in which substances have facts as aspects (facts are 'about' substances and in no sense elements of them, whereas it is natural to speak of them as elements as well as as aspects of situations). The way in which agents, situations, and 'acts' or 'activities' are involved together in a closely interrelated way in the structure of causal explanation is explained in (*c*) below.

(*b*) *The intermeshing of situations or actualities with each other*

Situations and suchlike are typically not discrete in the way that things are discrete. For instance, existence at one time is inseparable from existence over a period including that time and inseparable situationally from existence at preceding times or, in the case of a coming into existence, from a background to the coming into existence concerned. And in a slightly analogous way situations in one place are inseparable from situations in adjacent places and in some larger space which includes the place concerned. In these ways it is exhibited that the internality of relations to situations so emphasized by Whitehead prejudices the discrete identity of situations or their capacity to be counted in a way in which the internality of relations does not prejudice the capacity of substances to be counted (legs, wings, and heads are obviously relational but still countable). The concept of substance is such that kinds of substance are typically identified under concepts according to which they can be counted, whereas in the case of the concept of situation the reverse holds: the concept of situation precludes kinds of situation from being normally identified under such concepts.

We have indicated that temporal and spatial situations are not discrete temporally or spatially. But situations, actions, and episodes are also liable not to be discrete in other ways.

Thus, firstly, to take a particular instance, the act or episode of remembering something (bringing it to mind: what would be described as the second act of a potentiality in Aristotelian tradition) incorporates the asset or habit of having the knowledge concerned in possession available to be remembered (what would be described as the first act of a potentiality concerned), and this latter in turn incorporates the possession of the capacity to acquire and hold knowledge (the underlying potentiality involved).

But, secondly, the possession of any such underlying potentiality as a state of affairs or situation incorporates the existence of whatever it is that is of such a nature as to be capable of such possession, e.g. of the animal which is capable of knowledge, or of the world which is in a state capable of giving rise to such and such a situation. And, in general, any situation of potentiality or pregnancy presupposes a background of actualities which are conditions of or constitutive of this situation of potentiality. Any actualities (acts or actions, episodes or occurrences) which presuppose potentialities, and any potentialities, all presuppose some actualities.

In my last remark I appear to be relying upon an understanding of the notion of "actuality" which I was in the course of explaining. But my explanation proceeded by means of giving examples and was intended to exhibit the need for some more general notion than that of "situation", covering 'second acts' which are commonly episodic and not naturally described as 'situations', and not limited to 'first acts'. And the 'actualities' which ground potentialities are not thought of as 'actualities' in virtue of being realizations of other co-existent presupposed potentialities, in the way that bringing to mind is a realization of having in the mind as something known and knowledge is a realization of the capacity to know. Thus we do not think of an animal's existence as being in the relevant sense an actuality in virtue of its being a realization of some prior potentiality, e.g. Aristotelian matter, but in virtue of its being in some way positive or 'actual' in itself.

What has now come to light is an absolutely critical point. The notions of "actuality" and of the "real" and "positive" should not be introduced by means of the distinction between actuality and potentiality. Rather the notion of potentiality enters with the recognition that some actualities are temporal and arise from a causal background which is in efficient causal potentiality to them.

We are now in a position to observe, thirdly, that situations and suchlike ('actualities') are typically not discrete in a further way. Nothing static or attributed to a point-instant of time exists except as constituted by its setting in a context involving activities or episodes over a period including that instant. An animal is alive not in virtue of merely statically being in a state called 'being alive' but in virtue of exercising the activities of life over a period in a way appropriate to being alive (which includes being asleep at some times and allows of being damaged so that many capacities are for a period not exercised): living consists as much in so-called second acts (the activities of life) as in any state they presuppose; this is what is meant in common idiom when it is said that for an animal existence and life are 'dynamic' (the words "active" or "moving" would seem more exact).

The word "situation" connotes a state of affairs rather than an activity or set of activities, but situations include the going on of activities and cannot exist apart from the 'active' within them, and it is primarily for this reason that it is clear that we need some wider term than "situation", e.g. the word "actuality", to express the concept we are trying to identify. Moreover it is never a situation alone, but only an 'act' in the context of a situation, which produces any change. This is part of the structure of causal explanation to be made clear in (*c*) below.

The notion of "actuality" is polymorphous or analogical, not univocal or used mechanically according to one definition, and covers realities with different levels or degrees of 'positiveness', states of potentiality as well as actualizations of these states, none of them in typical kinds of case discrete from each other. Situations have the complex web of spatio-temporal inter-involvement mentioned; second and first acts and mere capacities have the interdependence explained; all count as 'actualities'.

(c) The key role of the notion of "act" or the "active": positive existence explained

But we do not need to view 'actualities', that is acts, activities, and episodes, along with situations, as an unstructured heap. Rather, inbuilt into what we have already said is the conception that the criterion of 'activeness', together with the realization that it is some actuality that renders 'positive' what we have called 'positive' existence and so-called 'positive facts', allows us to see actualities in a certain order. It is some general notion of "act" which allows us to describe the relations of capacities and first acts to second acts and of states or situations to actions or episodes. It is the same notion which allows us to pick out what brings something active into a situation: agents and situations do not, without 'acts' or 'activities', cause anything.

It is vital not to fudge the point here being made. We attribute 'reality' to primary substances or *supposita* and also to actualities, both situations and 'acts' or 'activities', because all have a role in normal cases of efficient causation. Only primary substances or *supposita* are causal agents, subjects of concrete causal predicates or of what Anscombe calls 'causal verbs'.[41] If a thing's existing and being of a nature is prior to its being 'active', a nature or situation is shaping or determinative of its act or activity and so natures or situations have some role in the causation concerned. But no primary substance acts without an 'act' or 'activity', and the mere possession of a nature by a substance or merely a situation involving substances is not enough. Both primary substances and actualities must be 'real' in order to 'cause', but they do not 'cause' in any unitary or univocal sense, but in the structured interrelation just indicated. (It will be noted that it is as causes in some way, and not just as effects, that actualities are candidates for any kind of 'reality'.)

At this stage, when we realize that the positive existence which we have attributed to substances (including that positive existence which they have in virtue of accidents—relations among them—in the way explained in Section 2) is what gives positive character to actualities, we must concede first that if anything is to be conceived of as an actuality it must be primarily this.

This has the implication that in speaking of 'the positive existence' of a substance we may intend to refer to what we have called an actuality, and not be signifying a fact as what is expressed in a true proposition. In brief it is as Aquinas held: we have to distinguish *esse* as an act of being from *esse* as what is asserted in the composition of a proposition (*Summa Theologica*, Ia, Q. 3, Art. 4, ad 2). The *esse* of a substance or *suppositum* as an 'act of being' will, of course, not consist in some mere fact, e.g. the aspect we referred to as a 'being of a certain nature', but in some structure involving interdependent situational and 'active' actualities, the latter typically involving facts of an accidental kind, those whose positive character is to be explained in terms of their indicating a fuller existence or life in their subjects (cf. Section 2 of this chapter).

Thus, upon examination, we have realized that every actuality can be conceived of as exemplifying the positive existence of substances. What we have come to must be either the conception that the positive existence of substances is positive precisely under the aspect of being an actuality or the conception that actualities are actualities in virtue of being elements in something else, something designated as 'the positive existence of substances'. If we

[41] G. E. M. Anscombe, 'Causality and Determination', p. 137.

took the latter view, in an attempt to make the primacy of substances yet more unqualified and isolated, we would have the oddity that actualities are what they are in virtue of being elements in what appears to be a mere fact or else leave a need for an explanation of the word "positive" in the phrase 'the positive existence of substances', having closed the door on the possibility of such explanation. (For this reason we set aside the theory that the set of actualities referred to as the positive existence of substances constitutes a distinct sub-set—a proper sub-set to use the language of set theory—within the set of all actualities, prior to all others, primary in the hierarchical intermeshing structure which actualities taken together make up.)

Granted then that the positive existence of substances is an actuality and not just a fact, the theoretician's question arises as to the relation between the fact of existence concerned and the actuality. The fact of existence *qua* fact is an aspect, we have said, of a situation or actuality rather than itself a situation or actuality. The answer must lie in our understanding of the way a span of true propositions asserting facts picks out a span of situations or actualities and their elements and makes possible the differentiation of these situations or actualities and their elements. Within the structure of non-discrete actualities, there are focuses, namely primary substances, which have reality but are not themselves actualities because picked out by proper logical subject or object-denoting expressions, not signified only by propositional expressions as is the case with actualities. The peculiarity of statements of the existence of substances is that they identify these focuses. What they state *qua* fact is a structural aspect of actuality, rather than an extra element in actuality. Such then is my present answer to this question of the theoretician.

Reviewing now what our account has achieved, we can observe that it leaves no room for any application of the terms "situation", "actuality", or "positive existence" in connection with abstract objects and subject-matters. The logical presupposition of existence which arises even in connection with the abstract, and the distinction between affirmatively and negatively formulated propositions which arises with every subject-matter, are neither of them as such bases for the development of the ideas in this section. It is true that the primary expressions of the structural facts which constitute key aspects of the intermeshing web of actualities, namely the propositions marking out the positive existence of substances, the focuses of the whole system, are affirmative, but this is not of relevance to our present discussion.[42] What is of key importance is that it is the existence, *qua* actuality, of these substances or focuses, which is constitutive of 'what there is' or reality—not any facts, existential, nature-stating ('essential'), or accidental, *qua* facts.

(d) Criteria of identity and distinctness

Having now dealt with the questions that arise about the notion of "actuality" as far as we can within the scope of this book, we can summarize the resulting picture as follows.

We attribute reality to two sorts of things, primary substances or logical subjects proper of the basic kind (*supposita*) and actualities. The actualities form a sea or mesh of non-discrete things of discrepant interrelated kinds, situations, activities, episodes, and suchlike, and within this sea or mesh these primary subjects or *supposita* constitute focuses or centres, subjects of the propositions in which we pick out aspects of the elements of this sea or mesh, aspects which we call 'facts'.

Before filling out this general perspective, we must consider the problem of the criteria of identity and distinctness for such 'realities'.

The question arises as to when two things to which we attribute

[42] E. L. Mascall (*Existence and Analogy*, p. 50) tells us that 'existing, not being an essence, cannot be reached in the first act of the intellect, the abstraction of the concept, but only in the second act, the affirmation of the judgement (*compositio*, as St Thomas calls it)'. After drawing on Gilson's exposition (*Le Thomisme*, p. 44 n. 1, and p. 49), he goes on to say, 'Strictly speaking, I cannot "conceive-a-thing-as-existing", I can only affirm (or deny) that it exists. Or if I do include existence in a concept it is, to use the jargon of the commentators, only existence *ut signata*, never *ut exercita*.'

I believe that it is significant that actualities, which are as such positively existential, are only signified by true affirmative propositional expressions, and never by object-expressions, so that the concrete is only properly attained in speech by the proposition wherein what is signified is as such existence as exercised, not *ut signata*. But the precise statement of this and its interconnection with what Mascall observes is obscure to me. Mascall's point would seem to arise in connection with abstract as much as concrete subject-matters, but it does seem that truth and even the assertion of truth where certain affirmative existential propositions in regard to the concrete are concerned is strategic for the latching on to truth in regard to any matter.

reality are distinct, whether or not they are discrete from one another. Obviously, if they are both *supposita*, the answer will be straightforward, since we will have to say that they are distinct if and only if they are discrete; and they are discrete if they count as two As where "A" is a concept proper for counting. But if they are both actualities, or the one an actuality and the other a *suppositum*, the question is more difficult. What we say in regard to the use of the word "real" must be guided by considerations of efficient causal explanation since it is this that determines the application of the word both to primary substances or *supposita* and to actualities in the first place: it is this connection with efficient causal explanation which takes us beyond merely insisting that both are subjects of first-order attention, enquiry, and attribution. Therefore we must say that actualities are distinct in reality, and that an actuality and a *suppositum* are distinct in reality, if and only if they are distinct in respect of relation to efficient causal explanation in the way explained—and identical in reality if and only if not distinct in respect of relation to efficient causal explanation in this way.

The type of identity and distinction here concerned is not proper or strict identity, i.e. it does not satisfy the Leibnizian requirement that if A is identical to B the expressions "A" and "B" are inter-substitutable *salva veritate* in all non-'intentional' contexts. On the contrary, since expressions for actualities are propositional and expressions for substances or *supposita* are object-expressions, it is normal for them to be not inter-substitutable. As Aquinas would have put it, the *modus significandi* of the types of expression concerned is not the same, a point important for what one says about the Christian doctrine of the Trinity[43] (on which I comment below): there is no general inter-substitutability between grammatically 'concrete' and grammatically 'abstract' uses of the same 'term' (using the word "term" informally to cover variations in grammatical form or character, e.g. as between "living" and "life",

[43] See n. 44 below. The importance of this for the doctrine of the Trinity appears in many places in *S. Th.*, Ia, QQ. 39–40. Thus, in Q. 39, Art. 5, we are held back from inferring from "*Deus* begat *Deus*" (referring to the Persons concretely) to "The essence of God (*deitas*) begat the essence of God (*deitas*)" or "Wisdom begat Wisdom" (i.e. using abstract terms). And in Q. 40, Art. 1, we are not allowed to say either "God is paternity" (this property is in the essence of God 'only by mode of identity' and not in 'mode of signification') or "Paternity begets" (the 'mode of signification' of "begets" forbids such predication).

"good" and "goodness", "is white" and "(the colour) white").[44] (We meet with some notion of a real identity which, like this one, does not satisfy Leibniz's Law in other contexts having to do with explanation, e.g. we identify lightning with a discharge of electrons because what has to be explained about it for the purposes of physics is to be explained in these terms without being able to speak of both as, for instance, perceived—we do not see electrons.)

We can now return to the picture or perspective which this way of viewing things opens out, involving a contrast between the temporal and composite and, on the other hand, if it exists, the incomposite and non-temporal.

From this it follows that in the case of what is temporal, or contingent in the way that results from being temporal, the *suppositum* or substance is distinct from its act of existence so that the whole is composite, actuality distinct from focus, the compulsion to regard it as complex arising for reasons to do with efficient causal explanation; and its having a focus in this case goes with being temporal, exemplifying the distinction between 'being of a nature' and 'exercising a nature', first act and second act, having accidents and so forth. This temporal case is the case of the composite to which our thinking and imagination are accommodated: it is plausible that all cases of the composite must be like this, but I have not argued this.

In contrast, we have (if it exists) the incomposite, which as we have explained it is intrinsically underivative, and for which, as we shall discover, the only reason for supposing that it exists is that it is needed as First Cause as condition of the existence of the temporal. In God, as incomposite, there will be no real distinction between subject or *suppositum* and actuality. That is, in God there is no distinction between the situation of there being this *suppositum*, i.e. this *suppositum*'s being there, and this *suppositum*'s being active:

[44] In various cases inferences of the approximate linguistic form "A is B ; B is C ; therefore, A is C", i.e. as Aristotle would have put it in the *Categories* of the form "C is said of B ; B is said of A ; therefore, C is said of A" are valid. For instance, "Socrates is a man; man is an animal; therefore, Socrates is an animal" and "Man is a biped; the biped is an animal; therefore, Man is an animal". But, because of what Aristotle would have called homonymy and paronymy, the inferences "Socrates is white; white is a colour; therefore, Socrates is a colour" and "Socrates is courageous; courage is a virtue; therefore, Socrates is a virtue" fail. In Aquinas's terms, this failure would be because the terms "white" and "courage" occur with a change in their *modus significandi*, from concrete to abstract as between the first and second proposition in each case.

this *suppositum* is as such active, and its existing is as such an act. In God, there is no real distinction between *suppositum* and His existing as an act: the first cause is at once a *suppositum*/causal agent and an actuality. This exhibits the vital importance of distinguishing the *modus significandi* of expressions for *supposita* and expressions for actualities, so that they are not freely intersubstitutable, and at the same time insisting that this distinction in modes of speech does not compel a real distinction, i.e. a distinction positively affecting efficient causation.

Any theism requires that, in God, there should be focus as well as actuality. The peculiarity of Christian belief is to hold that, in God, there are three focuses or *supposita* but still only one actuality: on this we will comment shortly. But a unitarian, e.g. a Muslim, just as much as the Christian, will have to insist on the distinction between subject and actuality in God. His reason will be that it makes no sense to regard a *person* (whether as subject or object of personal properties, relations, and attitudes) as an actuality, i.e. as a thing of a sort only properly indicated by a complete proposition, and not properly indicated by a subject-expression proper. Allah lives, knows, and rejoices in Himself, knows and wills all things, has compassion and mercy towards creatures and is the object of their fear, love, and worship. But it is precisely to the point here that what requires the distinction between subject and actuality is not efficient causation as such (though the idioms we use to express relations which have a causal aspect may be affected) but other relations of a 'person-orientated' kind.

It is part of the structure of causation that there should be a distinction, at least in our modes of speech, between agent and activity, between *suppositum* and actuality, and that both should be involved. But it is not required by this structure that this distinction, between agent and activity, in the cause should be a 'real' distinction in the sense explained. If the argument proposed in Chapter VI that temporal continuance needs a cause and this cause can only be incomposite is valid, then we will have proved in one act that there exists an incomposite thing and God as this incomposite thing is at once subject or *suppositum* and actuality.

As subject He is causal agent, causing temporal continuance, and also person and good according to the arguments of Chapters VIII and IX. But His incompositeness belongs to Him, according to our mode of speech, primarily as actuality, and the implications of this for the understanding of His infinity, not only in causal power, but in whatever other aspects His actuality may be realized to have, appear in Chapter IX.

(e) *Incompositeness as a form of first-level unity*

We must realize that 'incompositeness' is a form of 'unity' in the most common use of the word "unity" in the English language.

The temporal is, we have argued, composite: therefore, the incomposite is non-temporal and includes no real distinction between any 'being of a nature' and 'exercising a nature'. For instance, if God has love or understanding, we should realize that, if His existing were something distinct from and prior to them, it would be something situational or in the nature of a first act in relation to a second in relation to them. For, if His loving and understanding were realizations of a nature which was their background, the ground from which they sprang, then, for His existing in this nature to be distinct in the order of factual or causal explanation from these realizations of it, His existing would have to be a state (disposition or *dynamis*) related to these activities as a state is related to an episode. But this is impossible in a non-temporal being. Accordingly, God's existing, His being of the nature He is, His living, knowing, or understanding, and His loving will all be one and the same actuality.

We can now complete the picture. It will follow that in God, God and His existence, *suppositum* and its act of existing, nature, life, knowledge, and love will be none of them distinct in reality, but that all of them will constitute but one actuality, the only actuality with discrete existence. (God's existence is independent of all other existence and discrete in itself, while all other existence constitutes an intermeshing interdependent whole whose elements are neither discrete from each other nor independent of God's existence: rather, if God exists, the existence of creatures involves God's existence internal to it, creation being, as it is said, something real in the creature but not in God.[45])

This illustrates how incompositeness is a form of unity: what is incomposite is integrated within itself or, as it were, unified so as to contain in its existing no division or real distinction. This is "unity"

[45] *S. Th.*, Ia, Q. 45, Art. 3 and *De Potentia*, Q. 3, Art. 3.

as a first-level property or property of objects, not as a second-level property, like "being one in number" or "applying to just one object", of the classes of which objects are members or the concepts under which objects fall. (One has to be watchful of idiosyncrasies in the usage of such words as "unique" in English and *unicus* in Latin[46] and the tendency of "unity" in English to signify one of various first-level properties, e.g. the unity a chair has which a heap lacks, or which an animal has but a chair lacks, and so forth.)

We do not at this stage discuss how this first-level property may be related to the second-level property asserted in the statement "There is only one God" (we will do this in subsection (*f*), only after first considering the notion of "nature").

But it is at this stage, while we are dealing with the unity of God as a first-level property, that we should consider how the question of the unity of the Persons in God presents itself in Christian theology. It turns out that all the explanations of the theologians are directed towards, not the explanation of why there is only one God or the general proof of this, but the explanation of the first-level unity of God's actuality (and it is only incidentally that they explain how it is that there being three Persons does not involve tritheism). Their explanations *at this point*, as it were, leave it open that there might be more than one trinitarian-structured god or other gods not thus structured: these points are foreclosed in other ways in their thinking. That is, it is one thing to explain that the Father is *archē* in the Godhead and another thing to explain why there is no other such Father or *archē*. We leave consideration of unity as a second-level property, the fact that there is only one God, to Section 11.

We have remarked that God's personal nature involves that some distinction, not a 'real' distinction but only a distinction in logic or ways of speaking, will need to be made between God, *qua* subject, and God's existing, *qua* actuality and what it consists in. Christian Revelation proposes that there are three subjects in God, but that the existing of each of the subjects consists in the same actuality as

[46] In Allan Wolter's selection, *Duns Scotus: Philosophical Writings*, *unitas* is translated 'unicity', but Duns Scotus uses the term *unus*, not the term *unicus*. In Aquinas, *unicus* seems associated with uniqueness in regard to Persons or *supposita* as if to say "A is uniquely God" meant "A uniquely is God" and would be said of a Person, and would then have the false implication that there is only one Person (*S. Th.*, Ia, Q. 31, Art. 4.).

the existing of the others and that from the standpoint of causal explanation none of the subjects will be present or active without the other subjects being thereby present and active in the same divine activity: as a result, Christian doctrine involves again that there is no real distinction between subject and actuality in God.

The explanations given of this in Christian theology bear out that the unity insisted upon in the doctrine of the Trinity is this form of unity which we have called incompositeness.

Thus, in the Christian doctrine of the Trinity, as expressed by Athanasius and the Greek Fathers and recapitulated in Aquinas,[47] God *is* by there being three *hypostases*, *subsistentiae*, or *supposita* (primary substances or subjects of predication in the sense indicated in Aristotle's *Categories*). And it is because the Father is *archē*[48] and the Son and the Spirit proceed from Him and because this proceeding is involved in the very being of the Father that it is by one and the same actuality, not by three independent actualities, that the three exist. The Father's existing consists in the act of begetting the Son and breathing the Spirit, i.e. this is what His life consists in. The Son's existing consists in His being begotten or conceived by the Father (in St Augustine's conception in the Father's act of self-understanding wherein He conceives a perfect image of Himself, like Him in actuality and determinateness), and, thus constituted, exercising all the love of the Father. The Spirit's existing consists in His being (metaphorically) breathed by the Father with or through the Son (in Aquinas's conception, developing the thought of St Augustine, the Holy Spirit is God under the aspect of the fruit or goal towards which love tends or in which it rests in delight, constituted by and in, and inseparable from, the love of the Father[49]).

47 *S. Th.*, Ia, Q. 29, Art. 2.

48 *S. Th.*, Ia, Q. 33, Art. 1.

49 In St Thomas Aquinas's exposition of this understanding of the Holy Spirit as Love it is not that the Holy Spirit is the act of loving in God, e.g. the loving by the Father of the Son and by the Son of the Father, or God's act of loving however else conceived—any more than the Son is the act of understanding or knowing in God, e.g. the Father's act of self-understanding, or God's act of understanding however else conceived. Rather, by means of Revelation and not by any natural means of knowledge, we learn that, in God, and not only in man, *intelligere* involves or consists in something analogous to the expression of a concept or formation of an image or model, so that it involves or consists in the existence within the knower of some *likeness* of what is known or understood, with the implication that in God there must be an analogue (altogether perfect, and therefore, since in God there is no accidental division or diversity, as individual as the person known) of the likeness,

Accordingly, by the co-encompassingness, *perichoresis*, or 'circumincession' of the three, where the Father is in whatever respect, there the Son and the Spirit are present in Him; where the Son is in whatever respect so far as His divinity is concerned, there the Father and the Spirit are present in Him; and likewise in the case of the Spirit.

What is illustrated here is that, whereas it might have been imagined that the unity of God arose from His being analogous to a Platonic Form or Idea, one in number by the Identity of Indiscernibles applied to intellectual objects, and inwardly one by something analogous to the integratedness of a single intellectual conception, in fact the explanations of Christian theologians take a quite different form. What is central to these explanations is firstly the position of the Father as *archē* in the Godhead, so that everything outside God attributed to Him is naturally primarily attributed to the Father as a single Person, and secondly the co-encompassingness of the three Persons, so that by mutuality of in-presence whatever is concretely attributed to one Person in His divinity is under another aspect attributable to each of the others and the life and existence of the three is one actuality. It is here that the details of my account of the criteria of real identity, drawing upon Aquinas's notion of a distinction in *modus significandi* by Aquinas,[50] become most obviously crucial.

Such is the doctrine of the Trinity which it is no part of the concern of this book to argue for or support. The point of my present remarks is solely to indicate how the doctrine of the Trinity fits into the logical structure I have been explaining, and not to justify this doctrine.

concept, model, or image of what is known: hence, the Son is the (non-materially conceived and expressed, in God's eternal being) true and only Word of God, the express image of God. However, the loving which understanding of its intrinsic nature involves does not thus involve a new likeness to the lover: 'The object of intellect is in intellect *per similitudinem speciei*, but the object of will or love is in the will not by reproduction but as a goal is in the tendency to the goal' (Bernard Lonergan, *Verbum*, p. 203, where he refers to *SCG*, IV, 19: 4, 9).

[50] See nn. 43 and 44 above.

11. THE NOTION OF INTRINSIC POSSIBILITY: THE STATUS OF THE CONCEPTION OF "NATURES" AND THE BASIS FOR A CONCEPTION OF THE UNIQUENESS OF GOD

(a) The metaphysics of 'natures'

In analysing the concept of compositeness, we are (from the point of view of general philosophy) primarily analysing a feature shared by all objects which do not conform to the requirement of Wittgenstein's *Tractatus* that there should not be two ways in which subject-predicate statements about them *can* be false, one by the non-existence of the subject and the other by the truth of the negation of the predicate in regard to the subject. The problem is that in this phrase, '*can* be false in two ways', the "*can*" signifies intrinsic possibility or the absence of intrinsic impossibility; what is significant for us is the conditions of understanding this word "intrinsic". It is only incidentally that we get to appreciate what would be involved in being an object not of this kind, indeed the only such object.

What I now wish to explain more fully is the inseparability of four things which, taken together, have been considered to constitute the ontological structure picked out by the word "composite": (i) a distinction in reality between a thing and its existence; (ii) a distinction between 'nature' as a quasi-abstract object and the thing of that nature; (iii) a distinction between a thing's being of a nature and a thing's existing as a realization of that nature, illustrated in the temporal case in the way explained in Section 10; (iv) being not intrinsically underivative. I insist on (i), (ii), and (iv), but set aside (iii). I set aside (iii) because, I believe, it involves developing a special theory of a distinction between states (of being of this or that nature or disposition) and exercises of those states to apply only to non-temporal objects other than God: there seems no special merit in going out of one's way unnecessarily solely to cater for what seem to be mere figments of philosophical imagination.

I have explained what it is for a distinction to be a 'real' distinction, or 'distinction in reality', in such a way as to make (i) and (iv) inseparable. If a thing is incomposite, there can be no real distinction between the thing and its existing: and in this case the thing's existence is internally such as to leave no room for the possibility of its having a cause or having real contingency—that is, it is intrinsically underivative. And, conversely, if it is composite it is by reason of the internal conditions of the possibility of having a cause being satisfied.

In envisaging any possibility of a thing's having a cause we envisage the thing as possibly having an external ground of possibility. That is, then, we envisage the thing as of such a character internally as will leave it open that it might not exist or might not have existed: the internal requirements for there being a cause must be satisfied if there is to be even room for the possibility of a cause. In either case it is a prerequisite that there be a distinction between the thing along with its existence on the one hand and its nature, as a quasi-abstract object which might not be or might not have been actualized, on the other.

This nature as a quasi-abstract object in the case of what has a cause is part of the formal specification of the causal background to the thing *qua* external or prior ground of its possibility, and in this way has its quasi-abstract existence prior to the actual existence of the thing; and, since the total causal context of an effect must leave its nature determined (otherwise the supposition of causal connection is mere fantasy), as prior in this way it must be shaping of the effect. In brief, the situation of the possibility of what has a cause, a situation to which the effect's nature is internal as 'object', is prior to and shapes the actuality of the effect's being actual. That is, the nature's being internal as 'object' to the possibility concerned is prior to its being internal to the actuality, and so cannot be reducible to the actuality.

Let me explain why I set aside (iii).

It might be argued that, although the nature as a quasi-abstract object may specify the actuality formally, i.e. in the sense that its specification is integral to the specification of the actuality, it must also be in some other way prior in order to be properly described as "limiting": an actuality can only be limited really or from the point of view of efficient causation by something non-abstract and actual so that this priority requires, on this argument, some further accounting for. It would then be urged that there must be some actuality of the thing's being of the nature concerned which is determinative of the shape of the actuality of the thing's existing, *illustrated* by the way in which the situation of a temporal thing's being of a certain nature or kind is determinative of (though not discrete from) the actuality or actualities in which its existing is realized, i.e. the actuality or actualities expressive of this nature.[51]

[51] Further, it will then be better in the general case to say that the cause of the effect determines its nature by determining the quasi-situational state of being of the

And again we might argue that the internal requirements for there being a cause must also be satisfied if there is to be even room for the possibility of a cause: that is, just as the distinction between nature as a quasi-abstract object is required for the latter as well as the former so also with the distinction between the quasi-situational being of a nature and the realizations of existence according to that nature (the ontological structure of what has nothing internal preventing the having of a cause must be the same as it would be if the thing concerned did have a cause). On this argument anything not intrinsically underivative must involve a real distinction between its being of a certain nature and its existing.

A thing or *suppositum* is actual by acts or non-situational actualities, rather than by being of a certain nature or kind. For this latter is compatible with being the static background of capacity for acts, and is not as such what existing consists in: existing for an animal includes not only being an animal but activities of living as an animal, and it is these things that go on over time in virtue of being an animal rather than the being an animal which is attributable at each instant which make the animal actual; nothing static or of an instant has actuality except in virtue of its place in a context of non-static actualities. This suggests that, quite generally, if the nature of a thing is distinct and prior in the way indicated under (ii), then being of that nature is not an act of the *suppositum*; yet the acts of the *suppositum* are exercises or expressions of its nature; therefore the *suppositum's* being of the nature concerned must be prior to and shaping of its acts. Thus, if the nature of a thing as a quasi-abstract object is prior in the way indicated under (ii), it might seem that there must be an actuality (situational in character) of being of that nature prior to the non-situational actuality of exercising that nature, the two actualities being non-discrete in the way explained earlier.

But my opinion is that such an analysis is only required in the case of the temporal and that my argument does not depend upon its extension to the case of effects that are not temporal (anyway probably a mere fancy of philosophers, who explore every logical possibility without regard to its conceptual coherence in other respects).

My view is that we can identify a respect, *viz.* temporal continuance, in which it is evident that the cause of the temporal cannot itself be temporal and in the case of which the root of this impotence is compositeness, including the feature of a nature distinct from the existence of the thing concerned limiting its actions; on this basis I will argue that any cause of temporal continuance must be incomposite. It seems obvious that the incomposite must be intrinsically underivative but the question is

nature concerned than by directly determining the actuality or actualities expressive of this nature. For in the typical temporal case the act of causing is realized in the actuality of the coming to be of a thing of a certain nature and is not itself involved in the later continuance either of the state of being of a certain nature or in the actuality or actualities expressive of this nature.

how to articulate this: I have made the connection scarcely disputable by making positive relevance to efficient causal explanation a criterion of the 'reality' of the distinction between a thing and its existence but have then been compelled to resort to very abstract considerations about 'natures', considerations difficult to appreciate except by reference to temporal examples.

It is at this stage that it becomes important to note how abstract or general our definition of compositeness was. If we build too much into our notion of compositeness, e.g. so that the composite is by definition temporal or mundane, it becomes obvious that the cause of the temporal cannot be composite, i.e. must be incomposite, but problematic how it is to be shown that it is one or that it is intrinsically underivative. And we will show that in any case the root of the kind of impotence we are concerned with lies deeper than temporality. Therefore, while we may be of the opinion that being composite carries with it the whole ontological structure of states underlying temporally distinct acts, and that the supposition of the possibility of composite beings whose nature is expressed in an act or acts distinct from it but which are not temporal is not only academic but unintelligible or absurd, it remains that our argument does not depend upon this opinion. It may well seem that whatever is not intrinsically underivative (because not composite), i.e. whatever somehow even 'could' have a cause (because it has the structure of an effect), must be temporal: but we do not draw upon any such contention, even though Chapter II and Section 7 of this chapter invite it.

(b) That there can be only one incomposite thing, only one God

(i) The problem of the uniqueness of God: "God" as both a singular and a general term

We explained earlier how incompositeness was a form of unity as a first-level property. There remains the question as to how any inference is to be made from the first-level unity of God to the unity of God as a second-level property, a problem to which, as we observed earlier, Christian expositions of the unity of three Persons within one God were not intended to contribute.

Let me consider first how our treatment of natures resolves this problem and secondly how our philosophical treatment thus opened out contributes to Revealed theology. To infer that there is only one God rather than some other number from the concept of God as incomposite would seem to most philosophers as gross a mistake as to infer that there is at least one God rather than none from the concept of God as incomposite. Both would seem to involve an 'ontologism' of abstract essences as objectionable to theologians as to philosophers.

Yet it is useful to notice how tempting such an ontologism will always be to developed forms of monotheism.

For the religions rooted in Jewish monotheism the name or term "God" serves at one and the same time linguistically as a kind-term, grammatically comparable to other kind-terms such as angel, man, dog, horse, and fish, indicating a kind of being in regard to which one asks "How many . . . are there?", receiving the answers "one God" and "many men", and on the other hand a singular term, subject of singular statements such as "God created the world", "God is good", and suchlike, linguistically of the same grammatical form as "Bill built the house" and "Catherine de Medici was wicked".

The question arises how such an apparent violation of the canons both of logic and of syntax can be justified or in any way understood. If, in reporting a day's activities, a man were to say that he found the number four or the colour red or the species man behind the curtain, we would not accept this even as part of a tall story or fairy tale. Rhetoric and long periods of habituation within an ancient framework of worship will not suffice. Some gilding of the pill must always be incorporated into any such setting of rhetoric or cult, if the language of religion is to be integrated in any way whatsoever with the rest of human discourse. Indeed, some integration is involved even if it is imagined that it is the language of religion that has the mastery and which forms or re-forms and re-shapes the language of life.

Allegedly, the favourite method of gilding among some Christian and perhaps among some heterodox Jewish thinkers has been a Platonic conception whereby ultimate reality is to be attributed primarily to the Ideas or Forms, e.g. the Good in Itself of which all lesser goods are participations or likenesses, one archetypal fatherhood and one archetypal kingship, in which all lesser forms of fatherhood and kingship are participations or imitations, and so likewise in parallel cases, the basic conception being that the very use that men make of the general terms "good", "father", "king", and so forth, presupposes an implicit insight by the intellect into these archetypes (the meaningfulness of these terms presupposing the actuality of these archetypes). However, this Platonic gilding whereby God becomes the ultimate object of reference named by a privileged primary set of general terms seems to represent a mistaken view of how general terms have meaning and, indeed, thus conceived, to embody an idolatry of general concepts, as if the human intellect in its day-to-day linguistic workings was in its every act *ipso facto* directed towards the infinite.

Against this background let us make clear how what we have said about the nature of God in subsection (*a*) opens the way to a proof that there can be only one God which is quite free from ontologism.

(ii) The inference that, if there is a God, there is only one God

We must now consider the inference that can be made from the first-level statement that a certain thing, e.g. God, is incomposite, to the second-level statement that there is only one thing of the kind concerned—an inference from unity as a first-level property to unity as a second-level property, from a statement about an object to a statement (in Frege's terms) about a concept. Some such inference will work if an incomposite thing actually exists. This will make logicians suspicious of the coherence of the notion of "incompositeness" but it has to be conceded that the inference depends, not on introducing the technical term "incomposite", but only upon the significance of such expressions as "intrinsically" and "for internal reasons" needed in explaining the notion of intrinsic underivativeness, self-existence, or being *a se* as much as in explaining the notion of incompositeness.

There is a temptation to suppose that there is only a merely extrinsic necessity in the existence and singleness of God. On this view the necessity would arise solely from the absence of any external or prior ground underlying God leaving His non-existence possible, and no such ground leaving possible the multiple existence of beings like Him—no anterior ground leaving it open whether no God, one God, or many gods should arise. The necessity of His existence and oneness would be merely *de facto* because there would be no inner reason in Him or His existence why there was no such ground so that the absence of such ground would be merely *de facto*.

We have argued that if anything (e.g. God) were internally such as thus to be merely extrinsically necessarily underivative there would be a distinction between it with its existence and a quasi-abstract nature of this thing, and that, in the case of any incomposite thing, there is no such distinction to be made. Therefore, in the case of an incomposite thing (and we shall argue that God as cause of the temporal is incomposite), the nature of the kind of being concerned, instead of being a quasi-abstract object, must be in the logical sense a concrete or individual reality, nothing other than the thing itself or its existence (the actuality of its existing). That is, in other cases when we say that something is "such that so-and-so" or "internally such that so-and-so" this will be true in virtue of something general about it such as might have been exemplified in the case of other things or some character it has which might not have been exemplified by anything; but in this case when we say such things they will be true in virtue of nothing other than the reality (the incomposite thing or its existence) itself, not anything general nor anything which might not have been exemplified at all.

But if God were such as to be necessarily unique or one in number only by extrinsic necessity, then again there would be a distinction between God with His existence and a quasi-abstract nature of this God, since God, His existence, and nature would together be such as not internally to exclude the possibility of there being a different number of gods. But there is no such distinction, and therefore, for the same reason that, if God exists, He is intrinsically underivative, we can say also that, if God exists, there cannot be any other number of gods than just one: there could not be none and there could not be more than one.

The matter must be stated carefully. Firstly, we do not first treat God's nature as quasi-abstract and say that in virtue of the internal properties of this nature (thus conceived) there could not be any number of gods other than one: it is only that, if an incomposite thing exists (such as God is suggested to be), the nature of this incomposite thing will be perfectly individual and no different from the incomposite thing itself (God) or from its existence, and it will be in virtue of the lack of real internal distinctions in this thing and its existence (i.e. for internal reasons appertaining to this logically individual thing) that there can be no different number of Gods. Secondly, we do not presume a priori that any incomposite thing exists or that the notion that it does is conceptually coherent: we wait upon a posteriori argument that it exists, e.g. argument that it is a causal condition of the continuance in existence of the temporal.

We do not confuse properties of objects with properties of concepts, and we recognize that in order to assert or deny the number one in the case of God, so as to say that there is only one God or to question this, we have to use the term "God" as a concept-term. What we do is infer by a non-constructive *reductio ad absurdum* argument from there being something incomposite or intrinsically underivative to there being just one such incomposite or intrinsically underivative thing. In using the term "God" as a concept-term, we do not have to use it in such a way as to imply that it expresses a character which might have been exemplified some number of times different from one: the term "incomposite" does not express any such character. If, using the term "God" as a singular term, we say "God is incomposite" we can infer "There is something incomposite", "There is just one incomposite thing", and "There is no other incomposite thing than God", and so long as we are wary of words like "thing" we make no mistakes and do not fall foul of (for instance) the doctrine that, in God, there are three Persons or *supposita* but only one thing or reality, a doctrine we have already encountered in discussing incompositeness as a form of first-level unity.

But the non-constructive character of the philosophical approach means that no idolatry has been set up: for instance, there has been no suggestion that the root of there being but one God is in the nature of God, conceived of in a Platonic or rationalist way—it is not in virtue of any quasi-abstract nature or essence that there is a God or that there is only one God. We cannot even say that it is in virtue of the individuality of the nature in the case of God that there is only one God, as if individuality was one property among others or the key property of a conceptualisable or partly conceptualisable essence: even this would be to make a false concession to the ontologism which would allow the inference, first of existence, from the conceptualisable content of what existence is being asserted or denied of, and then of existence just once.

If we make a distinction between proving that there is only one God and explaining why there is only one God or stating or identifying the root of there being only one God, then the point of regarding the unity of God, like His existence, as intrinsic to Him is precisely to deny that this unity, any more than this existence, has any root distinct from or in any way prior to Himself: but He is no more cause of His unity than He is cause of His existence. The second-level fact, the fact 'about concepts', that there is only one *archē*, only one Father, only one God, like the fact that there is a God at all (or anything in real existence at all) is without 'explanation'. It is not to be 'explained' either in terms of some distinguishable feature within God or in terms of something outside God; but it does not lack explanation as if it 'merely happened' to be the case, as it were 'by chance', or as if there 'could' have been an explanation but was none; there is no room for an explanation, because it is 'intrinsic' to God's actuality. The puzzles of this 'intrinsic-ness' it has been the concern of this section to explore.

(iii) The role of our philosophical account of God's uniqueness within theology

In making clear that, in God, nature is nothing other than the reality itself, and is not something in any way abstract or general, we need to explain the syntactical propriety of the same term "God" serving both as a singular term and as the name of a nature, the nature identified but not expressed by "God" as a concept-term. Revealed Theology needs philosophy at this point according to the principle *Fides quaerens intellectum*: without it faith appears nonsense.

If what the theologian puts forward does indeed enshrine knowledge, how is what he says to be interpreted? If what he says is based upon the authority of some Scriptures as teaching from God, the question follows as to how the syntax of this divine speech in regard to God and His nature is to be understood in a non-idolatrous sense.

The philosopher, if he follows the account offered in the next chapter, suggests that any being to which a 'nature' stood as in any sense prior, not only would not be self-existent or intrinsically underivative (*a se*), but also would be just as impotent as temporal beings to cause and maintain existence unqualifiedly and unconditionally. Therefore, if man can know by metaphysical reflection (whether inarticulately 'felt' by the ordinary man or critically articulated by the philosophical writer) that the temporal continuance in existence of the stuff of the universe requires a cause, and a cause of this kind, then, by the same line of a posteriori reflection, man can also know that the nature of this cause must be as individual or particular as the activity or actuality in which its existence consists. That is, he thereby knows that the name of the First Cause, if it or he receive a name, will at one and the same time function as a kind-term and principle of counting, indicating a nature, and a singular term indicating a unique object or actuality. The importance of the a posteriori character of this reasoning will be further amplified in Chapter IX.

Yet in this the philosopher will not have reached to any content-containing conception of the First Cause. He has recognized its existence, its relational character as cause, and, by inference, its possession of whatever it takes in the way of active power for something to exercise such causality, negatively its incompositeness or freedom from any complexity relevant to efficient causation and thence, by inference, to the negative knowledge of God's uniqueness, negative in the sense already explained.

From the point of view of the theologian, the philosopher has served only the limited function of making it plain 'how, from the things He has made, the invisible things of God, His power and deity, may be known' and exhibiting a posteriori how it is that the First Cause must be by nature internally one and unique, so that the name "God" which indicates His ineffable nature and being should be both a kind-term and a singular term, without the appreciation of this framework embodying any contentful or constructive, positive knowledge.

VI

The Lessons of History: The Structure of Causal Argument to God's Existence: The Conceptions of Primary and Secondary Causality

My purpose in this chapter is to consider the systematic problems of identifying the appropriate structure for causal arguments for God's existence. My purpose in this chapter is not to propose any particular argument for God's existence in its final version. Rather my purpose is to review the materials thrown up by history germane to the right construction of such an argument, and to present these materials in what seems to me an order and perspective helpful to this end.

These historical materials can be seen in two perspectives; the first in their immediate relation to our project, and the second in the light they cast upon the whole philosophical consideration of the problems of 'explanation'.

In relation to the first, I take as my starting-point the need to adopt an a posteriori approach in any attempt to demonstrate God's existence. In Section 1 I analyse the impotence of mundane or so-called 'secondary' causes to cause temporal existence, and show that, if any such cause is needed, it cannot be mundane, but must be 'incomposite' (and therewith by our argument in Chapter V intrinsically underivative and unique): in other words, I identify the hole (if it is a hole) which, if it is to be filled, can only be filled by the notion of a 'primary' cause. Then in Section 2 I open out the conception that there has to be such a cause if any mundane causing is to be understood as itself a case of 'real' causal action: I explain how this 'primary' cause has to uphold existence, not just as a 'state', but as something which is as such 'active' or 'dynamic'; from this it follows that all such mundane causing is 'secondary', underpinned by the continuing action of a 'first' cause immediate to it. In Section 3 I indicate how the understanding operates in thus applying the single concept of cause in two such different types of judgement, judgements as to the need of a cause—in one case the need of a mundane cause and in the other the need of a primary cause—judgements which are interdependent, but which do not arise by means of argument by analogy from the one case to the other. I conclude the chapter by observing how any such position as I have opened out, insisting on the need of a cause as a condition of the existence of the effect, depends upon maintaining the priority of first-order over second-order judgements.

The plan of this chapter can also be seen in a way geared to the problems of 'explanation' and 'causation' within general philosophy.

In drawing lessons from history, I begin by marking the limits to explanation in terms of the natural and the mundane and trying to diagnose the cause of these limitations (Section 1). I then proceed to note the parting of the ways which arises in response to recognition of these limitations, in effect presenting us with a choice between a structured metaphysics, implicitly theistic, in which a non-temporal cause underpins mundane causes, and systems of metaphysics in which a non-realist view of explanation is taken (Section 2). In these non-realist systems, able to draw axioms only from logic and epistemology, all basic principles in explanation have to be taken in extreme generality (Section 2(d)), thus closing the door on the option in the history of philosophy of developing a theory of concepts to be applied analogously in different cases and not drawing upon such principles (Section 3). In these non-realist systems, it is characteristic always to give priority to epistemology over metaphysics and natural knowledge, to second-order questions and principles over first-order ones: but, historically, the establishment of the structured and theistic metaphysics I described has been commonly based upon the opposite presumption (Section 4); and I leave it to Chapter VII to show that the latter approach is the only proper one.

PART I. THE IMPOTENCE OF THE TEMPORAL

1. THE FIRST LESSON OF HISTORY: THE REASON WHY THE TEMPORAL IS IMPOTENT TO GIVE EXISTENCE AS THE KEY QUESTION: IDENTIFYING THE PLACE WHERE 'PRIMARY' CAUSALITY MIGHT ENTER IN

(a) The relevant notion of impotence to give existence

The first datum for our consideration of arguments towards God's existence is that we are at no time going to presume the possession by man of an adequate concept of God. We do not even presume him to possess a concept of God adequate to give him insight sufficient for him to judge such a being as God to be possible, adequate for man from the concept alone (by some logical trick) to infer to His actual existence.[1]

Likewise, we have no insight beforehand into the nature of God that makes it plain that He does have, or could have, or how He has, power to communicate existence to other beings. Nor do we have any insight as to how it is possible for there to be a being who exists of Himself, and, as it is said, is self-existing, or, in older terminology, has the property of aseity ('from-Himselfness') without this involving the self-contradictory or viciously circular notion of self-causation instanced when God is described as 'a cause of Himself' (*causa sui*)..[2]

[1] Human insight into the *possibility* of God's existence, or at least of the existence of a being of such a kind as European philosophical theology has assumed God to be (at least from Augustine onwards), was presumed without query by Duns Scotus, Descartes, Leibniz, and perhaps Anselm earlier, as well as Hartshorne later. It is this presumption, and not only the inference from possibility to actuality, which before all else needs to be questioned. But the latter itself involves a simply invalid inference drawing upon the premiss of a *de dicto* necessity in the proposition that God, under a certain description, exists (the context of 'reference' to God being here what is fashionably described as 'referentially opaque' or 'intentional') to reach to a *de re* assertion of His actual existence (within which the reference is 'referentially transparent'). The form of argument which depends upon these two objectionable premisses appears certainly in Duns Scotus, Leibniz, Hartshorne, Malcolm, and possibly in Anselm. I discuss both presumptions as they occur in Duns Scotus in Chapter VII, Section 1(*d*).

[2] Aquinas was aware, as Descartes made as if not to be, of the logical viciousness of explaining God's self-existence or aseity in terms of any notion of self-causation (*S. Th.*, Ia, Q. 3, Art. 4).

Rather, we proceed backwards.

What is plain to us is our own impotence and the impotence of any spatial, temporal, or mundane thing, or any system of such things, to be itself the cause of its own continuation in existence, 'unqualifiedly' and 'unconditionally' (I explain and distinguish these terms below), and likewise our, and their, impotence to be such a cause in respect of any other such thing or system of things. This we argued in Chapter I, leaving methodological queries to Chapter VII; but at any stage it always remains a case of making plain, or rebutting objections to, an argument which, once properly stated, is obvious in itself to the pre-critical or 'ordinary' man; therefore, our basic argument remains as stated in Chapter I.

Then later, upon reflection, we inquire into the ultimate cause or root of such impotence, and, it being once granted that the mundane is thus impotent, we inquire as to whether some other—non-mundane—cause of continuance is necessary.

When I speak of being a cause of continuation in existence *unqualifiedly*, I mean the cause of going on existing at all, rather than not existing, as opposed to being the cause of continuing in this state rather than that state, and I mean a cause of continuation in existence, not just in some, but in every, respect, e.g. even in respect of the continuation in existence of the underlying matter or stuff of which the thing might be supposed to be formed.[3] By speaking of continuation in existence, I mean to exclude consideration of mere coming into existence, which is the inception of something whose continuation after inception is normally assumed, and mean to concentrate attention on this continuation, which is not as such typically the immediate object of mundane causation at all.[4]

When I speak of something which causes existence *unconditionally*, I mean a cause which, without causally presupposing anything else,

[3] Cf. Aristotle's *Physics*, V, 1, and Aquinas, *De Potentia*, Q. 3, Art. 1. Note also Aquinas, *De Potentia*, Q. 3, Art. 5, and *S. Th.*, Ia, Q. 44, Art. 1, for the contrast between being a cause of being as such and in general and being the cause of anything more limited. This latter contrast is indeed relevant to the three arguments to God's existence listed in each place but is not the contrast I am marking in the text by the phrase 'cause of continuation in existence *unqualifiedly*'. These three arguments correspond to Aquinas's Fourth Way (*S. Th.*, Ia, Q. 2, Art. 3) whereas the argument of this book corresponds to the First Way.

[4] See Aquinas, *De Potentia*, Q. 5, Art. 1, and *S. Th.*, Ia, Q. 104, Art. 1.

gives existence: that is a cause which does not, in giving existence, rely upon or causally presuppose as already given or established either some system of nature involving regularities over time or the continued existing of any stuff with some persistent nature. Rather this unconditionality is a matter of causing perpetuation without causal presupposition.[5]

Strictly speaking, the unconditionality or absence of causal presupposition is a modal qualification on the causing or giving of the existence concerned, whereas the 'unqualifiedly' is a qualification or specification on the existence, i.e. on what is given.

Of course, if God upholds a thing in existence, the previous existence of the thing is logically presupposed, but not causally, since in His action He in no way uses this previous existence and in no way relies or depends upon it as something internal to His action. The difference between creation and upholding is precisely this, that in creation what is created comes into existence from nothing, while in upholding, what is upheld is something existing before, which is being given continuance in existence. But that what already exists should be the 'object' with respect to which God's action is specified in no way makes it an efficient causal precondition of this action and in no way compromises the immediacy of His action and presence.

For a mundane object to act in exercise of its nature and dispositions, its existence with this nature and these dispositions has to be already established and stands as an efficient causal precondition of action in exercise of them. True, a thing cannot possess a certain nature or disposition merely statically, since these things are as such principles of action, 'dynamic', as we say: therefore it is a logical condition of possessing a nature or disposition that it be typically in exercise; but it is not a causal pre-condition. Likewise, for God to uphold a thing in existence in its nature logically involves His upholding it through some exercise of this nature; but again this exercise and the continued upholding through it are not causal conditions of any previous and presupposed upholding. Hence, the upholding of things remains causally unconditional unlike any imagined synergism of God with created agents in their activity.[6]

[5] This is the primary concern of *S. Th.*, Ia, Q. 45, Art. 5 (cf. *De Potentia*, Q. 3, Art. 4).

[6] It has, for instance, been held that those human actions which intrinsically

It is a corollary of the reality of time that the past and present can have no causal role in that continuance of the substance of things—the continuance of the stuff of the world and the regularities or 'nature' on which its continuity depends—which is involved in the future's becoming present. It follows from this that the same power is involved in the causing of this continuance of things (if any cause is needed) as would be involved in the causing of the coming to be of things from nothing. As it were, if the continuance of the world is the result of the continuance of a divine brush stroke, then the same power is exercised at each stage in the continuance of the brush stroke as would have been exercised in the beginning of the brush stroke—if the brush stroke had a beginning. That is, so far as the active power required to secure the continuance of things as from now is concerned, this power can draw on nothing in the matter now present or previous to this juncture; and therefore, so far as the type of active power involved is concerned, the existence of things from this juncture is like a new beginning (the previously existing matter and its active powers are for the purpose in hand irrelevant).

It has been argued that any being with the active power to create out of nothing must be omnipotent, with a self-sufficiency in itself making it free in respect of whether it creates or not and free in respect of what it creates. What it is important to note at this point is that whatever omnipotence or freedom may be inferred from the power to create out of nothing follow equally from the power to uphold in existence. Moreover, acceptance of the existence of natural contingency (freedom and chance) and of the reality of time with the real contingency whereby the future does not yet exist make plain the presence of the same omnipotence and freedom at each juncture in space and time, the universe not being a real or natural unity or 'thing'. And, as we shall see in Chapter IX, it is reflection on the pre-conditions of the power to cause continuance in existence, rather than consideration of the case of creation out of nothing in isolation, which reveals how it is that omnipotence and freedom can be inferred from the power to give temporal existence.

Of course, if the nature or character of something is such as

depend upon grace, e.g. all actions which come from the supernatural virtue called 'charity', involve such synergism. I do not comment on this view: my concern is only to exclude such a view either of God's upholding activity or (see Section 2) of what is involved in created action in general.

logically to involve a previous history of some appropriate kind[7] (as, for instance, states of virtue and knowledge have, I believe rightly, been supposed to involve certain conditions as to their acquisition), then God could not give existence to a thing with this nature or character from nothing, since this would involve a contradiction. Therefore, Duns Scotus was wrong if he thought that the omnipotence of God as known to Revelation extends to the ability *always* to bring about any situation from no background or without intermediates what God normally achieves through some natural prior history—since God possesses and could possess no such omnipotence.[8] There are no limits to God's power in regard to *types* of thing and situation, except in so far as some types of situation have it in-built to them that things involved in them have certain types of prior history; but, as to *individuals*, there are limits to what can be or count as a continuation from now of the same individual with its individual history as these, the individual and its history, stand as now already established.

Indeed then, the upholding which God does precisely is an upholding not a creating, and therefore an upholding of something with a prior history, for which it will be normal that some of its attributes presuppose this prior history in some respect. But the implications of this are quite limited: nothing in this prior history is determinative of the future of the thing concerned or even of its having any future; in regard to each thing or natural group of things the only limits set on how its future may unfold are the limits set by what is required for its continuation to be a continuation of the same thing or group of things. Rather what matters for our general argument is that nothing in this prior history, not even the existence in order to serve in the Aristotelian sense as 'matter' to provide continuity, is efficiently causative of its having any future: the prior existence and history does nothing to perpetuate the thing—so far as practicality is concerned, it might just as well not have been. Continuance into the future may not need a cause—this is a common position amongst philosophers and the educated—but what is more clear is that the past and present things of nature, even imagined as a totality, cannot supply it.

[7] Cora Diamond, in her paper 'The Interchangeability of Machines', discusses this kind of property in her account of what she calls 'H–predicates'.

[8] It is an elaborate question to what extent Duns Scotus thought this, cf. Etienne Gilson, *Jean Duns Scot*, pp. 360 ff., *et al.*

(b) The problem of identifying the middle term: that is of identifying the root of the impotence in question

Our own impotence to cause the continuance of existence unqualifiedly and unconditionally and the impotence of any temporal or successively existing thing (i.e. thing that exists through a successive series of states) or system of such things in this regard is plain. What is not plain is the reason why there should be such impotence, or why the fact of such impotence should be so plain to us.

We often have this problem in philosophy. Aristotle described it as the problem of identifying the middle term.

For instance, we say that, in order to be actual, a thing has to be possible, and that in the relevant sense, in order for it to be possible, it has to have a background of pre-conditions (if you like 'a causal background') making it possible. We have here an argument, of a sort needing more careful statement, for the need of a cause, an argument which depends upon a middle term, the notion of possibility, a notion not free from obscurity and philosophical dispute.

Again, in order to prove lots of things in mathematics we have to proceed upon the assumption that something has the property of being continuous, but in order to generate any conclusions we need a criterion of continuity. It is fair to say that no non-problematic definition of continuity has been offered.[9] However, this is how mathematicians have proceeded, precisely via a definition of continuity. In many cases, it is intuitively obvious that lines and surfaces will have certain properties. Then we realize that these

[9] Cf. W. and M. Kneale, *The Development of Logic*, pp. 400 f., advert to, as already indicated in the note appended to subsection 1(*b*) of Chapter II, some of the difficulties arising from the customary mathematical definitions of continuity. The notion of a non-zero length shorter than any finite length (a so-called "infinitesimal") as deployed in the seventeenth century at least conformed to the intuition that lines are not composed of points, e.g. as expressed by Newton describing a line as the path of a moving point, against the background of Aristotle's explanation of how it was that Zeno's paradoxes arose (*viz.* from mistakenly supposing lines to be composed of atoms called points, whereas in fact the notion point involves even a further degree of idealization beyond that required for the notion of line). The attempt to develop a mathematics in which points appear only as a construction upon lines, divides in a line or in some other kind of continuum, appears never to have been made in modern times, owing to the practical serviceability of theories resolving the continuous into the discontinuous so that even in intuitionistic mathematics spreads and species turn out to be spreads or species of discontinuous entities.

properties are consequent on what we call 'continuity'; but we cannot demonstrate that lines and surfaces must have these properties, until we light upon an adequate analysis or definition of this middle term, the term "continuous".

Then, in a parallel way, in the present case, we are clear that mundane things are impotent in a certain respect. They have this 'property'. What is obscure to us is the ground of this property and what makes them thus impotent. Once we are clear what makes creatures impotent to give or uphold existence, as it were, out of nothing or without presupposition, we would be clear what property had to belong to a God if He was to be able thus to cause existence, if indeed any such unqualified cause of existence is needed.

(c) *Three candidates for being the desired middle term*

For example, one popular approach would suggest that the middle term involved in our use is the term "finite". According to this approach, the key insight which man needs to gain is the insight that whatever is finite is impotent to give existence. It would then be supposedly evident or quickly demonstrable that temporal (or spatio-temporal, or mundane, or creaturely) things are finite; and it would be supposed that it was because of this, their finitude, that they were impotent to give existence. Clearly, such an approach leaves three problems: (i) what is this 'finitude' (picked out by the suggested middle term)? (ii) why should temporal (or spatio-temporal, or mundane, or creaturely) things be of necessity finite? (iii) how does finitude involve and imply this impotence?

The terms "finite" and "infinite" in effect enter into the discussion as cyphers, meaning whatever their users want them to mean, or perhaps possessed of no genuine meaning at all, but surviving in discourse by association of ideas and rhetorical use. In rhetoric, we fondly insinuate that the infinite in a certain respect is greater in that respect than the finite so that, for example, "infinitely wise" is close in meaning to "perfectly wise", whereas the word "infinite" in the religious context is supposed to exclude as inappropriate any quasi-mathematical notions of measure, number, or quantity at all. In Aquinas, to be infinitely F is to be F not in this or that respect but without qualification, where F is a term not implying limitation: but for Aquinas, the terms "infinite" and "finite" are secondary terms indicating something consequent on a conceptually prior or more basic distinction between creator and creatures; if, with this definition of infinity, Aquinas had made these terms primary in his unfolding of theology, he would indeed have been through and through a Platonist.

Duns Scotus makes this distinction between the infinite and the finite the fundamental differential between God and creatures, but escapes such Platonism merely by insisting that neither this nor any other definite account of the meanings of the terms "infinite" and "finite" is accessible to the created mind. For Duns Scotus, the indeterminacy of the notion of the infinite leaves room for mystery, and is what prevents the rooting of Revealed theology in natural theology and makes it dependent upon Revelation.[10]

Until the thirteenth century, the theological usage of the terms "finite" and "infinite" was quite variable. Early Christian Fathers sometimes speak of the Father as Infinite, and of the Son as Finite as making the Father known. Early in the thirteenth century God was sometimes described as finite, because not vague or indefinite in Himself, and in no way like the Aristotelian matter or incomplete potential infinite (e.g. a never complete unending series of magnitudes, 1, ½, ¼, ⅛, 1/16, 1/32 . . .); but He was also commonly described as infinite because incomprehensible. As I have remarked, in Aquinas, at least in the case of the most general positive terms, the word "infinitely" is close in meaning to 'without qualification', to the Latin *simpliciter*, or to the Greek *haplōs*, rather than the Greek '*apeiron*'. If Gilson is right, it is, in effect, Duns Scotus who founded the modern use of the term.[11]

By contrast, what we may label the 'rationalist' approach, supremely exhibited in Leibniz and Wolff, and as such that approach to which Kant's discussions may be considered a reaction, but persisting in many modern expositions of an imagined Thomism, in effect substitutes the term "contingent" for the term "finite". In this approach we find an exactly parallel set of problems: (i) what is this 'contingency', i.e. how is the relevant sense of the word "contingency" to be explained? (ii) if the contingency is 'natural' or 'real', how is it plain that the Universe is thus contingent? (iii) how is it plain that 'contingency' in whatever sense the word is taken, involves or implies impotence to give

10 See Chapter I n. 6.

11 See Chapter I n. 6.

existence? e.g. how is it plain that God cannot concede the power to create out of nothing, or to uphold in existence, to a deputy?

Thirdly we have Aquinas's own approach wherein the relevant middle term is not "finite", nor "contingent", but "composite". This approach suggests in turn three problems: (i) the capital problem, fudged by most so-called Thomists, of what this compositeness is meant to be understood to be; (ii) what turns out to be no difficulty at all, the problem of demonstrating that ordinary things must be composite; and (iii) the problem of grasping why this 'compositeness' should involve and imply impotence to give existence. Prima facie, if only we could locate a satisfactory explanation of what 'compositeness' (allegedly a 'compositeness' even of 'existence with essence') is, it looks as if we would have on our hands a better candidate than "contingency" for the middle term we are seeking. For, on the face of it, contingency is difficult to envisage as the ground of impotence to give existence, whereas the supposed compositeness might seem to constitute the ground both of contingency and of this impotence.

One may remark that the word "*simplicitas*" used by Aquinas can be misleading as an expression of the incompositeness or radical unity which Aquinas has in view: this consists rather in the absence of any distinction relevant to causal explanation, than in any unity imagined to reside in a Platonic idea. The relevant notion of "compositeness" (or "non-simplicity") is the one made plain in Chapter V, Section 9, against the background of Wittgenstein's position in the *Tractatus*, whereby for real contingency to exist there has to be a distinction, rooted in how things and their relations are in themselves, between the subject which is said to exist and its existing. And it was considerations of efficient causal explanation, rather than of any other sort of explanation, that established the relevant kind of 'reality' in making this distinction. Our discussions in Chapter V suggested that incompositeness, or unity in this sense relevant to causal explanation (as it were, unifiedness or integratedness rather than uniqueness), might be of key importance in understanding the difference between God and creatures, and in offering a proper basis for a so-called "cosmological" argument from the existence of creatures to God.

It is the view I am arguing in this book that this term "composite" thus understood is indeed the middle term which is needed for a cosmological argument to be carried through. But this requires to be upheld against systematic epistemological or methodological objections in Chapter VII and brought into structurally clarified form as in our final chapter, Chapter X, before it can count as having been established.

Such then are three of the candidates for serving as this middle term we are seeking, which history has thrown up. There are others, but these three will suffice as examples to illustrate the structure of our problem.

(d) Resulting structure of argument

When philosophers or mathematicians wish to prove something, an ultimate conclusion or theorem, what is commonly necessary is that they should identify some intermediate conclusion, i.e. what specialists call a 'lemma', and prove this first. It is commonly the case, that, once the lemma is identified, ways will lie ready to hand to prove it and the derivation of the ultimate conclusion will be straightforward or obvious.

Suppose, then, that our middle term is "*M*". The structure of our argument to God's existence will then be as follows:

I. (Lemma). If the existence of ordinary things has a cause in the respect picked out, it is not anything temporal because temporal things are impotent to cause existence in this respect (the reality of time makes this plain to us), and it is not anything with the property *M* since it is the property *M* which makes temporal things thus impotent. (The question is as to why temporal and ordinary things are impotent in this way. The reason, we will say, lies in their being *M*: this is the middle term "*M*" which we seek.)

II. The existence of ordinary things, in the respect picked out, has a cause.

III. Therefore the cause of the existence of ordinary things, in the respect picked out, does not have the property *M*.

IV. To be a cause, and yet to lack the property *M*, implies being God: (*a*) it implies being incomposite, intrinsically underivative (*a se*), immediate as cause to the existence it upholds and unique; (*b*) it implies being possessed of all the attributes required for a thing to be appropriately called 'God', e.g. being personal.

In this book, the key property of ordinary things is taken to be their temporality. That the existence, in the respect picked out, of temporal things needs a cause (i.e. Proposition II) argued to on the basis that, in the context of the real contingency of the future and in general of the reality of time, the existence of natures requires reliable continuance over time. Proposition III is an immediate consequence of Propositions I and II.

Proposition IV requires further argument of a kind depending on what the middle term "*M*" turns out to be. If the term "*M*" is the term "composite", it turns out that the argument appears more self-evident. The fact that whatever does not have the property signified by "*M*" is intrinsically, not just *de facto*, underivative, i.e. the fact of its self-existence or aseity (so that it cannot itself have a cause but is in traditional terms *a se*), was built into the explanation of incompositeness in Chapter V. And its uniqueness is straightforwardly provable in the way there indicated and developed in Chapter IX, without any need for appeal to the principle of the identity of indiscernibles or to the unity of the Universe.

Accordingly, the primary problem is that of identifying the middle term "*M*", justifying the suggestion that it is the term "composite", and establishing the lemma in question, Proposition I.

It is possible for the ordinary man to be clear that some conclusion can be proceeded to, and be clear of this even in advance of having achieved such a philosophical task as I have indicated. The certainty of the possibility of a deduction or at least the existence of knowledge on the basis of the assumption of the possibility of a deduction or inference does not stand at risk here, any more than it stands at risk anywhere else where the most thorny problems about causal principles are concerned, the problems of their status, range of application, or validation.

That is, our own presentation may be defective at any of several points but yet the whole line of argument retain cogency. We have set up a programme each of whose points needs to be considered in its own right, in case a better account may be offered. Firstly, different accounts of compositeness may be offered and this with either of two different types of effect, the one making the reason why the incomposite must be intrinsically underivative less clear and the other making the reason why the composite should be considered the root of the existential impotence of the temporal more obscure: the explanation of each of these connections is open to more discussion. Secondly, there is our account of why the temporal needs a cause in the respect picked out, rather than having no such cause. The soundness of our conviction that an argument of the general structure outlined is cogent does not stand or fall with the satisfactoriness of what we have to say in these different respects: in regard to each separately it has to be asked whether and how my approach needs to be or can be improved.

However, after these preliminary remarks, it is not for us to fight shy of the problems in hand. Firstly, what is it about creatures that makes them in the relevant sense impotent? To this inquiry I now turn—looking for a reason for what seems 'obvious', beginning by digging amongst past suggestions as to this.

(e) Review of earlier approaches to the question as to the root of this impotence

In this inquiry as to what makes ordinary things in the relevant way impotent to give existence, one's first temptation is to offer as one's explanation the supposed difficulties involved in action at a distance, the difficulties made clear in Chapter III, Section 2(*b*). Wherever causation of existence is concerned, a cause has to be in some sense intimate to, or not at a distance from or external to, what it causes. For example, it must be that the externality of one time to another and of one spatial thing to another, and of the nature of things considered at one time to the nature of things considered at later times, carefully stated, likewise sufficiently involve this unwanted externality between cause and effect in many ordinary cases. But one thing cannot be internal or close to a second thing except insofar as the second already exists, and insofar as it already exists, the need for a cause of existence has passed. Therefore it cannot be that the problem is rightly stated in terms of the externality of creatures to each other, undoubted fact though this be.

If one wanted an explanation of the relevant kind of internality, one would have to explain it as the internality involved in one thing's being unmediatedly the cause of another thing's continuance in existence. Accordingly we can dispense with further discussions of what philosophical difficulties there are in relation to action at a spatial distance, the difficulties which led Faraday to bring field theory into existence, since it seemed to him absurd to suppose that a thing could act where it was absent.[12] (These difficulties at an earlier time and since his time have only been overcome by regarding the relevant kind of action at a distance, whether electrical, magnetic, or gravitational, as an exemplification of some regularity or law, grounded in the system of nature. In this way there would be no unconditional or presuppositionless action or power to act involved here, but rather the conception almost that Nature itself was the primary agent or *responsable*.) One may note that parts and natures are in obvious senses internal to and prior to wholes and to acts, but it is puzzling how anything can be prior to the existence of a thing and at the same time 'internal' to it—until one mentions the giving of existence by God.

A second type of approach has been to observe that, in the ordinary cases of causation which we consider, there is some sort of threeness involved. There is the actor or agent, the action, and the effect or upshot: and the action has some sort of externality to both cause and effect, as it were, lying in the middle. In practice, this is connected with using hands, material tools, and suchlike, i.e., in general, instruments. Accordingly, one meets arguments that creatures cannot give existence because they cannot be identical to their action, as if God were identical to His action: and this seems to be a mis-statement of the situation, prejudicial to God's freedom in creation. So far as the action is considered materially, the action *is* the very existing of the mundane, supposedly created things. So far as the existence is considered formally, the action should not be considered a thing at all.

Yet there would seem to be something right about the instinct that there needs, in the case of creatures, to be something intermediate, involving something quasi-instrumental, so that creation, if it were supposed deputed to a creature, would have to

[12] On Faraday's arguments against action at a distance, see Mary Hesse, *Forces and Fields*, pp. 197–206, 222–4, 291 f., and her references to Faraday.

have as well as a formal aspect, a material aspect (i.e. an aspect to do with the object, e.g. elements of the action pertaining to the use of faculties, limbs, and instruments, etc.).[13] This betrays the fact that all these instruments would have to be within the mundane world, coeval with the things whose causation of existence is in question and so, for the purpose in hand, the giving of existence, useless.

The other instinct behind this type of argument is the instinct that, in order to give existence to a thing, the cause of existence has, as it were, to put its shoulder, its whole 'umph' into the act. And it is obscure how any creature might be able to put the whole of itself into any of its actions.[14] (It is not easy to give precision to this argument!)

It must be realized that the moment we introduce the idea of an instrument or a bodily part as a means, we introduce the notion of something coeval with that object whose existence is to be explained, and lying with it together within one order of nature. Now, whenever a thing operates with instruments or with parts of itself, e.g. limbs or hands, it operates according to nature or its nature, but in this case according to a nature coeval with the existence of the things whose continuance in existence through time is to be explained.

Now, whenever the operations are thus coeval, it appears plausible to argue that the nature of the alleged cause would need also to be thus conceived. It is then plausible to suggest that the nature thus given will be given as part of what also sets the character of the context of the operations of the thing. That is, if the operations of the thing are within the world in the way explained, then the thing's nature will be conceived of as part of the order of the world, and it is difficult to see how one can avoid considering the thing with its nature as part of the order of the world. This would be true even if it occupy an extraordinarily noble position, somewhat as the Sun assumes a rather noble role in ancient conceptions of nature. In any such case, it will seem to follow that

[13] Aquinas explains how, in any creaturely action of bringing something about, there has to be something quasi-instrumental whereby the creature achieves its effect: *De Potentia*, Q. 3, Art. 4, and *S. Th.*, Ia, Q. 45, Art. 5.

[14] Duns Scotus bases his view that only God can act by means of His whole substance so that God alone can create on the fact that He alone is infinite, at least according to Étienne Gilson's exposition (e.g. *Jean Duns Scot*, pp. 351–5).

the thing's nature and operations are of a piece with, coeval with, the world and, for similar reasons to those that we met with in regard to instruments, that they are, like instruments, also impotent to give existence.

Thirdly, we gain here some inkling of a related kind of argument: *viz.* some argument that, if a thing is derivative in any respect at all, it must be within a field of other things within which it is placed with its nature, its nature allowing it to be operated upon as well as to operate, and derivative in *esse* (in being or existence) if derivative in or contextual in nature.

In this regard, one can argue that if a thing is derivative at all in any respect, e.g. in the determination of its accidents (accidents are indicative of the nature which they presuppose), then it is derivative in its *esse*, i.e. the existence whereby alone this nature subsists or has effect. If it is derivative in existence or *esse* from something *x*, and has an operation directed to an effect (an operation according to its nature) constituting this effect in existence, then, insofar as its nature depends on *x*, so its capacity to cause this effect will depend on *x* and so the very existence of the effect itself will depend ultimately on *x*. When I say 'depend ultimately', I do not mean the kind of dependence which is compatible with *x*'s having ceased to exist or ceased to act in the relevant respect, but I have in mind an action of *x* contemporaneous with the causal activity flowing from *x*'s activity.

There are other arguments: for instance, the key argument found in Aquinas (emphasized by Étienne Gilson) that the gap between being and absolute or unqualified non-being is infinite and requires infinite power if one is to get from one to the other.[15] That is, whereas, granted a framework of nature which might be presupposed in its causal activity, the finite power might achieve a finite effect, by contrast, in order to create things out of nothing or to uphold them in existence without relying upon any of the presuppositions which natural things, presupposing nature, rely upon, an infinite power is required. The point appears to be that, whereas one can specify the task set to the potter, the task which presents itself to him in moulding the clay and making the pot and what is required of him in order to achieve this task of making a pot, if one is not allowed to presuppose the existence of matter and

[15]. *De Potentia*, Q. 3, Art. 1, obj. 3 and ad 3, and Art. 4, and *S. Th.*, Ia, Q. 45, Art. 5. Cf. Étienne Gilson, *Being and Some Philosophers*, pp. 90–1.

its properties or the rest of any such context, then there is no limit, indeed no definition, no specification to be given of the object of the power which is required.[16]

However, how one reasons from infinity in respect of that in regard to which a thing has power ('object' of power) to infinity of power and how thence to infinity of being (*esse*) is obscure to me. The argument will depend upon taking Aquinas's, rather than Duns Scotus's, view of being: for Duns Scotus infinity or finitude turn out to be always in respect of some essence or property (a being is infinite in the sense that its essence is infinite) and not in respect of being as such, and the gap between the being of a finite entity and its non-being is as finite as its essence makes it.

Although, therefore, this argument of St Thomas may not succeed in strictly proving its proposed conclusion, nonetheless, it does suffice to give a real reply to the general objection, inspired by Kierkegaard,[17] that any argument from creation to God must be invalid because it involves an inference from the finite to the infinite. The argument also implicitly makes appeal to some of the considerations to which we have already alluded, *viz.* the non-reliance on matter, on instruments, on tools, on hands, or upon any means, so that God, without means, simply, or, as it was traditionally said, merely by a fiat, is able to bring things about.

(f) Our own approach

We must now stand back from these older arguments and make a statement of our own approach.

The argument from externality in time is instructive because as we have seen it is the relative place and relative time of two things that typically determine and limit what possibilities of causal relationship there are between them. And we are looking analogously for what determines the absence of a certain possibility of causal relation, *viz.* the possibility of one thing unconditionally and unqualifiedly giving continuance of existence to another.

What we need is some notion of being 'mad' or 'chimerical' to suppose such that, for instance, it is a chimerical suggestion that my moving of my fingers without intermediate mechanism could move

[16] The power to create cannot be communicated to a minister or deputy (cf. refs. in n. 5 above).

[17] Cf. Chapter I n. 10.

the stars: it is, let us say, chimerical because there is nothing in what the moving of my fingers is (the nature of this action) which could ground such an active power. An active power has to be grounded in the constitution or nature of that of which it is the power (i.e. in something internal to it).

Then we can say that it is chimerical to suppose that what is at one time can act directly at another time or on what is at another time. There is no sense in supposing an action or what happens at one time to be done by something operating at another time when there are no intermediaries.

Now what is composite in the cases we are accustomed to has an ontological structure involving distinct but non-discrete actualities, a nature which is as it were a situation or state and, distinct from it, acts or exercises of this nature associated with different times (because, being contrary to each other, they cannot be exercised at the same time). It is therefore temporal in the broad sense we indicated (the sense involving distinctions of order, but not requiring continuity) and it will be chimerical to suppose that it can directly cause continuance. Either this would involve acting from within one time or temporal order at one point of that time in order directly to act within the same order at a different point of that time, or else it would involve acting (if, contrary to what we suggested in Chapter V, there could be several times, temporal systems or 'worlds' of interrelated causes and effects) from within one time upon another. The latter hypothesis strikes one as even more chimerical than the former, because it is less easy to postulate intermediate mechanisms since *ex hypothesi* there is no over-arching 'world' for them to belong to (although in any case such postulates are contrary to the hypothesis of direct action). Only if something was by nature non-temporal could it be, without this difficulty arising, internal to and directly acting in a way bestraddling times, whether times within one temporal order or different temporal orders.

However, it is not because of its temporality that what is composite is impotent to give existence, but because, if it is temporal, its nature does not extend to this but is in this respect limiting or circumscribing. If we ask why its nature is thus limiting so that the attribution of certain active powers to the thing is chimerical, then the answer would seem to be that it is because its nature is a given which shapes the realization of its existence in act or acts and is not simply identifiable with its existence: this existence does not consist in the nature alone, whether conceived of as something quasi-abstract or conceived of as quasi-situational (that which in the case of anything temporal is conceived of as something static, attributable to instants or moments in virtue of holding over periods as a standing ground for an act or acts), but consists in this nature only taken together with some typical realization or expression of it in act or acts. That is, the ultimate cause of this impotence is this aspect of compositeness.

We have to look at the cases of the existentially impotent to which we are accustomed, *viz.* the temporal cases, and reflect upon what it is that makes them thus impotent. We then find that it is this role of 'nature' as limiting acts. Its having a nature is prior to the act or acts expressive of the nature, even to its so-called 'immanent' or non-external acts (thinking and understanding seem to be immanent acts). What we understand by this is illustrated by the temporal cases in which the nature is expressed in at least two non-compatible acts, associated with different times because incompatible.

We have to see why the root of this existential impotence should not be identified as temporality itself.

If we were to prove that the composite must always express itself in at least two acts associated with different times, we could say that the composite was as such temporal and that this was part of the ontological structure given with being composite. But still we would have to say that it is not separateness of times as such that is the root of the impotence but the localization in virtue of which times cannot be leapt over. A thing may be, in a bodily way or otherwise, at a place ('be present in being' at the place) by being immediate to its actions at that place, but if its nature debars it from acting at the place it is, as it were, localized so as not to be itself at the place and so not available to act internally to things at the place: hence the limiting character of nature is still what is crucial. In this way the root of the kind of impotence we are concerned with lies deeper than temporality. We said 'the limiting character of a nature', not 'the limited character of a nature'. We cannot usefully say that it is things' limitedness or finiteness in nature or being which limits their acts, because what it is for a thing to be finite in nature or being can only be explained in terms of the thing's having a nature distinct from its existence in the way just explained.

Because the expression "finite in nature or being" is meaningless until thus explained, to introduce it adds nothing. Meanwhile, we cannot say that it is their finiteness in powers which makes them existentially impotent because this is to explain impotence by itself.

Nor is the contingency of things the root of their existential impotence. If they are contingent, they will be composite, but their contingency is a matter of their background and their relation to it, not of how they are in themselves. But it is how they are in themselves, e.g. their compositeness, and their relations to what is coeval with them which determines what limitations there may be in their active powers, including their powers to use what is coeval with them.

That the ultimate cause of the impotence we are concerned with is the distinction of nature from existence so that nature plays a limiting role, limiting acts, appears in another way, namely that always, when in the case of some supposed external action it appears chimerical to attribute such an action to the object, the only way of reducing this appearance is by supposing some instrumental activity to take place within the context of some system or mechanism of Nature. The requirement of instrumental activity is another aspect of the ontological structure picked out by the word "composite". The activity (i.e. the agent's being in action) is distinct from the nature of the agent (i.e. it is distinct from the situation of the agent's having this nature) and distinct from its effect; it thereby has to have a distinct existence (which we identified in Chapter V as an element in the existing of the agent) and nature or character of its own; and this character must be such as to make it relevant causally to its effect if it is to be correctly described as 'causal', and this involves being instrumental.

It is obvious that the temporal is in the relevant way impotent and the notion of the temporal is very familiar. Therefore it may seem strange to root this impotence in something so obscure and abstract as 'compositeness'. But it is not mysterious that a thing should be limited in its actions by its nature—so that some actions should be chimerical to ascribe to it, e.g. ones which are not in any way instrumental to the achievement of their object and which overleap times without mediation. What is surprising is that there should be anything not thus limited, because its nature is nothing different from its existence so as to be able to limit its realizations: we have no reason apart from argument from its being actual (and arguments supporting this conclusion) for supposing any such thing to be even possible.

There is no doubt as to the relevant impotence of the temporal and it is clear enough that the properties of being temporal in the broad sense and being composite are inseparable. What we have argued is that, upon reflective examination, it is being composite that is best considered to be the root of the impotence concerned. This is not just a trite formula to solve a problem or answer a question, but picks out the ontological structure which is determinative—the ontological structure of a nature whose distinctness, whether as quasi-abstract or as quasi-situational, from the realizations or acts (or realization or act) of the existence of the thing means that this nature limits these realizations (or this realization). This is not an area in which a less abstract or different type of proof or discussion is to be expected—although the importance of the general character of our account of compositeness for the strategy of our general argument deserve to be re-iterated.[18]

The architecture of causal proof of God's existence requires that the cause of the temporal be shown both to be one (unique in nature) and to be, not just *de facto* underivative, but intrinsically underivative (that is, for internal reasons, incapable of being caused), so that there is no room for a regress in causes.

There is, in my view, no direct argument from non-finiteness, from necessity (non-contingency), or from being intrinsically underivative to being unique. In each case the argument has to pass via some other notion, such as the notion of incompositeness. Nor, in my view, is there any direct argument from non-finiteness or from necessity to being intrinsically underivative. But, by contrast, being incomposite and being intrinsically underivative are inseparable and imply being unique; and, in my view, argued above, only the incomposite could be cause of temporal continuance since compositeness is the root of impotence in this respect.

[18] In Chapter V, we noted that one might argue that, if the nature of a given thing was something quasi-abstract, then while the quasi-abstract nature formally specified the acts it would be some quasi-situational actuality which efficiently limited them, and then, if this argument were valid, it would be important that the existence of the thing could not be identified with this quasi-situational actuality but only with this together with the expressions of this nature which would be what are limited by it, thus limiting its existence. For the reasons given in Section 11 of that chapter, I do not think I need this elaboration for my argument.

PART II. THE NEED OF A CAUSE: THE STRUCTURE OF METAPHYSICS

2. THE SECOND LESSON FROM HISTORY: THE STRUCTURE OF METAPHYSICS AND THE CONCEPTION OF A PRIMARY CAUSALITY WHICH UNDERPINS AND DOES NOT ABOLISH SECONDARY CAUSALITY

In Section 1 we have shown that the reason why the temporal is impotent to cause continuance in existence unconditionally and unqualifiedly is that it is composite. It followed from this that, if the temporal has a cause in the respect picked out, this cause must be incomposite, from which, we suggested, it followed that it was intrinsically underivative and unique in nature.

There now remains to us the capital question as to whether and why the temporal needs a cause in the respect picked out.

The answering of this question is achieved by bringing it together with the question as to the reality of time and the real contingency of the future.

Both the appreciation of the impotence of the temporal to ground its own continuance and the appreciation of the need for this continuance to be causally grounded arise for us from the insight into the reality of time and, as it were, the real 'precariousness' of the future. It is only in the recognition of the reality of time and the real contingency of the future that we in one act identify (i) the respect in which the temporal, because composite, is impotent and (ii) the respect in which the temporal needs a cause as a condition of its existence. What (i) debars the past and present from a role in the efficient causation of temporal continuance into the future, rendering them impotent in this respect, is precisely what reveals (ii) the need of a cause for this continuance. As to (i), the clarity with which it appears chimerical to suppose continuance into the future to be caused by something with a temporal nature depends upon recognition that the future is so really contingent as to be in no way given with or internal to what already exists or is already established. But it is the connection with (ii) which is important for our immediate argument and the problem of the structure of metaphysics.

To sketch our argument as to (ii): it is the combination of

recognition of this kind of contingency with a recognition that temporal natures cannot exist at all except in a context allowing their successive expression over time (they can have no instantaneous existence and must require some continuance according to nature in order to exist at all) that compels acceptance that continuance into the future must be caused. Given the metaphysical contingency of the future, a trans-temporal ground giving trans-temporal existence to natures (upholding natural things in existence through time so that their natures can have effect) is needed if natural causation and natural contingency or necessity are to exist. In this way so-called secondary or natural causation has its reality grounded in the primary causality exercised immediately to every time and place by the First Cause.

Let us consider the general strategy of our argument.

We have already observed in an earlier chapter the possibility of an argument that the existence of real causation and real causal connection in nature involves the reality of time, i.e. a real contingency in respect of the future, transcending physics but belonging to metaphysics, not to epistemology or logic. Now we have the question as to whether this existence of causation within the realm of nature, set thus within a temporal framework, may depend upon the existence of a first cause upholding things in their natures, i.e. upon whether this natural causation may not be secondary in the traditional sense. And the view we urge is that there cannot be this real contingency in respect of the future, a real possibility of there being a future, without there being a causal background out of whose powers the future may come, a causal background of which the natural sciences give no account.

This then is the general strategy of our argument from the reality of time and the reality of causation within a temporal setting, to the existence of a trans-temporal cause of the whole Universe (this Universe taken with its order and constituted with its temporal framework).

The answering of our question as to the need of a trans-temporal cause of the general continuance of things in existence is thus achieved by bringing it together with the question as to the reality of time and the real contingency of the future.

Yet the latter question appears to be a question about the analysis of our concepts of natural causation: we argued that the reality of causal agency and action stood or fell with the reality of time, and that the realist position involved attributing a metaphysical contingency to the future. But the former question, as to the cause of temporal continuance, seems to take us beyond the sphere of the natural, and into the spheres of cosmology and theology. Therefore, to bring the two questions together is to raise the problem of the structure of metaphysics.

Let us first consider this question of the structure of metaphysics in general before considering the conception of 'primary causality' and the role of a 'first cause' in more detail.

(a) *The possibility of an Aristotelian view of the structure of metaphysics*

Against this background we can now view more clearly the second lesson to be learnt from history, one of those strategic points at which history repeatedly shows a division of the ways in the approach to philosophical questions.

The question is as to whether (i) the shape of metaphysics is geared, as Aristotle and Aquinas conceived it, to the structural relations among the realities that exist; or whether (ii) it has a shape indifferent to such structural relations, occupying itself with principles which apply in a uniform manner to everything, in the way that the principles of formal logic apply uniformly wherever they apply at all, as rationalists such as Leibniz and Wolff (possibly anticipated by Duns Scotus) seem to have supposed; or whether (iii) it has a shape set by epistemology or anthropology, Kant substituting epistemology for the psychological anthropology of Locke and Hume.

The materials thrown up by history include the suggestion of Aristotle that 'first philosophy' is at one and the same time the science of being as such and a science which has to do with the divine.[19] Against this there is a prejudice that general causal principles lie in the province of a general metaphysics which has no particular concern with any particular kinds of being as such, let alone with a certain unique being, God.[20] It goes with this that it is

[19] Aristotle, *Metaphysics*, Epsilon, I, 1026a 7–32.

[20] This view is nicely exhibited in some of the discussions (e.g. by G. E. L. Owen) of the dating of different elements in Aristotle's works, in which it is supposed that the association of first philosophy with a science centrally embracing the divine represents a early stratum in Aristotle's thought, united only by superficial stitching

commonly imagined that such general causal principles are pure in a Kantian sense, importing nothing in regard to particular empirically known or, as it happens, in fact actual kinds of being, completely general and univocal (i.e. with no variety in their interpretation or mode of application), just as the rationalists supposed.

In this second way, the materials of history suggest to us the questions which we require to examine with some rigour in our next chapter, while giving us at this stage the opportunity, useful for the purposes of exposition, of exhibiting the nature of the problem with which we have to contend. For there is no need for the genuine empiricist to concede so much to the rationalist. It is more plausible that there is a spread and variety in the causal principles we rely upon in thinking about, and augmenting our knowledge of, ordinary things. Each such lesser principle within such a spread may be supposed to be the object of knowledge just as easily as—indeed more easily than—any single supposed perfectly general causal principle (cf. Chapter VII, Section 2). In the way I shall indicate in Chapter VII, Section 1(*c*), such lesser principles are known in their actual application, in an interplay between their actual, first-order, use and second-order reflection upon this use.[21] So indeed, in my account, knowledge of them, or the rightness of relying upon them, does not wait upon any alleged recognition of cosmological or theological presuppositions.

(Indeed, in my account, it appears that knowledge of the applications of types of causal reasoning is prior to any statement of the principles exemplified in such reasoning. Therefore it is better to speak of a spread and variety in the types of instance of causal reasoning, each exemplifying certain very informal principles, than to think of causal principles as being, as such, objects of knowledge prior to appreciations of the need of a cause in the individual instances. Rather causal principles of greater or less generality are

with the evidently later conceived perspective of (e.g.) *Metaphysics*, Gamma, 1, and Zeta, 1, which is supposedly more sophisticated and almost modern in its quasi-linguistic and (except for the categorial structure which it takes as given) ontologically neutral and uncommitted character.

[21] This has been made abundantly plain to me, not only by the writings of more ancient philosophers, but also by the examples and arguments of my colleague Michael Partridge, both before and after my delivery of 'Two Kinds of Critical Philosophy' to the *Société Belge de Logique et de la Philosophie de Science* in 1972.

known by intuitive induction from the appreciation of the need of a cause in this or that particular type of instance. But this will become clearer later.)

Yet it may in fact be impossible to sustain the view that the concepts in terms of which causal relations are stated are coherent, so that the causal principles exemplified do have application, without conceding that these statements and principles presuppose, for instance, the reality of time, the unity of the Universe, and the existence of God. In this way, it may turn out, and indeed does turn out, that the general metaphysics which takes cognizance of these lesser causal principles incorporates also a recognition of these underpinning conceptions. (I prefer the word "underpinning" to the word "over-arching", since the relation of these underpinning conceptions to the ordinary causal principles is rather that of a ground or root to a consequent general necessity, than a law of thought or being, inducing necessity, as it were, derivatively in less general laws of thought or being.)

In this way, general metaphysics will turn out to be structured in a way quite different from general logic. It will not be centred on a small number of principles uniformly applicable without any content, in no way reflecting the structure of reality. Nor will it be, as Kant supposed, pure or formal in this way except insofar as it is supposedly shaped to the conditions of possible human knowledge, sensory and empirical. Rather, it will attend to a range of types of instance of the need of a cause arising in the case of each different kind of temporal object or actuality, along with its environment, in the way appropriate to it, grasping each such instance or type of instance as set within the context of an integrated system of causal relations, the context of whose realization is the reality of time, the natures active within which are actualized or energised by God. In this way His causal activity will be seen as giving active power to every other kind of cause, and giving it its character as a cause. And it will be because of this real relation (and not merely because of formal parallelisms) that the concept of cause as one unified concept is applicable in such systematically diverse ways.

The writer writes using a pen. The pen makes marks on the paper. The man grasps the table with his hand. His fingers grasp or exercise the force of a vice upon the table. The pen and the hand exercise causal power in a real but degenerate sense, degenerate compared with the prior sense in which the writer or the man exercises causal power. Analogously the activity of the First Cause is involved in every exercise of causality by mundane things: in what sense this is so will be our concern in subsections (*b*) and (*c*) below; in Section 3 I shall indicate how it is that one concept, the concept of "cause" can be used in such different, albeit interdependent, applications.

A caveat

The problem of the independence of the issues of the reality of time and of whether the temporal needs a cause has two aspects, a first-order aspect and a second-order one. It is vital that these should not be confused and that we keep the first-order aspect firmly in our sights, that is, the first-order question as to whether the possibility of causal relations among temporal things, and indeed their actuality inasmuch as the existence of causal relations is intrinsic to their nature, is grounded upon a metaphysically prior act of causation of their existence, with their natures, by God. We have to avoid confusing this question with the second-order question as to the extent to which the considerations which might be offered to justify the acceptance of principles of causality among creatures are independent of the considerations which would be offered to justify the principle that temporal existence needs a non-temporal cause.

We must as I have said distinguish these first-order questions as to whether the exercise of divine primary causality is a causal condition of the existence of temporal causal agents and of their actual exercise of causal agency from a second-order question. This is the second-order question as to the relation between, on the one hand, our knowledge of, or right to proceed upon, the causal principles we rely upon in ordinary life and in the development of science, and, on the other hand, any knowledge or insight that we may have into the need and character of a cause of temporal existence as such. That is to say, it is not plain that it is right to suppose that our knowledge of general causal principles is independent of insight into the real contingency of the future and that this insight may be attended by the possibility of insight into the actuality of a cause of temporal continuance. But what we will insist upon is that, in the order of knowledge, our realization of the real contingency of the future and the actuality of real causal action in nature comes first and knowledge of any primary divine causation comes only second.

Thus, although knowledge of principles of causality as we apply them ordinarily to ordinary things does not genetically depend upon but rather commonly precedes appreciation that nature needs a trans-temporal cause, it remains that a reflective or scientific knowledge of these principles may not be thus independent.

(b) Background to the consideration of the question of the structure of metaphysics: the conception of God's primary causality

It is important to observe that our main arguments to the existence and personhood of God proceed simply on the general basis that the continued existing of things in their natures depends upon a first cause of continuance, and therefore only on the basis of their existence as primary substances, without utilizing any such general principle as that every motion, or every actualization of a potentiality, has to have a cause such as we find in Aquinas's Prime Mover argument, his 'first and more manifest way'.

Nonetheless, our conclusion implies (I) the conception of God as upholding these substances in all their positive states, *qua* states, and (II) the conception of God upholding the existence of these substances, not just as a state, but as something active, so their natures are upheld not only in stasis but as dynamic. (Strictly speaking, the continuance of natures in existence is to be understood as a continuing in existence of the things with their natures, and the existence of a thing does not consist in the situation of its being of a certain nature. Since the temporal situation of thing's being of a certain nature is not *qua* static or *qua* situation the adequate realization of the thing's existence, giving the thing existence involves not only establishing or maintaining this situation but maintaining the thing not only as of this nature but as active according to and in virtue of this nature: the acts have to be 'in virtue of' the nature of the thing in order to be acts of the thing, but the situational nature is given existence in giving existence to the thing —we must beware of hypostatising or reifying natures).

That is, the activity of this trans-temporal cause will involve two aspects. In the first place what it causes has the aspect of a continuance of the states, substantial and accidental, of things. In the second place its activity has the aspect of underpinning the expression of these states episodically in time, since it cannot give continuance to natures, first or second, in stasis only, but only (if they are to be natures) in action also. But it does not achieve this underpinning by adding an extra push or premotion to every motion, as if synergistically bringing natures or dispositions from stasis or potentiality into act in order to give them the character of natures or dispositions (which are as such effective principles of action); it is the created subject or *suppositum* which brings its nature or dispositions into act by acting. Rather the First Cause achieves this in the fact that it gives to the created existence which it causes the character, not of a mere state or situation, but of an activity - something in itself active.

The character of God's action is better seen in this second aspect because what is maintained is not a mere fact or state of existing but an 'act' of existing: that is, what is maintained is precisely 'existing' as an activity or something 'active', something whose character as an 'act' is seen in its being 'dynamic', i.e. as something whose vigour is exhibited in the acts which are its expressions—the acts do not add vigour but exhibit it.

This second aspect of God's role as First Cause of continuance connects up closely with the idea that real causation requires that the thing put its shoulder or 'umph' into its action and that it would be capable of putting its shoulder or 'umph' into its action only insofar as the maintenance of its nature as something active (not as a mere fact, state, or situation), which is what its existence consists in, was itself the action of some prior, perhaps prime, mover—i.e. the action of this prior or prime mover specified object-wise, that is according to the content of the action or materially, following Anscombe's principle that 'the agent does what happens',[22] introduced in Chapter I and developed in Chapter VIII, Section 7.

The upshot of this second aspect of God's role as First Cause of continuance thus turns out to be this: that God, as the prime mover, actualizes, or, as it were, energises, the existence and *thereby* the nature of the individual thing. This act of immediate and effective or 'real' causation, wherein God does all the continuance of substances (it is the object of His action), and as it were in virtue of His immediacy and His radical unity (so-called 'simplicity') putting His whole being, energy, or 'umph' into His action, would be, in this perspective, the ground of the possibility and indeed the

[22] See Chapter I n. 16.

actuality of creatures exercising 'real causation': that is, of the causation whereby a creature can, in virtue of its nature and in exercise of its powers, have an immediate and effective relation to its effect.

But logically this internal causal involvement of God in all positive existing whatsoever will only be the activity of God alone in relation to the continuance of the existence of the first substances in their dynamic natures and dispositions—upon which our general argument rests—not in relation to all the particular actions expressive of this existence. In regard to these actions, we must insist that, of secondary causal action, the secondary causal agent is the only true agent, so that in general, in the exercise of their natures by mundane causal agents, although there remains a dependence upon God, what is involved is not an action upon His part but a underpinning in that He has to maintain natures, not just as states possessed at each of many times, but as dynamic, i.e. as actualities active in actions, actions which typically involve activity through time and therefore the continuance of the existence of the first substances involved through time.

Yet, in this conception, God is immediate to the existence and natures which He upholds, as maintaining them as 'active' (not that He maintains them as possibly active and adds 'activeness' to them, but that He maintains them as what they are, 'active' actualities), and this activeness of this existence of mundane things with these natures is expressed in, and achieves natural second realization in, their actions, typically their causal actions. This provides the basis for the argument from the existence of motion to a Prime Mover. In this way, the activity of God is involved in the activity of mundane causes: there is no gap between His activity and their existence (this existence being something active or dynamic in the sense explained) and no gap between this existence and their action, even though this action does not thereby become His or thereby presuppose any extra action on His part, some extra beyond His constituting their natures precisely as dynamic.

In this way, we have indeed reached the conception of God as the Prime Mover and First Cause, prime and first in that without His first action (establishing in each case a thing's existence as the dynamic actuality which it is) there would be no later movers or causes. He is first, not as being first in a long series from whose later members He is separated by intermediates so that the actions of the later members depend upon His action externally, but as being immediate to every action (whether conceived of as a moving or as a causing). Thus we remove the ambiguity from such formulae as 'this cannot go on to infinity, because then there would be no first mover and consequently no other mover' and 'if in efficient causes it is possible to go on to infinity, there will be no first efficient cause and neither will there be an ultimate effect nor any intermediate efficient causes': this disambiguation was achieved by Aquinas himself in his giving as reason that 'subsequent movers move [other things] only inasmuch as they are put in motion by the first mover, as the staff moves [another thing] only because it is put in motion by the hand'.[23]

In this sense, we may incidentally remark, there will be an immediate causal involvement of God in all positive existing whatsoever, not just in substances having substantial existence. Thus, all positive temporal existing whatsoever, not only existing 'in a substantial nature', but even the existing which consists in the positive existing of substances in some accidental respect (i.e. the 'existence' of 'accidents', cf. Chapter V, Section 2), will depend immediately upon the activity of God. For it to be the 'same thing' which is given continued existence more is required than continuation as of a thing of the same species or general kind: some continuity in so-called 'accidental' properties and relations is also necessary.

Not only is existing in a nature something whose character as an act is seen in its expressions but aspects of its character as an act are seen in its expressions. The character of the act of existing in its positive aspects can only reflect the character of God: but the character of the act of existing is what appears in its expressions. This connection, more carefully stated, is the basis for an argument that God pre-possesses all the 'perfections' to be found among creatures, an argument underlying Aquinas's Fourth Way, and anticipated in the formulations of Chapter V, Section 8.

I make these last incidental remarks because they seem important

[23] Aquinas, *S. Th.*, Ia, Q. 2, Art. 3. In Q. 46, Art. 2, ad 7., he compares God's causal action to the action of a hand in moving a stick to move a stone or of a blacksmith in using hammers to achieve his ends (a succession of hammers being needed over many years as a succession of fathers was needed to generate children over the ages).

for generalizing our argument beyond the existence of God as First Cause of the continuance of temporal existence, and beyond His personhood to, for instance, His goodness, along the lines suggested in Chapter IX.

(c) *The avoidance of a misunderstanding: the reality of 'secondary' causation*

Aquinas and Duns Scotus have been contrasted, Aquinas as holding that God alone is cause of existence as such while for Duns Scotus there is no difference between causing a thing and causing its existence.[24] And it is indeed vital for Aquinas as well as for the argument of this book that secondary causes are not causes of their effects in respect of their existence as such rather than their coming to exist, and not causes in respect of being beings rather than of being of this or that nature, and in general not causes (as we expressed it) 'unqualifiedly' or 'unconditionally' of continuing existence. All this goes with the insistence that God works in the operations of nature, and that the continuance of things depends upon Him.[25]

But it is necessary to retain a balance. It is part of the same view, of Aquinas, that this working and upholding is not a creating.[26] It does not involve setting aside the causal action of creatures according to their natures, but rather precisely an upholding of them in their natures, so that they can act according to these natures. In this way the views of Malebranche and Berkeley are rejected beforehand in arguing against such mediaeval Arabs as Avicebron in Aquinas's key discussion in *De Potentia*, Question 3, Article 7.[27] Therefore, in a certain sense, existence is the object of secondary causation even for Aquinas. And that God should cause things to exist in their trans-temporal natures is vital to the continuity or connection of time, and so integral to the reality of time insisted upon in this book.

The secondary cause is always something whose nature is temporal, extended in time, and its causal action is therefore, in the

[24] Gilson, *Jean Duns Scot*, p. 90.

[25] Cf. *De Potentia*, Q. 3, Art. 7 and Q. 5, Art. 1, and *S. Th.*, Ia, Q. 104, Art. 1.

[26] Cf. *De Potentia*, Q. 3, Art. 8, and *S. Th.*, Ia, Q. 45, Art. 8 and Q. 104, Art. 2.

[27] Étienne Gilson's discussion in *The Unity of Philosophical Experience*, pp. 45–8, is instructive in this connection.

view being explained, dependent on God's maintaining the cause in question, with its nature, in existence. And, on this view, secondary causes never act without the presupposition of, and reliance on, some order of nature within the framework of which they act, and the continued existence of the stuff or substance of the world, of which this nature and order are properties. True also, this presupposition of the continuance of the stuff of the world and the order of nature carries with it the fact that no secondary cause can act as agent in isolation from a context in which other natures are actively as well as passively operative. Nonetheless, it remains that a particular secondary cause acts within a context which is already set in the same sense that the continued existence and powers of the secondary cause are set, and within this context its action can be decisive as to the existence (doubtless, after coming into existence) or non-existence of some particular effect. What the secondary cause acts thus decisively to bring about is existence with nature, never existence alone or nature alone.

The phrase 'cause of existence as such' is ambiguous. In the sense explained much earlier, God alone, or something supra- or extra-mundane, is unconditionally causative of existence without qualification. Also, formally the secondary cause causes being this or that rather than just causing being a being. But it remains that the object of secondary causation is still existence as actuality, and not anything else or less; in this sense it is existence as such.

In this connection, we need lastly to note that the identification of any effect and, indeed, the point of any action of bringing into existence, if taken within any larger view, depends on the continuance of the product or state brought about, and not merely its inception; but the secondary cause does nothing to maintain the stuff and order of the Universe, on which this continuance centrally depends, so that although the intention of an agent may extend to such continuance and be exhibited, for example, in protective actions later on, as well as in the bringing into existence, it remains that the bringing into existence does not include a maintaining in existence. Even so it does remain that the object brought about, that whose inception we are discussing, is existence, i.e. even secondary causes do genuinely initiate existence or bring things into existence. Secondary causes do not bring about a 'bringing about', but rather existence or something real.

We are now in a position to put forward for examination an

historical surmise. Let us grant that for Duns Scotus all causation was indeed regarded as being a causation of existence. However, Duns Scotus explains the existence concerned in a way compatible with thinking that it consists in nothing more than the actual completeness of an essence, i.e. the essence's being complete in all its individual determinations and 'this-ness'. [28] Such a way of thinking whereby nothing, apart from such determinateness, is required for an essence really distinct from God to exist, so that existence is a mode, degree or accident of essence, at least in the case of creatures, is traceable in Avicenna and such rationalists as Leibniz. And any such way of thinking exhibits that the mere form of words 'causation is of real existence' is not sufficient to secure realism in respect either of existence or of the notion of real causation.

In their reaction against rationalism and conceptual idealism, Hume and later anti-rationalists attempted to restore the notion of real existence as something extraneous to essence (but not accidental to essence as if essence was prior to it), but failed to restore the conceptions of active power and real causation: rather they fumblingly tried to connect real existence with 'objective experience'. In effect the conception that in causation what is brought about is some finite thing, in respect of its finite essence is compatible with a development in thought whereby causation comes to be envisaged as the succeeding of one essence or set of essences, quality or set of qualities, upon another in conformity to law, not involving any real action of the one upon the other, or any non-ideal existence in either. The attempt to found the notion of the real or the non-ideal on the notion of 'objective experience' is hopeless: unless it is connected with the access of knowledge to causal and explanatory relations existing or having validity prior to this knowledge, objectivity degenerates into inter-subjectivity and inter-subjectivity degenerates into accidental or conventional agreement.

Reflection upon these aspects of the history of philosophy suggests a certain question. To what extent is it possible to secure that the terms "exist" and "cause" are being taken in a realist sense and in such a way as to involve 'real' causal connections, without

[28] I put the view of Duns Scotus argued by Étienne Gilson in *Being and Some Philosophers*, ch. III, pp. 84–90.

doing so in some way which, incidentally rather than by design, but nonetheless surely, leaves the way open for an argument that an underpinning first cause of temporal existence and nature must exist. In this way, the possibility of a valid argument to a first cause may be inseparable from the acceptance of the reality of time and of causal agency in Nature.

To put the point differently: the first lesson we drew from history was that there are limits to what can be explained naturally or in terms of mundane and temporal causes; the second lesson is that the response to this can either be to offer a structured metaphysics wherein a real mundane 'secondary' causality is underpinned by a real divine 'primary' causality or to retreat to a 'descriptivist' or quasi-Humean view of explanation whereby there is no real causality, mundane or divine.

(d) The difference between Aristotelian and rationalist presumptions in regard to the character of metaphysical principles: the limited scope of individual causal principles

At this point it is appropriate to stand back and reflect. We made a contrast between a rationalist metaphysics attempting to build metaphysics on certain univocal general principles bestraddling all subject-matters uniformly and an Aristotelian metaphysics utilizing structurally interrelated sets of analogous principles or types of application of analogical concepts. It is now worth inquiring in a preliminary way what history suggests as to the causal principles actually required for a non-rationalist proof of God's existence.

The arguments drawn from tradition which I have rehearsed, albeit without always identifying their originators or those by whom they have been most notably expressed, are arguments directed towards establishing a certain intermediate step or lemma in our argument. This was in effect the lemma that, if the existence of a temporal substance requires a cause in respect of its coming to be out of nothing or its being upheld in existence independently of nature or without presupposing nature, then no such thing, i.e. no temporal, mundane, or ordinary thing, has any power thus to give existence to another such thing: mundane things are in the relevant way impotent. (In effect, our lemma says that, if temporal existence requires a cause, then this can only be non-temporal, non-mundane, non-ordinary . . . and so the question arises as to whether it is possible to prove that it can only be such as God has been believed to be.)

We have to examine what is being said to need a cause with an eye to discovering how broad the principles are which would have to be appealed to in order to prove that it does need a cause.

Now the antecedent of our lemma is that temporal existence needs a cause. Thereby it concerns the causation of existence that is in some way successive, complex, and involving causal relations, and therefore some type of 'real existence' to use the expression we used in Chapter V. This was the real existence lacking to abstract objects as well as to the fictitious, the mythical, the imaginary, and the ideal, but also lacking to certain things more nearly concrete: *viz.* things whose existence consisted in the absence of something else, such as ignorance or holes in the ground, and properties or accidents, even of a positive kind such as knowledge, which should be regarded as having positive existence only in the sense that they give a fuller life or existence in some accidental mode to some substance which possesses them.

Accordingly, the antecedent of our lemma is carefully devised. We are not saying that every happening requires a set of necessary and sufficient conditions so as to be entirely determined according to laws in the way in which a Humean or Kantian train of thought would lead one to look for, a view from which the modern physicist would divert one. Nor is one making the Aristotelian requirement that there be typically four kinds of cause, not only an efficient cause but also a final cause or end (so that monstrosity, chance, and suchlike things are only to be understood in terms of the notion of the presence of some kind of defect), a formal cause (however that be understood), and a material cause. (The material cause was the kind of cause which the Semitic based theologies, Jewish, Christian, and Muslim, regarded as dispensable in view of God's power to create out of nothing, while the final cause, along with formal causes except in the sense of laws, were what mechanists and empiricists have sought to do away with.)

Rather, the antecedent of our lemma incorporates a much more modest claim. Firstly, it posed the need only for an efficient cause in the Aristotelian sense, i.e. a substance with adequate power to bring about the effect: adequate, not sufficient in the mathematician's or logician's sense (guaranteeing the existence of the effect, making it so that the effect cannot not follow upon the cause), but in a sense compatible with it still being open to the exerciser of the power, either freely or as a result of some chance deficiency, not to exercise the power. Secondly, it posed the need of this kind of cause only in regard to the coming into existence or continuance of certain sorts of things, *viz.* certain of those things which in Aristotelian terms would classify as 'first substances' in the sense in which Aristotle in the *Categories* uses this phrase.[29] Thirdly, it posed the need of such a cause only in respect of their successive or continued or temporal existence.

This is not a final and precise specification of what causal principle or principles may need to be appealed to in argument to God's existence, but does provide a preliminary indication of the ways in which their scope is anyway very limited—separated by a great gulf from such implausible conjectures as the so-called law of sufficient reason.

3. THE THIRD LESSON FROM HISTORY: THE UNDERSTANDING'S MODE OF OPERATION EXHIBITED IN DIFFERENT INTERRELATED CASES OF REASONING TO A CAUSE

The third lesson which I believe is to be drawn from history is one which arises from the second. It is this: that what is basic to any operation of human understanding at all, but in particular basic to any well-founded metaphysical knowledge whatsoever, is the understanding of certain trans-generic concepts (transcending any definition, in somewhat the manner of the six so-called 'transcendentals', and often labelled 'analogical'). Such concepts in each case underly the acceptance of a structured set of analogous types of application, rather than any single universal principle univocally applied.

The notion that there is a spread and variety in the causal principles we rely upon in thinking about and augmenting our knowledge of ordinary things or in their types of application, is a notion which may appear to the reader to run flatly in contradiction, not only to the rationalist, but also to Aquinas.

The consideration of what Aquinas says presents puzzles which

[29] One of the uses of the word "substance" in Aristotle's *Metaphysics*, Zeta, 3, is of this kind.

suggest ways in which the problem of human knowledge of generality is to be resolved, the problem attacked in Chapter VII, Section 2. This consideration will serve as a convenient preliminary to the statement of what our own argument involves.

(a) Identifying 'causal principles' in Aquinas

In *Summa Theologica*, I. Q. 3, Art. 4, there is a nice distillation of three general principles to which, in that Article and very often elsewhere, Aquinas chooses to make appeal. We have to raise questions about the understanding and grounds of application of these principles which it is not clear that Aquinas himself raised, not foisting our own views upon him, but also not allowing it to be presumed that (writing long before the rationalists) he would interpret these principles in the manner of a rationalist (which would, of course, exclude our approach).

Firstly, there is a principle stating the need of a cause, a principle to the effect that, if the *esse* of a thing differs from its *essentia*, this *esse* must be caused either by something exterior or by its essential principles.[30] This is the principle that what is composite needs a cause to which we alluded at the start of Section 2.

Secondly, there is a principle relating *esse* to the conceptions of actuality and potentiality, to the effect that, since *esse* is the actuality of every form of nature, *esse* must be related to *essentia*, if the latter is other than the *esse*, as act (*actus*) to power (*potentia*).

Thirdly, there is what appears to be a Platonic principle, explained by means of examples, to the effect that just as that which has fire, but is not itself fire, is on fire (*ignitum*) by (*per*) participation, so that which has *esse* but is not *esse*, is an *ens* by participation, and that one could only describe as the first *ens* something which is *ens* essentially rather than by participation. This third principle would require statement in general terms in some such way as this: that which has something indicated by an abstract term, but is not that which is indicated by the abstract

[30] Aquinas enunciates first the more general principle that whatever is in a thing which is besides (*praeter*) its *essentia* (whatever it 'has' which it 'is not') must be caused either by the principles of that *essentia* or by something exterior. But he then takes as a particular case of this that, where a thing's *esse* is what it is concerned as being distinct from (*ab*) its *essentia*, then this must be caused either by the principles of that *essentia* or by something exterior—the former hypothesis being absurd, so that if *esse* and *essentia* are distinct the *esse* must be caused by something exterior.

term, has the adjectival expression with which the abstract term is paired applicable to it only in virtue of participation, and one can only describe something as the first or primary thing qualifiable by an adjective if it is essentially, rather than in virtue of participation, thus qualified.

(b) *Systematic problems in the consideration of these principles*

In reflecting upon the role of these three principles in Aquinas's thought and argument, we need (i) to observe firstly that there is no key place in metaphysics in which they are ever drawn upon separately or independently and that the nuances in the statement of each have significance for the understanding and application of the others. Then, secondly, we need (ii) to observe that none of these principles is an empirical principle in any sense catered for by Hume and his philosophical descendants. Thirdly, and above all, we need (iii) to take seriously the implications of what is in Thomistic thought commonplace, that the unlikeness of God and of divine causing to creatures and creaturely causing is greater than any likeness so that it could not be by any induction of the Humean kind or analogical extrapolation from observed cases to unobserved cases that these principles are to be understood or applied. Rather the support for these principles arises from reflection upon or 'intuitive induction'[31] from a variety of very diverse individual examples.

Let us examine each of these points in turn and then draw out some of the consequences of the third.

As to the first of the above observations, (i) above, although it is the consideration of the relation of actuality and potentiality, act and power, to which Aquinas gives pride of place in his presentation of arguments to God's existence, in that it is key to what he describes as 'the first and most manifest way', these considerations would have no consequence whatsoever independently of the presumption that, in the case of anything whose existence constitutes the actualizing of a potentiality, this actualizing must be the work of some cause, in Aquinas's way of speaking a 'mover'. The role of the third principle (apart from any heuristic

[31] For an explanation of the notion of "intuitive induction", see W. Kneale, *Probability and Induction*, pp. 30–7.

function it may have served in the development of Aquinas's own thought) appears to be solely that of introducing the presumption of the likeness of an effect to its cause—granted that the causes and effects being envisaged are not mere facts or events, but things or substances or actualities.

In this connection, the way in which Aquinas states the principle that a cause is needed has a certain background, namely the supposition that in the typical case a thing and its *esse* are distinct. Yet, to over-simplify, although those in the tradition of Plato and Avicenna have taken it that in no case, unless God, is a thing and its essence identical, the normal view has been closer to that of Averroës who interpreted Aristotle as if there was no real distinction between a thing and its *esse*. Therefore, in approaching the text of Aquinas we need to avoid being deceived by the apparently artless simplicity of his argument into supposing that neither art nor inspiration has played any part in the strategy of his argument. The conception that there is a real distinction between a thing and its *esse* was maintained by Aquinas in measured dissociation from the traditions represented respectively by Averroës and Avicenna.[32]

In order to understand the philosophical as distinct from the dogmatic basis for a view such as either Aquinas's or that of this book, one needs to grasp that where efficient causation is concerned (a matter which of its nature bears upon questions of contingency and necessity in the relevant senses, a fact betrayed by its bearing upon questions of the openness of the future and the fixedness of the past), there is a distinction to be made in the case of creatures, a distinction which has no place in the case of God.

Now in the tradition represented by Averroës all causation must be external to existence as such and relate only to coming to be within the context of a given order of nature, and in the tradition represented by Avicenna existence is almost an accident of a thing (the thing ever remaining as an essence in the mind of God). By contrast, in Aquinas a thing has no real existence as an essence merely but is inseparable from its real existence—causation has as its object real existence and essence inseparably. Yet, though a thing (with its essence) and its existence are inseparable, its

[32] The positions Aquinas is to be set in contrast with are well delineated by Étienne Gilson, *Being and Some Philosophers*, ch. II, III, and V.

existence is not (in creatures) identical to, or part of, its essence, and not only its coming into existence but its continuance in existence is contingent, and the statement of this requires a distinction between *esse* and subject of *esse* or *suppositum* as we explained in Chapter V. This is the background presupposed in the conception of the relation in creatures of existence to essence as analogous to other cases of the relation between actuality and potentiality.

As to my second observation, (ii) above, it draws attention to the need to take a critical look at what we mean by 'experience'. In experience, what we learn is indeed partly this, that any continuous change needs a cause, not only in respect of the state at the end being different from the state at the beginning of the change, but also in respect of the continuous movement (ever incomplete so long as it is still continuing) from the one state to the other. But we learn, it appears, more than this, namely, the need that anything that might yet come to be issues out of a background adequate to make its coming to be a real possibility, i.e. the issue of the workings of natures, or things, already actual.

Now, we do not *observe* (in any sense which a traditional empiricist would recognize) that whatever comes to be comes to be as the issue of the working of powers already actual. Supposing the stuff of the Universe to continue, the Newtonian physicist did not observe powers at work active in conserving momentum (inertia) or energy. Nor, for that matter, did Newton or his successors observe powers at work conserving the stuff of the Universe in its constant quantity, whether the stuff by mass or mass-energy. Therefore the sense in which the principle that the actualization of a potentiality needs a cause is not such as to make it an empirical proposition in any banal simple-minded way. It may be that we have empirical knowledge of the existence of substances with active powers whose exercise requires the reliable persistence in existence of the substances with their natures through some appreciable period of time. And this would be something absurd to suppose if we did not have empirical knowledge of the exercise of these active powers in real causal action and relation. However, this empirical knowledge is one of which Locke abstained from giving an account, and whose existence in the case of material substances Berkeley and Hume denied. We are therefore supposing, presuming, or (in Chapter III) arguing for the existence of empirical knowledge of the sort not allowed for by the empiricists. But the things which we thus take to be empirically knowable also knowably have pre-conditions. If nothing gave continued existence to the dynamic nature of empirically known substances, they would have no continuance, let alone action. Yet it remains that the sense in which Aquinas's principle is empirical is not one which would satisfy the so-called 'empiricists'.

However, this observation needs to be set within the context of the third observation, (iii) above, which is the key one if we are to get into a position to gain a general perspective on how the generality of these principles is to be understood. Our third observation deserves recapitulation. It was this: just as the unlikeness of God to creatures is greater than any likeness, even though anything positive in the nature of creatures arises from likeness to God, likewise the unlikeness of Divine causing to creaturely causing is greater than any likeness, even though there is no mode of creaturely causality which does not owe its efficacy to some likeness to Divine causality. Therefore it is not by any argument by induction or analogy from observed cases to even remotely similar unobserved cases that we extrapolate some imagined principle that whatever is actual and/or ordered must have a cause of its being actual or ordered.

Rather, in order for it to be coherent for us to assert (what we certainly know) that within the world, in many types of case, what begins to exist and is ordered has a cause which is also within the world, we need upon reflection to grant that something must make reliable the continuance through time of the system of nature, and the natures of things: otherwise there would be no Nature and no natures, but only observed regularity; but we know that there are natures more certainly than we know that there are perfect regularities.

Likewise, in considering the existence of things whose existence is not identical with what constitutes them (as it were, their substance or stuff), i.e. their *essentia*, as an example of an actuality, involving the actualization of a potentiality, it is not just that Aquinas or those whose views parallel his are using the terms "actuality" and "potentiality" analogously, albeit even more extravagantly than Aristotle, but that the types of application of the principle that the actualizing of a potentiality has to be by the action of something actual are related by one application underpinning others (underpinning the empirical types of case) rather than by the generalization of an empirical principle.

The distance between Aquinas and the so-called 'empiricists' is yet more plain in the case of the third principle. In natural causation, animals or plants in each case in reproduction generate an animal or plant of the same kind; in the work of artists, the product is supposedly like the model in the artists's mind; and in the execution of human plans, there is typically some likeness between the effect and what the agent intended. Yet this seems an inadequate basis, if the thinking were inductive or by analogy along the lines suggested by Bacon, Hume, or Butler, for any advance to a grandiose metaphysical principle that effects must resemble their causes. Rather, the acceptance of any such grandiose principle (more carefully formulated) depends on some quite different way of using examples, e.g. in such a way as to give rise to the understanding that anything caused which has positive existence will have positive existence only in so far as it is related in some constructive or enriching way to the world of causal agents, of substances, and could be in some way a realization or *entelecheia* of existing natures. This is not the place to develop further an account of Aquinas's, or our interpretation, or knowledge of the truth, of any such principle, the most problematic perhaps of the three. I have only wished to make plain that it is even more obvious in the case of the third principle than in the case of the other two that examples are being drawn upon directly to give rise to an understanding which underlies any generalization, in a way which is entirely foreign to modern empiricist tradition as established by Locke and Hume.

In other words, such thinkers conceive that we have an understanding of such conceptions or notions as "actuality" and "potentiality", "causes" or "things with natures so as to exercise active power" (constituting a "causal background") and "effects", "the thing to which likeness is had" and "the thing exhibiting likeness". And they conceive us to have such an understanding as enables us, and indeed requires us, to put these notions into application at each of many levels, or in a variety of ways and types of case. The understanding is rooted in experience of the actual, but does not operate by first allowing itself to be captured in some empirical generalization, and then vainly seeking to extrapolate on this basis: it is not thus capturable. Rather the consideration of examples establishes the understanding of some concept of analogical generality in such a way as to ground its application in a variety of very diverse although interrelated instances. This is the third respect which I wish to pick out in which the materials thrown up by history suggest to us a perspective needing to be opened out in Chapter VII when we discuss systematic epistemological problems: in Section 2 of that chapter we shall establish the primacy of judgements in regard to particular examples, in the case of the application of these trans-generic concepts the concepts being applied analogously in the diverse cases.

(c) *What our own argument involves in the light of these problems*

In order to see the implications of what we have just been saying, let us consider the example most germane to the argument of this book.

In brief, our understanding of the concept of "cause" is such as to allow us to appreciate a community between the need which particular things have to have a cause in order to begin to exist and the need that the continuance of things has to have a cause in order thus to continue, and to appreciate that we cannot accept the one need without accepting the other. We do not first grasp the necessity of some one principle and then separately, in virtue of the acceptance of this principle, register the necessity of each of two types of application of it. Rather, in virtue of our understanding of the concept of "cause", we appreciate the necessity of each of two applications of it and appreciate the interdependence of these applications. No principle covering both applications enters in at this stage. (Likewise no recognition of an 'analogy' between the two cases enters in at this stage. In virtue of the understanding of a single concept—in the case concerned the concept of "cause"—we apply the concept analogously in different judgements, applications which are interdependent: we do not argue 'by analogy' from one judgement to another as if we identified the analogy prior to rather than in the act of judgement. The recognition and description of the analogy is the work of reflection and involves the exercise of the same concept.)

What needs a cause in the one case is typically the beginning to exist of an ordered whole (either a substance or a substance as positively qualified by some property or relation) and the cause explains this against the background of the presupposed continuance of things with their natures and nature. What needs a cause in the other case is the continuance of this presupposed background, appreciated as a continuance which is 'reliable', 'given', 'non-chance', or 'according to nature' (without being 'caused' by nature). In both cases what is appreciated as needing a cause has a temporal aspect and is in some way a complex which we have to conceive as given as involving some sort of unity or 'prehension' marked by such terms as "ordered", "reliable", and so forth: the function of such terms seems to be little more than to indicate that the 'unity' or 'prehension' concerned is such as to require a cause.

Thus, upon reflection after judgement in regard to the particular cases, we can discern some analogy between the effects being considered, something in common between the diverse things discerned to need a cause and in virtue of which they are so discerned. That is, we are able to state an 'analogous' generalization as a summary of the conclusions reached in a spread of particular judgements applying the concept of "cause" by a kind of piecemeal induction: we might say that the temporal and order-involving complex needs a cause (I do not need in this book the principle that the 'composite' or the 'not intrinsically underivative' needs a cause). But the function of such generalization is solely in second-order debate in order to rebut the allegation that there is no community between the different cases in which it is judged that a cause is needed, and the community noted is identifiable only in 'analogical', not univocal, terms—besides being a community between effects, not between ways of causing.

4. THE IRRELEVANCE OF THE CONSIDERATION THAT EXISTENCE IS NOT PART OF THE CONCEPT OF ANYTHING: THE QUESTION TURNS ON THAT OF THE PRIORITY OF FIRST-ORDER INQUIRIES

We have so far undertaken a somewhat external, albeit analytical and critical, review of the materials thrown up by history. We need now to pick out certain key elements which are relevant in certain particular ways to the perspectives to be opened out in Chapter VII.

In the discussions of creation, both in the *Summa Theologica* and in the earlier *De Potentia*,[33] Aquinas raises the objection that whatever is possible to the understanding must be possible in

[33] *S. Th.*, Ia, Q. 44, Art. 1, obj. 1, and *De Potentia*, Q. 3, Art. 5, obj. 1.

reality, and that therefore, since neither the existence of a thing nor its having a cause belongs to the concept of a thing, it must be possible for the thing to exist in reality without a cause. One meets the principle involved, albeit in a watered down and more dubious form, in Hume, who holds that whatever is imaginable is not absolutely impossible in reality.

Hume is here expressing in an empiricist version (substituting images for concepts) the principle found in Descartes and Leibniz that what is possible, granted clear and distinct ideas (concepts), in the human understanding must be possible in reality (this was the basis of Descartes's argument that it must be possible for man to exist as a soul without a body). This principle is open to the objection that the human mode of possession of concepts is not such as to give such an adequate understanding of anything as to provide the basis for any positive judgements as to possibility. Rather these must be based on judgements of likeness to cases of known actuality or on other extensions of the argument *ab esse ad posse*. Human concepts are adequate only to ground various kinds of judgement as to conceptual or logical impossibility, not possibility.

But Aquinas does not enter into any of these considerations whatsoever. His method of refutation of St Anselm's ontological argument[34] has put him into the position of agreeing beforehand with Kant that existence is external to the concept of a thing (something more clearly stated and grasped by Kant than by Hume). And his response to the objection is laconically brief: simply this, that, existence being external to the concept of a thing, the question is not of the possibility of the concept but of the possibility of existence, and a cause is needed for the existence of the thing, not for the concept of the thing. This argument issues from his way of viewing existence as the form of all forms, or that which renders actual anything that is actual, even forms, and as itself in a certain way not properly speaking a form or concept at all. (It must be noted that, if existence in the relevant sense, or of the relevant kind, is a transcendental, then this argument implies that the concept of cause will also be a transcendental, or, of course, if the use of the noun "cause" is founded in the use of the conjunction "because" in the way that the use of the noun "condition" is

[34] *S. Th.*, Ia, Q. 2, Art. 1, resp. and ad 1.

founded in the use of the conjunction "if", then indeed the term "cause" will not be a concept-term at all, properly speaking.)

The significance of this response on the part of Aquinas is in the first place this: that it refuses to concede any priority to second-order inquiries in regard to concepts over first-order inquiries about existents and their causes, an absolutely key point in relation to our argument in Chapter VII, Section l. This is a first respect in which the materials thrown up by history indicate the character of the problems which we need to deal with in Chapter VII. The systematic problems belong to epistemology.

It must be part of our general systematic epistemological concern, to be dealt with in Chapter VII, (i) how epistemological problems as to the status of the general principle that a cause is needed, whatever the range or application of this principle and however exactly it is to be stated, are to be resolved, and, subsidiarily or secondly, along with this, (ii) in what sort of way the antecedent of the intermediate step or lemma that I stated (our lemma amounted to this: 'if the temporal needs a cause of continuance, this cause cannot be temporal or in any other way composite') is to be justified and in what sort of way any argument that might be used to back up this antecedent might have the back-up offered to it itself grounded.

These are the supreme problems in epistemology, problems whose solutions as proposed by the rationalist appeared to concede too unlimited an adequacy to the human understanding. As dealt with by Kant they appear to lead one into various unwanted and ultimately incoherent forms of idealism. In other hands they appear to have generated various ultimately conventionalist systems, whether these conventionalist systems be dressed up under the guise of the view that all thought is culture-determined or in some way historically bound or bound to the consensus of the 'scientific community' or rooted in a 'form of life', or not so disguised. However, to escape these unsatisfactory ways of resolving this problem as to how some perfectly ordinary knowledge, which we undoubtedly do have and which undoubtedly does qualify as knowledge, can possibly be had by us and how, *post factum*, to justify the recognition of it as knowledge is not easy. Any genuine justification must be according to principles which are not *ad hoc* and therefore restricted in statement in some merely arbitrary way, e.g. in a way carefully devised so as to allow the derivation only of knowledge within some pre-limited sphere, some sphere of statements which it has been decided beforehand are the ones to be declared justified. That is, it must be according to principles attractive in themselves in their details, and not carefully calculated simply to preclude any risk of openness to conclusions leading in a metaphysical, ethical, or theological direction.

How thus to set epistemology on a non-arbitrary basis is not clear. This will be our concern in Chapter VII, Section l. It will involve an interplay between first-order judgement and second-order reflection, but without the first-order ever becoming derivative from the second-order as such.

With this chapter, the kernel of our argument has now been presented and its main strategy made plain. The reader who wished to turn a blind eye to the systematic epistemological problems confronting such argument and forgo the pain and difficulty of working out a critical approach to the establishment of God's personhood, fullness of power, and goodness might pass directly to Chapter X. But, as we have seen, the epistemological hurdle is the main one to be crossed in any argument to God's existence, and the question of the character of His personhood and goodness is determinative of the shape of any development of theology. Therefore it is to these matters that I now turn in Chapters VII to IX.

Note: My Approach to Aquinas's Prime Mover Argument

My argument is close to the prime mover argument of Aquinas (the 'First Way'), properly understood.

The prime mover argument as presented in the *Summa Theologica*, Ia., Q. 2, Art. 3, and in the *Summa Contra Gentes*, I, 13, is couched in the idioms of Aristotelian philosophy so that unsurprisingly it is often misinterpreted along excessively narrow Aristotelian lines, e.g. by Kenny in *The Five Ways*. However, although couched in Aristotelian idioms it is so presented as to be suseptible of a non-Aristotelian interpretation, an interpretation inviting a close marriage with the Fourth Way. The appropriateness of a non-Aristotelian interpretation becomes plain when we discover in Q. 3 of the *Summa Theologica* that the prime mover argument shows that in God there is no potentiality (Q. 3, Art. 1) and this is interpreted as showing that in Him essence does not differ from existence (Q. 3, Art. 4), as if in all things other than God continued existence was the actualization of a potentiality. The rightness of this line of interpretation is borne out by his own preferred statements of arguments that everything needs God as creator to give it being (*Summa Theologica*, Ia, Q. 44, Art. 1, and *De Potentia*, Q. 3, Art. 5). The same ways of thinking about existence are most fully articulated in *Summa Theologica*, Ia, Q. 45, Art. 4, and Q. 104, Art. 1 (corresponding to *De Potentia*, Q. 3, Art. 4, and Q. 5, Art. 1, and also evidenced in the reply to Objection 7 in *Summa Theologica*, Ia, Q. 47, Art. 2).

The question then arises whether, if Aquinas has moved from the sphere of merely physical argument, where Kenny would place him, he has, at the opposite extremes, succumbed to ontologism or rationalism. As to ontologism, identifying God with a concept, he is at no risk. It is sufficient to note the distinction he makes between existence as what is implied or presupposed in any predication (it is this with which we are concerned in the proposition "God exists") and existence as an act or actuality, and to take note of the distinctions he makes in respect of types of identity in discussing the Trinity.

In considering his relationship to rationalism, we have to raise questions which he never himself addressed. He does not discuss the role of analogous concepts of reasoning, does not use the word "analogy" in connection with the word "cause", and always states causal principles with a greater generality than he needs for his arguments, but also with a greater informality. What is plain is (*a*), that his causal principles never have the univocal character or generality of the principle of sufficient reason and (*b*), his tendency to present them by means of examples.

VII

The Systematic Epistemological Queries

The suggestion that it is possible to prove the existence of God as cause seems to many wild, implausible, and objectionable for four kinds of systematic reason.

Firstly, it is presumed that the possibility of any such demonstration has been so decisively subjected to criticism by Hume and Kant as to be capable of no revival. Second-order epistemological consideration of the systematic limits of human knowledge, qualifying the status and restricting the range of application of any possible causal principles, is supposed to establish this.

Accordingly, after a short review of the structure of the historical background of our concern in this chapter, I shall turn in Section 1 to the fundamental question of method confronting all philosophers, and indeed all men, as to the relation between second-order epistemological and methodological inquiry and first-order inquiry into reality itself.[1] Is second-order inquiry primary so that all access to reality is mediated through it, or is it merely parasitic upon a presupposed access to reality as it is in itself? Can knowledge of reality be self-reflective and critical without being indirect?

The result of this investigation in Section 1 will be to establish the standing of concepts of "substance" and "cause" in metaphysics.

Secondly, it is presumed that the principles of causality which we learn to apply in connection with ordinary or mundane things, either of the way in which we get to know them, or of their meaning, are useless in any argument to God's existence: either the way in which we get to know them is such as to render such extrapolation illegitimate; or thus to apply them would involve using such terms as "substance" and "cause" in a non-univocal way, i.e. not disciplined by any definite set of criteria of application.

[1] Roughly, I count an inquiry as being second-order when any appropriate response to it will implicitly involve the embedding of one sentence, clause, or proposition within the larger whole, the response itself where what is embedded figures as in some sense an object of thought or knowledge, rather than an object of desire or pleasure. cf. David Braine, 'The Nature of Knowledge', pp. 44–50.

What is involved here is the whole question of the way general principles enter into the getting of human knowledge, and in Section 2 I shall show how it is that what is basic to human knowledge, in the respect concerned, is a certain kind of understanding of concepts, an understanding of which knowledge of general principles is a fruit, but which is more primitively or fundamentally exhibited in the handling of examples.

Thirdly, it is presumed that to intrude the concepts of "life" and "personhood" into general metaphysics and cosmology represents the survival of a childish anthropocentrism, exalting a temporarily dominant animal species inhabiting a minor member of a small part of but one of a multiplicity of galaxies, along with the perspectives set by the interest of this species, to a master role in judging the nature of the Universe as a whole. It is represented that only a highly uncritical approach, or else some sleight of hand in argument, can achieve this.

Therefore, in Chapter VIII, we shall seek out and identify the only possible or only appropriate or correct basis for bringing the concepts of "life" and "personhood" into the centre of a truly critical metaphysics.

Fourthly, it is presumed that so far as human understanding and reason appear to succeed in bringing God within the ambit of metaphysics, it thereby, to that extent, brings God within the orbit of human comprehension, and sets up an idol with which the religions of Jew, Christian, and Muslim should have no truck.

I leave to Chapter IX the task of considering to what extent the structures of thought about God in metaphysics correspond to the structures of thought to be discovered in systems of systematic Revealed theology, and the extent to which proofs in respect of God contribute to theological understanding or so-called 'theological science'.

INTRODUCTION: BACKGROUND TO THESE SYSTEMATIC QUERIES

Any causal argument for God's existence has to draw upon whatever understanding man has of relevant causal concepts. It is a commonplace of the history of philosophy, signalled first in Aristotle's review[2] of the senses of the words *aitia* ("cause") and

[2] For Aristotle's reviews of the senses of *aitia*/*aition*, see *Physics*, II, 7,

archē ("principle" in a wide sense embracing "root" or "starting-point", as well as "cause" and "first premiss"), and obvious from the discussions of 'cause' from the time of Hume onwards, that the notion of "cause" and the questions of the status and range of application of causal principles are central to several of the most vexed problem areas in philosophy. To prove the existence of God within the framework of modern thinking, in which it is part of the meaning of the word "God" that God is personal, it is necessary to become clear as to the proper status in general metaphysics of such concepts as "personhood" and "life".

The human tendency in first thinking about these problems has normally been to begin with an uncritical empiricism, seeming to embody naturalism and anthropomorphism. Conformably with this, the philosophers' starting-point, at least in the West in modern times, has been, strangely within any reflective context, to presume that viable approaches (if any) will be mere variants upon such naturalism and anthropomorphism. Teleology is presumed to be immanent in Nature, so that there is little reserve in thinking of Nature as having ends, and parts of Nature as having functions, without there being any thought of an external mind or designer.[3] Whenever an ordered whole has a beginning in time it is presumed obvious that the whole concerned must have a temporally prior cause. Ordered wholes, if they are ordered in respect of their nature and have a beginning in time, are commonly the workings of human life, intelligence, and personal action. Accordingly, the causing of the Universe to exist is presumed to be the work of some analogous (perhaps, in Hume's words, 'remotely analogous'[4]) life, intelligence, and personal action.

How there can be aim or function without an understanding mind underlying it, how the Universe (within which Time is comprised) can have an, as it were, temporally prior cause, and how one can argue from examples of causation by personal action within the Universe to the nature of a cause in the unlike case of the causation of the Universe as a whole, are questions to which this

Metaphysics, Alpha, 3–10, and Delta, 2, and *Posterior Analytics*, II, 11. For his use of the word *archē*, see *Metaphysics*, Beta.

[3] Aristotle writes in *Physics*, II, 1 and 8, on immanent order without arguing it from transcendent designer in writing about 'nature'.

[4] Hume, *Dialogues Concerning Natural Religion*, Philo's concluding concession in Part XII.

uncritical naturalism and anthropomorphism provide no answer. The questions of Philo, Hume's *alter ego* in the *Dialogues Concerning Natural Religion*, echo difficulties emphasized by the early Christian Fathers and scholastics as to how God could be in any way 'like' His creatures, and have barred the way to any warm espousal of Butler's way of analogy. Indeed, the propensity of modern apologists for religion to justify the ways of God to man, e.g. in suggesting that if we were planning the Universe ourselves we could devise no better training-ground for virtue than our present scene of evil, have seemed to the religious presumptuous, and to the atheist morally offensive. In this way, the last twists of uncritical empiricism and anthropomorphism have seemed less plausible or natural and more futile than their predecessors.

Within the setting of naturalism and anthropomorphism in ancient Greece, there arose the intellectualism forever associated with the name of Plato. This never envisaged an absolute disparity between the human intellect and other intellects, although it envisaged the human intellect as subject to temporary forgetfulness and blindness, needing to draw upon the shadows given through the senses in order to revive its memories or mitigate this blindness. Whereas, to the materialist, 'life' and 'personhood' appear to be but local phenomena within the Universe, in what we call Platonism the intellect is presumed pre-eminent: the supposed capacity to know, think about, and 'contain' all things is taken to dwarf the grandeur of the natural Universe which comprises only some of all the things the mind can know. The material and natural are thought of as either less real than the intellectual or as deriving their reality from their being caused by some intellect and from their likeness to some model formed in this intellect.

This way of thinking has long survived the death of particular Platonic theories of Ideas or Forms and of reminiscence, and is present whenever man's capacity to reflect, know, and choose is taken without argument to dwarf anything material. Aristotle, and the Aristotelian tradition after him, presumed the human mind to contain at least potentially all things,[5] conceived the celestial spheres to be moved by intellectual desire, and celestial bodies to be moved by intelligences greater than the human. The same presumption about the human mind took new life in Hegel (albeit that

[5] Aristotle, *De Anima*, III, 4–5.

within the Hegelian tradition Man may have sometimes come to be conceived of collectively). But the same presumption constantly recurs more modestly in the presumptions of preachers and apologists, whether of religious or secular world views.

This is the background from within which pre-critical rationalism came to appear to be a natural mode of thought. What is special to pre-critical rationalism is the presumption that human concepts are clear and distinct, i.e. that they are for certain general purposes adequate.

Out of this background it comes to be presumed not only that human beings can identify certain cases of conceptual impossibility, e.g. the case of formal logical contradiction, but also that they can make sure and reliable judgements as to the absence of any kind of conceptual impossibility—not only the absence of logical contradiction, but also the absence of impossibility arising from any of a multiplicity of other roots of conceptual incoherence, or of any impossibility arising from an accidental combination of logical and other factors. Thus, because human beings can conceive of a thinking being without a body, it is supposed to follow that there is no incoherence in the concept of such a being even for God with His perfect grasp of concepts, and so no obstacle to God's creating and hence to the existing of such a being.[6]

For something to be necessary is for its opposite (its contradictory) to be impossible and for something to be impossible is for its opposite to be necessary. The problems of proving the presence of a conceptual impossibility and of proving the presence of a conceptual necessity lie together, and are not often realized by philosophers (as distinct from logicians) to be in quite a different situation from the problem of proving the presence of conceptual possibility. To prove conceptual impossibility it suffices to produce just one ground of impossibility of the right sort (logical, imaginative, where sensory properties are concerned, and other kinds of ground in other cases as indicated in Section 2 below): the existence of one such ground is sufficient. Likewise, to prove conceptual necessity it suffices to produce just one ground showing the opposite to be conceptually impossible. But to prove conceptual possibility is like proving non-existence, and we have no way of proving the non-existence of even logical impossibility within the

[6] Descartes, *Meditations*, VI.

field of natural arithmetic, except for the proof from truth or actuality, let alone of proving the non-existence of just any kind of conceptual impossibility whatsoever where more open fields are concerned.[7]

Again, when it comes to principles relevant to causation, the rationalist has not hesitated to generalize from Aristotelian causal principles, such as from that 'any substance which comes into existence has a cause' to 'every happening or state of affairs has, not just an Aristotelian efficient cause with active power adequate to produce it, but sufficient (i.e. determinative) reason for obtaining or existing'.[8] The concept of "cause" or "reason" is supposed to be so clear and distinct as to make it evident that the Universe, considered as an ordered whole constituting a single object, would need a cause or reason for being, independently of whether it had a beginning or of any considerations of time.

Within the context of this pre-critical rationalism it is presumed that human beings have such an exhaustive grasp of all the possibilities that they can judge from human concepts alone that, if a substance is not material or physical, it cannot be other than intellectual and personal. Likewise, if an event or state of affairs is not by chance or necessity, it is presumed conceptually evident that it must be the issue of free action. The cause of the Universe, as being external or independent of the Universe (whereas for its operation to be by chance or necessity would involve its being integrated within the Universe as within one system or order), must therefore be intellectual and free acting in its operation, its operation of causing. Thus from the formal structure set up by a negative, "not material or physical", or "not by chance or necessity", it is presumed that some positive and determinate conclusion can be inferred.

[7] W. Kneale argues in *Probability and Induction* that there can be necessities the reasons for which we do not have insight into (pp. 71–2, 77–89), there being evident examples which suggest that mathematics includes such (pp. 79–80). He says 'In any sense of the word "conceive" which is relevant to the argument, an ability to conceive the contradictory of a supposed law of nature does not disprove the suggestion that the supposed law is a principle of necessitation', and then adduces his argument from mathematics.

[8] For the importance of distinguishing these principles, the more limited or Aristotelian and the more unrestricted or rationalist, see Chapter VI, Section 2(*d*), and, on the possibility of envisaging a spread of analogous applications of concepts, see Chapters VI, 3, and VII, 2(*b*). See also n. 19 and subsidiarily nn. 10, 34 and 36 below, in regard to the presumption that 'being is intelligible' which we may consider a variant of the principle of sufficient reason.

Or again, the rationalist might attempt to argue that the causing of the Universe involves the teleological and 'intentional' structures characteristic of life and intellectuality (cf. Chapter VIII, Section 2(*a*) and (*b*), below). He will then infer from these formal characteristics of this operation of causing the Universe to the applicability of such positive concepts as life and understanding to the cause which acts.

As against this pre-critical rationalism we must insist that it is illegitimate to presume that any understanding of "cause" thus completely independent of the temporal is possible solely upon the basis of some merely formal structure, whether that set up by a mere negative, or whether by a richer kind. For the human intellect legitimately to reach to supply any positive or determinate content, such as the concepts of "intellect", "life", or "person" involve, the conditions of human knowledge require some less formal approach, involving some further resort to experience of the actual. Or, at least, such would seem to be the weakest and most elemental constraint which any critical or self-reflective approach ought to place upon rationalism.

However, instead of resorting to consideration of the actual, modern philosophy has concentrated rather upon a direct consideration of the conditions of being critical or self-reflective in respect of knowledge itself. In this way it has attempted to put the resolution of certain second-order questions first, before the consideration of any first-order questions.

The attempt to make a second-order inquiry primary has taken two principal forms: firstly, a form in which what is considered is the conditions of thinking, questioning, and reflection in general; secondly, a form in which it is presumed that what is concerned is the pre-conditions of empirically or sensorily based knowledge and inquiry.

Thus, in working out his Promethean strategy, Descartes attempted to represent knowledge of God as the author of human nature and the divine veracity as the basis of human knowledge: Berkeley appropriated Descartes's view of mental substance and causation without the same epistemological preoccupation with the material world, but offered the same basis for accepting the rest of human knowledge, so far as it concerned the empirical, albeit with a re-interpretation of the content of this empirical knowledge along mentalist and idealist lines. Prima facie, an approach that depends on identifying the pre-conditions or presupposed 'horizons' apart from which empirical knowledge cannot exist is bound to have, built into it, tendencies to solipsism or conventionalism, or, at the least, restrictions in its conception of what might possibly be justified under the heading of the so-called 'empirical'. Kant introduced the 'transcendental' approach in the sense of that approach which considers the pre-conditions of human knowledge, and in addition presumed that this approach would, in the case of human knowledge, reach only so far as the 'empirical'.

Next, by various remarkable moves, representative of a kind of *ju-jitsu* (in which one allows one's opponent to do all the work), Maréchal and his succesor transcendental Thomists,[9] such as Frs. Rahner, Lonergan and Coreth, along with the Anglo-Saxons, Germain Grisez[10] and Hugo Meynell,[11] attempt to represent the existence of God as an epistemological presupposition of human knowledge, in some cases because it is allegedly presupposed by any thinking, questioning, or self-reflection whatsoever (a Cartesian approach, imitative of St Augustine) and in some cases even as the pre-condition of sensible or empirical knowledge (an approach more true to Kant).

Of recent times, the attempt initiated by Descartes and Kant to make philosophy hinge upon epistemology has been replaced by an attempt to make it hinge upon the 'philosophy of language'. This may only involve insisting upon the key place of the philosophy of language in all reflective philosophical thinking, from which it may serve to reinforce recognition of the primacy of the first-order. But it can come, as in Dummett,[12] to embody a new Cartesianism

[9] E. L. Mascall in *The Openness of Being*, chs. 4 and 5, gives a clear introduction to the work of the transcendental Thomists with extensive references. The inspiration of trancendental Thomism is expressed in this remark of Fr. J. Maréchal: 'Kantian agnosticism is not only not irrefutable; it can actually be overcome while starting from its own principles' (*Le Point de départ de la métaphysique*, vol. v, ch. 1). Since writing the body of this book there has come to my notice Étienne Gilson's *Thomist Realism and the Critique of Knowledge* with its notable critique of Maréchal in ch. 5.

[10] Germain Grisez in *Beyond the New Theism* makes his version of the cosmological argument depend upon the requirement of respect for "rationality norms", pp. 76–82 and proposition 2 of pp. 82–4.

[11] Hugo Meynell, 'The Intelligibility of the Universe'.

[12] Michael Dummett, 'Can Analytical Philosophy be Systematic, and Ought it to Be?', reprinted in *Truth and Other Enigmas*.

wherein philosophy of language holds the place which epistemology held for Descartes and Kant. What results is then a new kind of transcendental idealism with all the difficulties which Kant envisaged in respect of the conceiving and knowing of the real as it is in itself. Berkeley, in his thought in relation to the material world and in relation to abstract objects such as numbers, may be conceived of as the precursor of these modern linguistic idealists. Michael Dummett, perhaps inspired by Berkeley, has intimated that he envisages the possibility of some argument from the philosophy of language wherein theism will appear to be a pre-condition, and therefore consequence, of "realism", in the sense in which he defines the term.[13]

Kant spoke of himself as instituting a Copernican revolution in philosophy. In fact, Descartes and Kant are both guilty of variations in a gymnastic which is more analogous to a Ptolemaic revival of geocentrism. The effect of making an epistemological theory about man the basis for the rest of philosophy is to erect as a principle a certain kind of anthropocentrism, wherein man is the measure of all things and, as Kant had it, rather than the forms of knowledge being shaped to the objects of knowledge, the objects of knowledge are shaped and limited by the forms of knowledge, or by the faculties of human understanding and sensibility. Our first task in this chapter is to show that thinking is shaped by the real, rather than vice versa, and that therefore this whole transcendental approach wherein this second-order inquiry is made primary is non-viable and self-destructive. Accordingly, the implausible devices whereby theism is supported upon such transcendental argument are misconceived in starting-point. The transcendental approach to theism, had it worked, would have put the person or subject at the centre and left it no mystery why God should be conceived of in personal terms - but, this approach being abortive, some other basis is needed (if indeed there is any to be found) for introducing the concepts of "life" and "personhood" into the centre of metaphysics.

Incredulity in respect of the capacity of the transcendental

[13] In his published work this idea appears in the Introduction in *Truth and Other Enigmas*, p. xxxix, where he confesses to having suspected that, while anti-realism is ultimately incoherent, realism is tenable only on theistic grounds, mentioning an earlier unpublished paper (doubtless the Part II, almost delivered in my hearing in 1964, whose Part I had gripped an audience for 100 minutes) in connection with the strengths of Berkeley's strategy, which he regards as prefiguring his own thought.

approach to justify any theoretical knowledge has engendered a propensity among philosophers as well as non-philosophers to exalt *praxis* (or action) above theory or science. This only reinforces the anthropocentric or at least geocentric character of the thinking concerned, and can do even less than the Kantian approach to provide a basis for theological knowledge: indeed, it does not even leave room for faith as an intellectual commitment. Rather it leaves room for religion only as a certain kind of ritualism or legalism, even if this ritualism in the form of a legalistic moralism extends its operations to other men and animals. Clearly, non-formalistic Jewish tradition, Christianity, some variants or adaptations of Islam such as Sufiism or Sikh religion, and much other human religion, have involved a union of knowledge and love: but, if the intellectual character of the 'knowledge'-aspect of the religion is undermined, the love, whether of God, of men, or of the rest of creation, becomes instinctive or conventional and loses its fullness of character as love.

Accordingly, I now turn to consider the relation of first-order inquiry in regard to the real to second-order inquiry in regard to thinking and knowledge and perhaps action. In this way, I shall show, by a rather more extended discussion of the key arguments of the thinkers concerned, principally Descartes and Kant, whom I discuss in Section 1(*b*) below, and secondarily those authors from Duns Scotus through to the transcendental Thomists, along with Grisez and Meynell, whom I discuss in Section 1(*d*), that the perspective in which I have briefly cast the history of thought in this Introduction is correct.

1. METHOD IN PHILOSOPHY: THE RELATION OF FIRST-ORDER TO SECOND-ORDER INQUIRY; ESTABLISHING THE CONCEPT OF SUBSTANCE IN METAPHYSICS

(a) *Our posture in approaching the question of method*

The fundamental question of method which we must consider first is the question of the order in which first and second-order inquiries ought to be taken. In philosophy, should we begin with second-order considerations of epistemological and methodological kinds and only afterwards allow ourselves to advance to any first-order conclusion, or should we regard the consideration of epistemology and methodology as parasitic upon an underlying level of first-order apprehension of the real? We must consider the question of method in philosophy in such a way as not to prejudge the issue.

In approaching this fundamental question, I am drawn, upon reflection, to consider that whatever total view one ends up with, it should include this: that it is not intellectually appropriate for any being less than God, if there be any God, to develop completely a priori a positive account of what exists (whether or not it be meaningful to speak of an 'a priori knowledge' of what exists on the part of God). All our knowledge, and all the knowledge of any possible creature (or if the only beings that exist are other than God, creaturely or finite, then of all beings) in regard to actual existence, is derivative upon experience of actually existing things, at least so that the actuality is ground to the knowledge, rather than vice versa. And since the kinds of possible things that exist cannot be, either in fact or according to any reason which human beings can give, limited to such as we actually know or to such as are like them, it is not right that the shape of a general metaphysics of things that exist, and of nature, should be set by, or limited by, theories about the human understanding.

It will follow from this that there will be no possibility on the basis of reasoning about what is involved in thinking, inquiry, and reflection as such (the approach initiated in modern times by Descartes), or in regard to the conditions of the possibility of empirical knowledge, understood as based in sensory experience within a spatio-temporal framework (the approach initiated by Kant), or upon the conditions of inter-subjectivity or of the co-operative use of language, to generate any conclusion as to what actually exists (with its nature as it is in itself and in its relations so far as these exist objectively), antecedently to any incidentally consequent possibility of knowledge of them. In so far as followers of such approaches have come to any realist conclusion, it is only by smuggling the conclusion in at the start, e.g. in some presumption that the thinker is a substance with identity persistent through time, various presumptions about causality, the presumption that space and time are real, or the presumption that physically realized human communication ('language') is real, or else by some sleight of argument. We find, I believe, such sleights of argument among various so-called transcendental Thomists, discussed below, such as Coreth, Lonergan, and Rahner, whose approaches usually turn out to be at root Cartesian rather than Kantian, although sometimes dressed up in Kantian or even Hegelian garments.

Rather, I take it that there is a world within which there are things that actually exist and operate according to the natures they have and with which we enter into interrelation, interrelation both physical and cognitive; and it is the character of the things that exist that shapes our use of what cognitive faculties we have, rather than vice versa. That is, there is, as Kant observed, a certain conformity between, on the one hand, our concepts and faculties of understanding and sensibility and, on the other hand, the objects of knowledge; but Kant was wrong to conclude that the explanation of this lay in the shaping of the objects to our concepts and faculties; rather it is our concepts and faculties that are shaped to the objects. In order to justify or render palatable this realist and Aristotelian viewpoint, one will need on the one hand to unpick the plausible arguments put up by Kant for his point of view, and, more importantly, on the positive side, exhibit how the openness or unrestrictedness of man's capacity to initiate new angles of inquiry renders him free because of, instead of imprisoned by, the scope and nature of his own faculties—free to obtain knowledge which is still genuinely critical.

On this view Spinoza[14] and Hegel[15] were right in their stance that epistemology could not generate any first-order knowledge without first drawing on some first-order premisses as tools or, as we might put it, 'grist to the mill'. Accordingly, Spinoza and Hegel were right to reject the Cartesian attempt to make second-order inquiry into the nature, conditions, and extent of knowledge prior to first-order inquiry into the truth about things known or to be known. As Hegel insisted, this criticism stands equally against Kant. Indeed, it stands as an indictment of most modern philosophy, although not of Bradley (for whom knowledge could not be arbitrary and for whom therefore the prima facie data of experience were to be presumed correct until some incoherence came to light

[14] Spinoza, in writing about the fourth mode of perception or knowledge, compares the perfecting of knowledge with the perfecting of workmanship by progressive improvements in the use of tools: no progress is made without reliance upon some tools or some native intellectual instruments (*Works*, pp. 8–12.)

[15] Hegel criticized Kant's way of making a criticism of the faculty of cognition a preliminary to its exercise as soon involving the misconception of already knowing before you know (*The Science of Logic*, ch. IV, pt. II, sect. 41).

and some possible explanation of the error lay to hand[16]), and not of so-called ordinary language philosophy and phenomenology in so far as these proceed genuinely a posteriori.

Historically, Descartes stood in the midst of a scepticism of the type represented in Montaigne and sought out some ground, standing upon which he could, Atlas-like, hold up the edifice of knowledge. His method was to put second-order inquiry and the solution of his methodological doubt into the first place in philosophy, but the effects of his adoption of this method in theory were mitigated by the survival in his thought of the notion of substance and the persistence of many scholastic and Franciscan-mediated modes of thinking,[17] often rooted ultimately in St Augustine. He was not in his thinking as thorough going in his mentalism as such thinkers as Leibniz; nor did he espouse any such radical rejection of the status in reality of relations,[18] or any such over-simplifying reduction of all causal principles to one (the so-called principle of sufficient reason).[19]

Leibniz espoused no such doctrine of the primacy of second-order or epistemological inquiry. Locke, by diverting attention from questions about epistemology as such to questions of what Kant described as the mere physiology of the understanding,[20] does not stand clearly in any camp, either in regard to the a priori status of causal principles or as to the primacy of second-order inquiry.

[16] In ch. VII, 'On Truth and Coherence', in *Essays on Truth and Reality*, F. H. Bradley makes clear that in his view coherence is violated by any principle of approach which would make whether or not something constitutes knowledge accidental. For instance, to ignore the privilege of commonly received perceptions, and to treat all consistent systems of imagined perceptions as on a par, would be to adopt such a principle. It is false to represent so-called coherence theories of truth as only requiring internal consistency.

[17] The importance of Franciscan and Augustinian influences upon Descartes is evidenced in many places, including Etienne Gilson's demonstration of the influence of Scotist tradition on Suarez and thence on Descartes (*Being and Some Philosophers*, chs. III and IV), as well as standard histories.

[18] Leibniz's rejection or supposed rejection of the reality of relations was a common position, presumed by Kant following Wolff. Kant presumes that the inner properties of a thing in itself would be non-relational, for instance giving this as a reason for saying that a thing in itself cannot be known through mere relations and therefore cannot be known by outer sense (*Critique of Pure Reason*, p. 87).

[19] The over-simplification embodied in the principle of sufficient reason is well treated of by Roger Verneaux in 'Note sur le principe de raison suffisante" in *La Crise de la raison dans la pensée contemporaine*, pp. 39–60, and, in different style, by Germain Grisez in *Beyond the New Theism*, pp. 69–76.

[20] See Chapter II n. 23.

Hume, following in the tradition of Descartes, Malebranche, and Berkeley, accepted all the sceptical conclusions which proceeded naturally from the adoption of their approach (albeit regarding ideas as faint images, the converse attitude to regarding percepts as vague concepts or as the mere objects of certain operations of spirit called perceptions), and did not himself make any of their concessions in respect of first-order fact, whether first-order principles of causality or first-order facts about perception. Yet Hume's proposal was merely to substitute a physiology[21] of the understanding for any epistemology whatsoever.

It is therefore only in Kant that the principle of the primacy of epistemology, and thereby of the second-order, comes to adulthood; true, he has to allow some appreciation of what we may regard as facts of a first-order kind about the human use of the senses within a spatio-temporal framework to shape his doctrine, but his doctrine itself relegates these to the status of the transcendentally merely ideal, albeit indubitably empirically real. Berkeley's doctrine made the nature of the human faculty of knowledge determinative of the status of material objects, relegating them to the status of the ideal; Kant relegated all the objects of our faculty of knowledge to this status. In him, the transcendental approach, of making human thought and knowledge a datum, and considering its pre-conditions, comes into its own.

We must therefore turn to address ourselves directly to the Kantian strategy, foreshadowed though it may have been in Descartes (Descartes himself developing certain ambivalent or analogous ideas expressed in St Augustine). Hegel and his idealist progeny, Bradley among them, may constitute the later fruits of the hubris in regard to the human intellect represented in Plato and surviving in Aristotle: but although they make thought and its framework in a way primary, they do not thereby intend to make the second-order or the epistemological primary. Hume and his sceptical descendants, the conventionalists among them, do not propose any way of rescuing science, as such, from the predicament in which Descartes placed it. Only Kant conceived of the possibility of making the very weakness of the human intellect, identified in his second-order inquiry as to the possible cause of any conformity

[21] The psychology which Hume substitutes for epistemology (see John Passmore, *Hume's Intentions*, especially chs. I and II) would count as a physiology of the understanding in Kant's sense.

between objects and concepts, the ground on which science itself was to be restored.

Can epistemology or second-order inquiry be thus or in any such way made primary both within philosophy and ultimately within human self-reflective or critical knowledge as a whole? Or is it wrong thus to conceive epistemology as exercising a mediatorial role between man and first-order knowledge?—that is, should one rather regard first-order knowledge (including, it may be suggested, knowledge of substances persisting through time and some of their actions with aspects of their natures) as primary, and second-order inquiry as parasitic upon it? Only after resolving these questions in subsection (*b*), will I proceed to consider how first-order knowledge can nonetheless be critical in subsection (*c*).

(*b*) *The escape from Kant and the restoration of substance*

I shall begin (i) by indicating the more obvious and natural-seeming arguments in refutation of the Kantian position and strategy. I then proceed (ii) to articulate the counter-position, chiefly represented in philosophical history by Aristotle and his descendants, ordinary language philosophers such as Ryle and Austin as well as Aquinas among them, and Hegel with his progeny (Marx as well as Bradley), and (iii) to summarize how the result of the refutation of Kant and the establishment of epistemological realism is the restoration of Aristotelian conceptions of substance and cause. I shall then proceed (iv) to indicate what is required for a more exact refutation of Kant. I delay until the next subsection, (*c*), the consideration of how this realism is still critical.

(i) To the ordinary man and to any philosopher who retains just a little of what Bertrand Russell called 'a robust sense of reality', the notion that all the objects of human knowledge are constituted or shaped by the human faculty of knowledge, the objects of our knowledge by our concepts, has always seemed slightly mad. On the face of it, the galaxies, the stars, the solar system, the earth, and even the lower animals would still all have existed, even if the human species had not come into existence; they did exist, before man came to exist; their present existence is independent of man, except in so far as he may exterminate some animals; accordingly, it seems just obviously false to suggest that either the existence of these things or what they are is constituted or shaped either by the actuality or by the possibility of sure knowledge of them.

True, human beings do exist and do have some knowledge of these things and so human knowledge of these things is possible, and it follows that these things are such as to be capable of being known by human beings with the faculties which human beings do have, if human beings with such faculties do come into existence. But this no more suggests that the actuality or possibility of human knowledge is in any way the cause of the existence or natures of the things known than the fact that, if "*p*" is true, then "it is true that it is true that it is true that *p*" will be true, shows that its being true that it is true that it is true about *p* is a cause or ground of the situation that *p*.

Moreover, it has always seemed both to the ordinary man and to the modern scientist that the advance of knowledge was a process of discovery (indeed, full of surprises) in its primary aspect. New phenomena are discovered, and, as a result, conceptual frameworks, far from having the whip hand, are recurrently in need of 'radical' re-shaping to cater for these new discoveries which, in a certain context, force upon man a re-casting of his science in key structural respects. Of course, the philosopher may note that the distinctions of "singular term", "predicate", and "relation", and perhaps other distinctions marked in Aristotle's *Categories*, must remain always inviolate. But it is tendentious to assume beforehand that these fundamental concepts are so limitedly understood, inflexibly applicable, and quasi-mechanical in use as to constitute a restriction upon human vision, rather than a platform leaving man open to the possibility always of initiating freshly varied and ever new types of query or investigation—therefore intellectually incapable of capture in any formalisation or rule. Neither philosophical view can be assumed without further discussion, but to presume the first is to load the scales of debate beforehand in a direction opposite to that which the shape of the progress of science would suggest.

However, the philosophers' objection to the Kantian conception that our concepts make, form, or shape the objects of our knowledge, not vice versa, has run along different lines. The philosophers have adverted to the necessary emptiness of any metaphysics or view (first-order) about reality itself *if* this is actually founded on second-order inquiry unassisted by any first-order information, and have marked the impossibility of second-order inquiry getting anywhere at all even in respect of the second-order, let alone in respect of the first-order, without drawing upon the first-order.

Something has to be conceded about the knower or thinker if any conclusion even about thought and knowledge is to be reached, and something about the senses if any conclusion about sensory knowledge is to be reached. This insight of Spinoza and Hegel into the self-defeating, self-nullifying nature of the approaches to be found in Descartes, in Kant, and later, exhibits that the results of any such approach must be in Aristotle's terms unscientific and merely contentious.

(ii) If, then, the Kantian view that it is our faculty of knowledge, our concepts and categories, which determine their objects and not vice versa is mistaken, one has no choice but to revert to the view that it is the objects which determine our concepts—albeit looking for a more sophisticated version of this view than the one Kant considered. We need not take it that objects and kinds of objects or natures establish themselves in our thought one by one, independently of each other, or in abstraction either from the situations and types of situation in which they occur or from the accommodation of our thought to the general structures of reality. As it is in the nature of human beings to be talking animals, it is also in their nature to be both argumentative and reflective: accordingly it is not just that human knowledge is such as to be articulated in statements whose parts are heterogeneous, but it is such as to involve the objects of thought coming into human knowledge, not one by one but in structured relations, corresponding to the relations within a complexly structured tapestry of propositions as well as to the relations suggested by consideration of logical and syntactical categories.

Against this background, first-order judgements about reality, including therefore judgements about relations (against which Leibniz and others had a misconceived prejudice), and the understanding embodied in first-order judgements, are needed, not only to give a topic to second-order discussion, but even to give content, direction or orientation, and determination to the questions raised in respect of this topic. Always second-order judgements and questionings are in a way ancillary, subsidiary, and parasitic.[22]

The starting point of human thinking is what Aristotle referred to as the seemings, common opinions, appearances, or experience (he used the relevant Greek terms interchangeably[23]). These constitute a structured array from which incoherent elements need to be weeded out, but always with some attempt to diagnose the source of mistake. Experience is not to be presumed to be embodied only in singular statements but also in general statements. And the common opinions or seemings include fruits of or embodyings of experience whose title to knowledge will turn out never to be brought into real question: the mere possibility of formulating, and therewith posing, a question, may open the way for some play of argument, but will never suffice by itself to establish the presence of a real question or doubt. (This is, I believe, very close to what Bradley, if rightly understood, actually thought, as opposed to what he is commonly represented as thinking.[24])

(iii) All this once granted, equally the most cursory and the most extended examination of the 'seemings' and 'experience', especially those embodied in the more salient distinctions marked in part structurally in our modes of speech, will reveal the primacy of notions of "substance", "causal agency", "nature", "time", and suchlike. In other words, some refinement of the structured system of Aristotle's first eight categories is a datum of experience—not just a reflection of a certain culture or of the structures of Indo-European languages[25] and not just the reflection of some universal syntax, nor of anything else which might be supposed to reflect man's nature but yet still be accidental to thought and to reality.[26] In this way the privileged position of the concepts of "substance" and "cause" is a 'given', not just because we are too limited and enclosed by our perspectives to be able coherently to question these

[22] David Braine, , pp. 49–50.

[23] See David Braine, 'The Nature of Knowledge', and G. E. L. Owen, '*Tithenai ta Phainomena*'.

[24] Cf. n. 16 above.

[25] Bernard Lonergan instances Aristotle's categories when distinguishing the categorial from the transcendental after remarking that categories vary with cultural variations, *Method in Theology*, p. 11. The idea that the distinctions marked in Aristotle's list are culturally limited seems to deserve little respect. The quasi-Cartesian background of his ideas about categories appears in his earlier work, *Insight*, p. 395.

[26] It seems to be Kant's view that this *might* be accidental.

concepts, but because they are 'givens' of experience itself as a structured whole, and therefore of knowledge, and, in virtue of the meaning of the terms "substance" and "cause" themselves, a meaning signifying something prior to knowledge, therefore givens, not just of knowledge, but also of reality.

(iv) We need now to turn to a more exact consideration of what is involved in the rejection of the initially intuitively plausible arguments put up by Kant for his point of view. These arguments appear in a small number of purple passages, more distinguished for their eloquence than for their precision. He presumes that, for a necessity in some principle to arise from something on the side of the object, the very concept which we form of the object would have to be such as to give rise analytically to the principle concerned.[27] However, his explanation of the notion of the "analytic" is either psychologistic, drawing upon the idea of one psychological content containing another and reminiscent of Locke's conception of a trifling proposition as one in which the predicate is part of the identity of the idea of the subject, or else quasi-logical, drawing upon the notion of 'that, the denial of which implies a contradiction' (but how cavalierly or how strictly are we to understand the words "imply" and "contradiction"?).[28]

If we review a wide variety of propositions, some relating to the relations between colours and some having to do with space, the number of its dimensions, its continuity, and so forth, some having to do with motion in its intimate connections with a continuous space and time, some having to do with time, its non-cyclical

[27] See Kant, *Critique of Pure Reason*, pp. 22 f., 85–7, 173–5.

[28] For Kant's unsatisfactory explanations of the terms "analytic" and "synthetic", see Kant, *Critique of Pure Reason*, pp. 48–55, and *Prolegomena*, pp. 16–22. He defines the analytic in terms of the containment of a predicate concept in a subject concept, a definition reminiscent of Locke's conception of the trifling, and leaving it plain that arithmetic and geometry are synthetic, but still ambiguous as to whether the definition is to be understood in logical or psychological terms. Meantime he presumes that, if the denial of a proposition would lead to a contradiction, then that proposition would be analytic: but, until one explores the ambiguities of "lead" and "contradiction", it is not clear whether what this strictly implies is a Leibnizian position which would make geometry analytic (in Leibniz's terms a 'truth of reason' or in Hume's terms a 'relation of ideas', since these correspond to the wider modern use of the term "analytic") or whether what it implies is rather the position of Frege, Husserl, and suchlike, which would make geometry synthetic (corresponding to Hume's recognition that being straight and being the shortest line between two points are different ideas, *Treatise*, bk. I, pt. II, sect. IV). A. J. Ayer in *Language, Truth and Logic*, ch. IV, emphasizes the presence in Kant of a twofold account without noticing the huge ambiguities in the second.

character and suchlike, some with the relations between solidity, spatiality, and causation, some having to do with the relations between time and causation, some having to do with the intrinsic evil in pain and harm or damage, and others with the good in thinking the truth, in knowledge, in rationality, and in many non-controversial aspects of well-being and its structures, it becomes extraordinarily implausible to question that there are many conceptually necessary propositions to deny which implies no such formal contradiction. At the same time it has long been evident that there have been many propositions which are formally necessary by the strictest standards, without its being psychologically necessary for us to assent to them, e.g. the infinity of the series of prime numbers, quite apart from Kant's own example of $5 + 7 = 12$.

Thus the presumption that any non-analytic necessity of thought must be either a subjective necessity born of habituation to the regularity of appearances, or else a necessity arising not from the side of the object but only from the side of our concepts and the character of our faculty of knowledge, and that any necessity from the side of the object will be analytic, is worth very little, being based on an uncritical use of the word "analytic". The presumption, that is, that any necessity arising from the side of the object will be either, on the one hand, a tautology or identical proposition, or, on the other hand, a general truth of logic, expressive of the formal aspects of objecthood, has been left baseless.

If there were an intellectual necessity arising from the side of the object that, for instance, an object's coming to exist or continuance must depend on a cause (we note that neither existence nor continued existence will be part of the concept of the object and any such a necessity would attach to a propositional fact rather than a concept), it would be otiose to demand an explanation of how this necessity arose or how it was possible for it to arise, if in this demand it were presupposed that all necessities are primarily "conceptual". But we have, because of the sophistical use of the word "analytic", the presumption that there can be no such intellectual necessity, so that any necessity from the side of the object must be empirical and (Kant following Hume) 'inductively' known in the Humean sense and so ultimately accidental so far as knowledge is concerned. And it is indeed obvious that, if we once allow ourselves to be pushed this far so as, for instance, to start explaining why the human faculty of knowledge is well shaped or well adapted to its objects in terms of the contrivings (conceived of as external or accidental to these objects) of evolution, nature, or God, we can be asked the basis for the relevant knowledge of evolution, nature, or God, and obvious that, set upon this path, we will not recover the situation.

However, in the meantime, there is in the background of Kant's thinking, not an argument, but a presupposition born of a picture. We are led to suppose that the conformity of concept and object is problematic because man is the possessor of limited intellectual faculties which in virtue of the fact of their being limited must be presumed to operate like a set of restricting spectacles introducing systematic limits on what man is able to know, systematic in such a way that, if we follow our leader and champion (Kant himself), we will know how much we do know or how much is, at least in principle, well in reach of our knowledge, but disarming in another respect in that naturally we can form no conception whatever (except in the most formal terms) of the sorts of thing we are or might be ignorant of. In this picture Kant is the heir of Locke.

But, opposed to the empiricism of Locke, we have, with an equal title to the name "empiricism", the view, traceable in Aristotle, many mediaevals, and Hegel, that the unformalisability, or lack of restriction to one image or model, of the most general questions or question-structures arising among human beings frees them from restriction in self-critically reviewing, querying, and constructing alternatives to their own procedures and presuppositions in thought. Since the word "concept" is liable to be deceptive, I speak rather of questions and question-structures. All speech presupposes the capacity to raise questions about existence, about the nature of things ("What is it?"), about the efficient cause of things ("Why?" in a first sense), about the good of things (the final cause, "Why?" in a second sense), about truth and reasons for assent ("Why?" in a third sense), and so forth.

In every area where the issue is capable of being put to the test it is evident that man's capacity for disciplined use of such question-structures so as to generate an increase of knowledge, and not merely to destructive or sceptical effect, outruns any formal definition or model, in each area the matter being demonstrable in a manner appropriate to that area. Thus in mathematics, we meet Gödel's rigorous proof that, if man has any general knowledge of the arithmetic of the natural numbers (i.e. this arithmetic is consistent), then man's capacity for arithmetical proof outruns that which any formal criterion of provability would allow. In physics, history has again and again seen earlier paradigms overthrown and today not even the notion of "mass" is bound to any sacrosanct model. There seems to be no case of a term as general as the terms "substance" or "cause" being limited in its disciplined or scientific application by a 'schema' in the way Kant envisaged,[29] but it belongs to a different book than the present to justify what I have said in full and illustrate it from a multiplicity of areas or to demonstrate from the philosophy of language that the opposite supposition (the Kantian) is incoherent. However, I have said enough to indicate the lines along which the establishment of an epistemological realism and the restoration of the concepts of "substance" and "cause" should proceed.

(c) *The vital need that metaphysics be critical*

It does not suffice merely to establish the primacy of first-order judgements and thereby to establish the necessity as well as the possibility that metaphysics be epistemologically realist; it remains still, nonetheless, absolutely vital that it be also critical. Second-order inquiry cannot have a mediatorial role whereby it constitutes the ground of the rest of knowledge, the sole route towards 'science'. Yet, it remains indispensable.

If any proposition man has to utter is to be genuinely an articulate and understood proposition, it has to be open to and potentially the survivor of some self-reflection, so that conclusions in respect to first-order questions can never be reached in isolation from the possibility of second-order methodological or epistemological criticism or inquiry.[30]

If one considers the multiplicity of things relevant to the consideration of the reasonableness of assent to, and even of a claim to knowledge of, some particular proposition, and along with these the further things relevant to their being believed or ranked as known, and the yet other things relevant to these further things, and so on unendingly, it becomes evident that it is a condition of the possibility of knowledge or of reasonable belief that it should

[29] See Chapter IV n. 9.
[30] David Braine, 'The Nature of Knowledge', pp. 50 f., 62.

not be required that there exist any such totality as 'all the data' relevant to the grounding of knowledge of a particular proposition. The possibility of knowledge or of reasonable belief depends upon there always being, as well as the 'data' upon which knowledge or belief is actually grounded, also some other 'data' or 'considerations' not actually drawn upon but available to be drawn upon in reflection upon the soundness or well-groundedness of the knowledge or belief.[31]

In the approach which I attributed to Aristotle in subsection (*b*), whereby metaphysics proceeds from the so-called 'seemings', 'common opinions', 'appearances', or 'experience', the same is true of ethics also and indeed every other part of philosophy as such. Accordingly, the bases of second-order inquiry are given among the same data, not separately from the rest of experience, but in their integrated relationship with it. The same structure which gives us such concepts as those of "substance", "nature", and "active power" or "causal agency" also gives us these in a context which places many controls upon their use.

There is a requirement that general first-order judgements, second-order considerations about constraints on knowledge and methodology, and views as to the mode of interplay between first- and second-order, should all together conform. This has implications to be considered later in regard to causation and personhood. But, the 'three elements' I have just mentioned are not normally presented separately in 'experience' or in the 'seemings'. It is not just that there is the obvious coming together of certain judgements as to causation (first-order in aspect) and explanation (second-order in aspect), but plainly the 'experience' or 'seemings' presented to a human being determine without room for dispute that remembered experiences should typically have the same status as present experiences and that experiences of others known through testimony or written record should typically have the same status as one's own experiences, so that, although the sources of error or deviation from the norms may differ in the instances mentioned, it remains that this sameness of status constitutes the norm. (Again, Bradley's tendency of thought comes close to this.)

And, within any metaphysics or epistemology appropriate or possible to human beings, wherever there is some species of

[31] I argued these points in my paper 'Two Kinds of Critical Philosophy'.

'experience' or 'seemings' which given precisely as 'one's-own' or 'present', there will be experiences or seemings of equal status, given precisely as 'witnessed' (by other person or record) or 'remembered'. 'Experiences' or 'seemings' are given not as 'provisional', 'probable', or 'merely seeming', so that from them some contrived probabilistic construction or inference is to be made, but as data[32]—although some need to be weeded out as incoherent and have their deviance explained.

It is of peculiar importance in any consideration of first-order principles about causation, since these are of peculiar relevance to this book, to avoid landing ourselves in a position in which we are entirely uncritical. So by some hook or crook, by some device, we need to make it possible for man to assent to a first-order conclusion in respect to causal principles without making the causal principle or its application thereby derivative from some second-order level theory, set of propositions, or approach, and yet still leave it within the range of second-order reflection and criticism: e.g. the exact mode of its statement, its exact range of application, and so forth, will need to be subject to criticism and second-order discussion or, in common parlance, reflection. So we need, not only to latch on to causal principles that are genuinely valid as well as plausible, but also to discover how it is possible for them to be subject to reflective examination and criticism without being thereby made in some way rationally based on, or derivative from, the second-order theory that might be drawn upon in such reflection and criticism.

There is no occasion in this book to consider the controls on every concept in metaphysics, but it is very necessary to consider in more detail the controls to be placed upon the use of the concepts of "causal agency" and of "personhood". This will be my main concern in subsection (*e*) and also in Chapter 8. Then in Section 2, I proceed to discuss the role and standing of general principles about

[32] If some qualifier such as "it is to be asserted that", "it is true that", "it is probable that", "it would seem that", or (mentioned in a different connection) "it is not the case that" were supposed to be understood, or even to be expressed, in every proposition whatsoever, such a qualifier would be meaningless, contributing nothing to the sense or force of the proposition. There is no warrant for attributing to Bradley the idea that "probably" or "seemingly" are thus to be understood in every proposition. (cf. Wittgenstein, *Tractatus*, 4.061–4.063, 4.442, and *Philosophical Investigations*, I, 22, 547–557, *et al.*)

causation and the way in which it is the consideration of examples which sets controls upon what we say.

(d) *The role of purely a priori arguments in any metaphysical approach to God*

We need now to assess not just the general question of the conditions of any metaphysics being critical, but more particularly the conditions of any proposed metaphysical approach to God being critical.

First, let us review the role of a priori argumentation as historically met with within natural theology. What is relevant to our consideration is not the a priori in just any sense, but the a priori in any sense which involves in principle prescinding from knowledge of the actual as somehow experienced and instead somehow proceeding from the noetic or non-phenomenal (e.g. the Ideas, conceived of as objects of intellect not imagination or sense), or from the supposed pre-conditions of human knowledge, to some conclusion as to the actuality of God.

I do not consider here arguments from the existence of truth to the existence of God to be found in St Augustine, St Anselm, and St Bonaventure, amongst others. This is because, in my view, these and many other such arguments are not unambiguous in the way they are to be taken, inasmuch as it is not clear in what way the notion of truth is being conceived in them, i.e. not clear to what extent the notion of "truth" is to be taken as a first-order and to what extent as a second-order notion in these arguments. The conception of 'truth' as an archetype and standard according to which both things in existence and modes of thought are to be conformed was a common one in those times. But this book is not the place to elaborate on these problematic and obscure arguments.

Accordingly, I will mention Duns Scotus as perhaps the first of those who produced an unmistakably a priori argument for God's existence, arguing from our knowledge of the possibility of existence in the created world to the actuality of God, an argument subject to three key weaknesses.[33] His argument (i) is lacking in any

[33] See Étienne Gilson, *Jean Duns Scot*, pp. 128–43, especially pp. 136 ff., and references. In *Duns Scotus: Philosophical Writings*, on pp. 42 f. with pp. 44 f. we find it argued that an independent power to produce something is possible on the

proper consideration of the ambiguities in the notion of "possibility" (the argument depends on the 'possibility' being 'real', not merely 'logical', a distinction which a rationalist may not be able to make) and (ii) takes no account of the limitations corresponding to this in the appropriateness of our arguing backwards from possibility to some real actuality as the pre-condition of this possibility: e.g. he does not explain how, from the logical possibility of something's having a cause with power to produce it, the real possibility of this follows. He also, at several points, (iii) assumes a certain adequacy in respect of our concepts both of the "actual" and the "possible", characteristic later of the rationalists. But such assumption is neither, in my view, right nor, in their discussions any more than in Duns Scotus, given appropriate justification.

In more recent times, again and again neo-Thomists whose understanding of St Thomas has been reshaped along Kantian lines under the influence of Father Maréchal, and who have subsidiarily been influenced by Hegel and by Heidegger although in more superficial ways, have urged that 'being' implies 'intelligibility', and have so understood this statement that any concession of the existence of real being, whether divine or temporal, will generate the conclusion that God exists. This approach is found in such as Coreth, Rahner, and Lonergan.[34] In Rahner's case, his thought from the start in his doctoral thesis, *Spirit in the World* (1936), has always been shaped by a consideration imagined to accord with Hegel. In *Spirit in the World* Rahner has the same problem as Hegel as to how to grant to what we normally conceive of as the purely material any reality at all. But Hegel's idealism takes the form, somewhat reminiscent of Leibniz, of almost making material things alive, of making it that the very being of material things consists in their realization or *entelecheia* in being known by others, so that the 'being present to itself' which Hegel regards as necessary for the existence of a material thing is something realized when, e.g., a human being takes cognizance of a material thing. By contrast

basis that some power to produce something is possible, and then on p. 46 it is argued that such an independent power to produce something would only be possible if it were actual (this summary exhibits the structure of the argument).

[34] See E. L. Mascall, *The Openness of Being*, chs. 4 and 5, for an introduction to these thinkers.

Rahner follows a more solipsistic course in his account of the knowledge which we have of material things, and regards our knowledge of material things as a mode of our own 'being present to ourselves'.[35] In this way he is in a way more idealistic and indeed, subsidiarily, if not more Cartesian, at heart less emancipated from Descartes than Hegel. But subsidiarily, as a result he has increasingly, since 1936, for the same reason found the same difficulty in giving an account of the existence of the purely material that Descartes, Berkeley, and Kant experienced, a difficulty greater than that in the pages of Hegel.

This difficulty as to why material being has to be intelligible is at least noticed by Rahner, and is what involves him in a flirtation with Hegel and in the end a more thoroughgoing Cartesian-formed idealism. Lonergan, after some intelligent exercises in the philosophy of science in the first half of his book *Insight*, in the second half merely adopts a Cartesian methodology (of examining the conditions of thinking, inquiry, and reflection) and then draws out the conclusions that being is intelligible and God exists like rabbits out of a hat; the rhetorical method seems to be that if you bury argument in enough prose, you need not flirt with Hegel but anything will pass.[36]

In all these thinkers what we find is the supposition that an argument to God's existence will depend upon the assertion that being as such is intelligible understood in an extraordinarily global sense, i.e. understood in suchwise as to assert no less than Leibniz's

[35] This position can be traced within *Spirit in the World*, pp. 68–77, 90, 132–5, and 337, and in *Theological Investigations*, vol. i, pp. 326–9.

[36] In *Insight* Bernard Lonergan appears to slip from the proposition that being is all that is known together with all that remains to be known (p. 350) to the propositions that it is what is to be known by intelligent grasp and reasonable affirmation (p. 445) and that it is what can be known whose affirmation is the same as the affirmation of the intrinsic intelligibility of being (pp. 500, 552). He insists that so far as human knowledge is concerned this 'definition' is of a second-order kind (pp. 350, 643) but presumes that from the 'intrinsic intelligibility of being' (pp. 499, 652, 655) it follows that there is such a thing as what being is or the idea of being or complete intelligibility (pp. 643–5, 673) from which it follows that there is a transcendent intelligence to which all being is as such intelligible (pp. 643, 672–3, 677, cf. 658). Surely, it is only required that those beings which are actually knowable should be intrinsically intelligible, not that all beings or beings as such should be so.

Less objectionable argumentation to a similar conclusion similar to the doctrine that 'being is intelligible' appears in Germain Grisez, *Beyond the New Theism*, pp. 67–82, but explicitly drawing upon an extra premiss that 'rationality norms' are prescriptive.

principle of sufficient reason. It is very curious to find theologians thus involved in making into a first premiss of their arguments what for many of their predecessors, including Aquinas, would have only been a consequence or a theological conclusion. For instance, in Aquinas's understanding, nothing occurs without point, so that, although within some lower order of causes a thing may appear to be purely by chance, nonetheless it comes within the scope of divine providence which is God's practical wisdom in respect of the world: in this sense, nothing, in his view, is absolutely by chance. But the only basis for this view is for him derivative on having already first concluded to the existence of God and reached to sundry conclusions about God's providence in the world, and in particular also about the extension of this providence, not only to the sphere of general structures and types of things, but also to individuals.

Hence, before any such general proposition as that every event has a cause, in the modern sort of interpretation, and before any such principle as the principle of sufficient reason could be asserted by a follower of Aquinas, he would first have to prove the existence of God and then envisage such a principle as a corollary or consequence of God's intelligence and wisdom and the world being His creation and under His government. The causal principles from which we are arguing to the existence of God are, as we saw in Chapter VI, much more limited in scope than any such principles implausibly defended by modern neo-Thomists, and much more confined than any principle of general conformity to law or any principle of the universal application of the law of sufficient reason as might emanate from Leibniz or Kant. Indeed, if the account proposed in this book is correct, then it is not that there is one principle of causality applied univocally in some cases to relations between creatures and in other cases to relations between God and creatures, but there are various principles of causality whose every application to relations between creatures involves the attribution of an underlying exercise of causality to God.

We are now in a position to apply the conclusions we reached in subsection (*c*) as to the conditions of any inquiry being at the same time genuinely critical and genuinely realist. Against this background we can set aside straightaway any attempt to base either general causal principles or those causal principles appealed to in an attempt to argue for God's existence in which the basis proposed for accepting these principles is of a purely second-order character, as if man completely turned in on himself could gain any knowledge of reality whatsoever. Accordingly, we can set aside the approaches whereby those neo-Thomists who have followed Father Maréchal, the so-called transcendental Thomists, have tried to utilize a Kantian approach in order to justify a conclusion in this regard. Indeed, if they could have justified anything in this way it would be a causal principle far too extravagant for even physicists to be happy with it, but certainly far broader than anything the intellect should feel any compulsion to accept. Equally we should set aside any attempt to make acceptance either of this or that general causal principle, or ultimately of the existence of God as a condition of the exercise, or the having been exercised, of human rationality.

That is to say, our inquiry ought not to be into what it is necessary to assent to in the course of argument in order to be accounted rational, about the validity of argument and the conditions of rationality and suchlike, which are second-order questions, but into the question of what is needed for a thing to exist This is a first-order question and it is not appropriate methodologically in philosophy to make our answer to the first-order question derivative upon an answer to the second-order question.

Nor can we even allow the more sophisticated approach found in some papers of Hugo Meynell which still import some second-order consideration as the basis for the acceptance of some first-order conclusion. Meynell's approach bases itself, not like the approach of Germain Grisez which rests on the consideration of what is required for man to be deemed rational,[37] but on a consideration of what character the world must have for the approach which we deem rational to be appropriate.[38] Those philosophers who are familiar with discussions of the so-called problem of induction will be used to the emphasis laid by some on the difference between the assertion that inductive procedures are rational or demanded by reason and the assertion that the Universe is such that such procedures are appropriate; but they will also be familiar with the failure of the second approach to do more than the first to solve the supposed, possibly misconceived, problem. The

[37] See n. 10 above.
[38] See n. 11 above.

division between different groups of transcendentalists, some more methodological or idealist, like Grisez, and some, in appearance, at least, more realist, like Meynell, makes no difference to the capacity to solve the problem of justifying causal arguments to God, any more than to solve any problems about induction. There is still the ultimate dependence upon second-order or epistemological judgement as to what is rational.

Therefore, we absolutely require, even and indeed especially in the case with which we are concerned, an approach which is realist and not involving any such mistaken second-order resort or transcendentalist gambit.

(e) *The basis of a truly critical metaphysical approach to God*

Our question, therefore, is of what is needed for a thing to exist, not any question about what would make other men right to deem us rational in our approach to such a question. Yet our approach must still be critical: it must not be that just any natural-seeming or rhetorically attractive argument passes. How?

Now, we had posed the proposition that the continuance in existence of temporal things as such requires a cause (I shall not repeat the detailed qualifications and explanations given earlier).

Thus, in effect, we argued as follows. Temporal existence as such involves a certain contingency and we have to realize and explain the difference between this type of contingency and the types of contingency discussed respectively by logicians, natural scientists, and epistemologists. That is, there is some metaphysical contingency distinct from any of those three or four other types (there are, I suspect, at least two very different types of contingency of interest to epistemology as such[39]). This kind of contingency, like the natural or physical contingency of concern to natural scientists, must be rooted in a certain kind of a causal background.[40] There is

[39] Epistemology has an interest in that kind of contingency (or possibility both ways) which arises when it is neither certain that *p* nor certain that not-*p* but possible that not-*p* and possible that *p*. Epistemology also has an interest in that kind of contingency which is present when it is neither necessary a priori that *p* nor necessary a priori that not-*p*. (It may also have an interest in a so-called contingency which is present when it is neither analytic that *p* nor analytic that not-*p*. But this does not involve an absence of certainty or necessity *tout court* but only the absence of a certainty or necessity whose roots are in some narrow sense 'logical'.)

[40] Argument to the effect that any 'real' possibility and therefore any 'real'

thus a kind of contingency cognizable to metaphysics, implicit in the ordinary use of notions such as "freedom" and "chance" and implicit in the distinction between the past and the future, and not adequately explained by natural science or taken cognizance of as such by natural science. This metaphysical contingency provides a basis for an argument for the need of existence of a cause which is still quite limited in the scope of the causal principles to which it makes appeal.

What we have here firstly, is not that the need of a cause in physical nature and the need of a trans-temporal cause giving continuance of existence over time and explaining the continuance of things are two entirely separate types of judgement, but rather that they are types of judgement in which we meet with the same structure of thought and argument in the way we explained at the end of Section 3 of Chapter VI. This structure has the effect of involving the notions of contingency and of causation or causal background as a pre-condition of the existence of contingency (albeit that there are two different kinds of contingency and causation here involved in the two cases, kinds of contingency and causation taken cognizance of in different spheres of human speculative knowledge and discussion, *viz.* metaphysics and the natural sciences).

Then, secondly, it needs further to be recognized that the type of contingency of which natural and human science takes cognizance presupposes the type of contingency of which metaphysics has to give an account. The natural and human sciences can at least to a large extent have their results formulated in a descriptive style in terms of the existence of various regularities, leaving many matters descriptively undetermined by the specifications of these regularities, so-called 'laws': the ordinary man and the metaphysician both recognize the indeterminacy as real. Likewise the sciences observe what present themselves as exercises of causal agency: the ordinary man and the metaphysician recognize this presentation as non-deceptive and discern that, inasmuch as such exercises of causal agency (i.e. actions) presuppose a reliably persisting non-momentary agent, they presuppose therewith an underlying causality, a cause of persistence as such. Thus an unconditional or genuine a

contingency must be based in a causal background, i.e. it is a matter of what the natures actually existing and operative in the world make or leave possible, adverted to in Chapter I, appears in David Braine, 'Varieties of Necessity', pt. III.

posteriori knowledge of mundane causal relations gives rise a posteriori to a knowledge of the primary causality of the First Cause, knowledge of it, not *qua* pre-condition of knowledge of secondary causality, but *qua* condition of the exercise of secondary causality itself, something internal to every exercise of secondary causality because involved in the continuance of the natures and existence of the things exercising secondary causality.

Now, if one proceeds in this kind of way, exploring the presuppositions involved in the notions of "real existence" and of "contingent existence", one is in one way still involved with the second-order, inasmuch as one is involved in conceptual inquiry and in dealing with certain particular concepts expressed by particular terms or words such as "cause", "contingent", etc. But one is not making the second-order primary in the sense of making some principle within second-order theory, call it epistemology or methodology as you like, the basis of some conclusion at a first-order level. Therefore it is not precluded that the conclusion that one reaches is primarily a first-order conclusion, although limited in scope by the very way in which one has approached it, and not attained uncritically despite not being derivative from second-order inquiry. It is a first-order conclusion because it is the conclusion that such and such, say "*x*", has a cause, that this cause establishes *x* in its nature, *viz.* as a thing whose actions involve persistence through time, and, while this cause gives reliability, dependence on this cause means that at any locus in space-time the future is open (because what is future to it does not yet exist) so that freedom and chance are real: it is not any second-order judgement as to the meanings or implications of such words as "cause".

For second-order inquiry to be made without qualification primary is for man in his thinking to adopt the position of the person who thinks that he could live by eating himself. It is not just that the second-order is never in fact primary, because something first-order always turns out to have been smuggled in in the justification or even inside the meaning of the statement of the second-order; or that something first-order must have thus been smuggled in if any first-order consequence is to follow. Nay, rather, it is more than this: let something be presented even under the guise of being second-order (if it is second-order or dependent on the second-order in reality, the situation will be worse), then you will be debarred from drawing any first-order consequence from it—the second-order is as such never a possible basis for the first-order.

2. HUMAN KNOWLEDGE OF GENERALITY: THE RELATION OF UNDERSTANDING TO EXAMPLES AND GENERAL PRINCIPLES IN JUDGEMENT

(a) *The basis of general knowledge consists in a possession of concepts which is exhibited primarily in judgements about particular cases, and only secondarily in knowledge of general principles or laws*

(i) *The problem stated*

It might seem that, although a particular example may be illustrative of some general principle, no conclusion can be drawn from such an example in respect of any other example without first establishing such a general principle. Such a principle might be concluded to by, on the one hand, mere extrapolation from a particular example, either by the method of induction or by an argument by analogy (as identified by Hume, Butler, and their contemporaries), or on the other hand by some more informal 'intuitive' induction.

Thus it might appear that, while examples might illustrate or suggest general principles, it is the general principles, and the understanding which grounds their acceptance, which do all the work, and not the examples, except perhaps accidentally, which determine philosophical judgement. However, this received commonplace of modern philosophy generates some problems. Why, for instance, is it that we are so often more certain about particular cases than about the formulation of watertight general principles which might determine judgement about particular cases?

One mistake that philosophers generally make is to address this sort of question in too general and amorphous a form. The remedy is to begin by considering some examples of particular concepts and associated general principles.

(ii) *Review of examples*

Take the paradigm case of intuitive induction presented with a paradigm example of a red object and a paradigm example of a green object. It would appear that the judgement that something cannot be red and green at the same time is intuitively obvious. But, on reflection, questions will arise as to the status of 'reddish green' or 'greenish red' objects, objects red on one side and green on another, and objects partly red and partly green (however configured); and there will also be questions about the workings of human perception which forbid the identification of even the spectral colours with a range of wavelengths of light and introduce consideration of lighting conditions and the contrast background. What will result from this inquiry will be an extremely cautious formulation of a general principle, involving many levels of qualification and stipulation of conditions of application. But it will disguise the original perceptual insight that red and green are contraries, even though this will still be presupposed in any such general principle.

In questions of this sort we are in an area where the empiricist principle that what is imaginable is possible,[41] and vice versa,[42] is, with some minor qualifications, valid. Though more free, imagination is determined in structure by perception, so that, if it is not possible to imagine something as being in a certain respect red and green at the same time, this limitation on imagination marks a limitation on perception. In other words the unimaginability of the state of an object's being red and green at the same time indicates that there must be some way in which their being contraries, so that in paradigm situations a thing cannot be red and green at the same time, is foundational for judgement. Judgement may proceed against the background of this insight in advance of any successful analysis attaining accurate formulation of a watertight general principle.

Thus, generalization has proceeded, in the sense of extension to other cases, on the basis of an insight that might ground a general principle but long in advance of any accurate formulation of the general principle concerned. Even when contrasting cases involving

[41] In Hume the principle that the imaginable is possible appears in many forms, e.g. nothing imaginable is absolutely impossible (*Treatise*, bk. I, pt. II, sect. II), what is clearly conceived may exist and separability of ideas implies separate existence (*Treatise*, bk. I, pt. IV, sect. V).

[42] The converse principle to that mentioned in n. 41 above also appears in Hume in many forms, e.g. what appears impossible must be impossible (*Treatise*, bk. I, pt. II, sect. I).

some restriction on the extension concerned are presented, discrimination (or judgement) comes into operation in advance of any precise formulation of general principles. What we have to observe is that some reason has been adduced, once the question has been raised and second-order inquiry set in train, for allowing judgement to proceed on the basis of examples in advance of any accurate formulation of general principles, namely in this case the relation between the faculties of perception and imagination, one or both being presupposed in our use of colour-words.

What we have in every one of the varied cases which present themselves to the philosopher for consideration is the acquisition, as the result of the consideration of one or a small number of particular cases, of a capacity to make judgements in a large range of other particular cases. This capacity is liable to be spoken of as having some 'root', 'ground', '*archē*', or 'principle' (*principium*), and in regard to each type of case the question arises as to how best to describe or think about this ground or principle.

The rationalist error consists in supposing that this ground or principle is always a general law-like proposition or statement of some kind: indeed this has become the normal meaning of the English word "principle". However, consideration of the case of colour-judgements showed that the capacity concerned, at least in that type of case, operated reliably in advance of the formulation of any law-like proposition. So, we find, for instance, at least in that type of case, the root from which the capacity for further judgement arises would better be described as the appreciation of some perceptual *Gestalt* or group of perceptual *Gestalten*. This same appreciation would also be that which made possible judgement or discrimination between different proposed formulations of any general law-like proposition or judgement.

The choice of the word "appreciation" marks the convenience of sometimes leaving it open to what extent it is 'sensibility' and to what extent it is 'understanding' which is, in the type of case concerned, determinative of the use of words or concepts concerned, or in what way the two may be involved inseparably together.

In the case of colours, it was plain that in a certain sense the imagination held sway, and the unimaginable is, as such, impossible. In the case of geometrical concepts, the imagination holds a much more limited sway: it might appear that, as a perceptual

Gestalt, straightness excluded a vantage point from which a straight line might be described as curved in such a way as to join itself again, and included a context within which a set of parallel straight lines remaining equidistant from each other was possible, and not solely an association of straightness of path with shortness of path; but in mathematics and in physics, we abandon being governed by any such perceptual *Gestalt*, and make the meaning of the word "straight", as used in these contexts, consist in something entirely defined in terms of shortness of distance or separation or in modern physics, equivalently, in terms of the path of a light signal taking least time between two points. In this and related ways the principles of physical geometry in certain respects pass outside the government of the imagination. When it comes to concepts of time and causality, the imagination—in that limited sense in which imagination has to do with the visual, rather than in any larger sense whereby it has to do either with modelling in an extended metaphorical sense or with supposing—appears entirely without competence.

Yet it is evident that circles in explanation or in temporal order are vicious, and it may be presumed that there are other principles relating to time and causality in which the understanding grounds judgements of 'impossibility' and of 'necessity' (what is impossible not to be, or inescapable), both in dealing with particular examples and in formulating general principles. But what the consideration of the perceptual cases as well as of other types of instance, including instances of practical moral and legal judgements, shows is clear. It is most obviously this, that there is no a priori reason, or reason to be taken for granted beforehand, for assuming that the understanding can only make judgement possible in regard to particular examples of temporal and causal relation by reaching first to the formulation of a general principle and then applying it to the particular examples.

The rule that we should be guided, as far as is compatible with coherence, by what presents itself as 'experience', 'the seemings', and 'common opinions' as these are enshrined in the underlying structures of speech, not only in respect of the more general first-order judgements, but also in respect of second-order judgements and the mode of interplay between first- and second-order considerations, leaves very little room for doubt in regard to what we should say in connection with the present issue. The structures involved in the exercise of sight and visual imagination are determined by the datum that the animal who sees and imagines is an animal who looks, moves about to look from different vantage points, touches and explores the world with sight and touch in large measure integrated. Accordingly, he deploys the concept of "substance", and thereby implicitly the notion of "causal agent", at the most primitive level of his making judgements about experience, the level upon which the shape and sense of all other judgements about experience depend. Thus the most basic and determinative judgements of experience involve exercises of the understanding of the concepts of "substance" and "cause" inseparably from any knowledge of particular instances whatsoever.

Therefore, there is no learning situation in which a judgement has been made in regard to a particular example of an empirical kind, in which the understanding of the concepts of "substance" and "cause" has not already in the same act been brought into exercise. Accordingly, whatsoever is integral to the possession of these concepts, even if it includes implicitly a capacity for generalizing, does not wait upon an induction from observed instances to unobserved instances for its legitimation. Nor does the capacity to make judgements about particular instances (a capacity, after all, exercised in the first exercise of judgement in a learning process) wait upon the formulation, whether by Humean or by any other species of induction, of a general principle.

The notion of "cause" brought thus early into play in empirical judgements is the notion of "causal agent". That is, the notion we are speaking of is not whatever is expressed by the interrogative "why?" and the conjunction "because", which turn one's attention not only to efficient causality but also to reasons for belief, action, and attachment, but the particular notion of "substance with active power" on which all other discourse about efficient causality pivots. The efficient causal agent can have its action, its exercise of active power, only in time. Temporal concepts inexorably accompany the application of causal concepts. And indeed, inasmuch as the human being who makes judgements of experience is a causal agent of the kind we call animal, his experience from being present becomes past, a possible object of memory. Accordingly all his judgements of experience are tensed, the use of the present tense being basic, marking the norm against which all other expressions of time and modality stand in contrast.

(iii) Conclusion

Thus we have shown firstly that the primary expression or realization of understanding, or having, a concept is in the judgements we make applying it in particular instances: the general (universal) judgements we make using these concepts represent only a secondary realization of this understanding. It is natural to see the general judgements we make using them as arising by intuitive induction[43] from the singular judgements about particular instances under the critical controls we mentioned. We have seen secondly that this must apply not only with more limited (e.g.) visual and spatial, but also with such general, concepts as those of "substance", "cause", and "time", so that general principles in regard to these will likewise be well based on intuitive induction, albeit always under some critical controls.

What we need to realize thirdly is something that immediately follows from these first points, namely that the possession of the most general concepts can ground a spread of principles related by analogy, as we might have expected since these concepts are not univocal.

(b) Because the knowledge possessed in virtue of having the most general concepts is expressed primarily in handling examples, it can be realized in a spread of analogous applications of principle, rather than in some single univocal principle

We need to approach the question of the shape of general metaphysics and of the way in which the concepts of "life" and "personhood" may enter into it via a consideration of the way knowledge of causality enters into the general framework of metaphysical knowledge.

The mistake of rationalists such as Leibniz was to suppose that a knowledge of causality was founded on certain perfectly general principles such as the principle of sufficient reason, and that such principles, and the concepts such as "cause" or "reason" in terms of which they were expressed, were known or possessed in one act. The propensity to view knowledge as founded on general principles

[43] For an explanation of the notion of intuitive induction, see W. Kneale, *Probability and Induction*, pp. 30–7.

so that the move from experience to science is a move towards axiomatization is found in Aristotle and, before him, Plato.[44] The conception that the concept of cause would be exercised in certain perfectly general principles was the basis of Duns Scotus's presumption that the concept of "cause" is univocal.[45] It is various perfectly general causal principles which become the subject of critical consideration by Hume and Kant. Accordingly, Leibniz is representative rather than atypical of the general tendency of Western philosophical thinking.

However, although whenever general concepts are applied, it is with some knowledge of the implications and possible settings of their application, and some capacity to judge of other respectively similar and dissimilar cases in which they would or would not have application so that they always involve an implicit generality, nonetheless knowledge of their applicability in a particular case does not wait upon or carry with it appreciation of either the formulation or the range of application of the principles governing such implications or of the concept concerned. We find this in morals where we can often judge the good of a particular end or object or judge the wrongness of a particular action without yet being able to formulate in a perfectly watertight way the general principle relating to all cases of the type concerned: e.g. in a particular case we may be clear that wrongful lying is involved, without being ready with a definition of "lie" and a principle expressing precisely all the conditions on which the wrongfulness of telling an untruth may depend. Likewise in science, it is evident that most generalizations are expressed in ways not intended to cover an indefinite range of contingencies not normally allowed for: e.g. the intrusion of a large foreign body into the solar system, the approach towards conditions in which singularities would arise, or, in laboratory experiments, the destruction of the apparatus due to accident or political action.

In the case of the concept of "cause" in the sense of "causal agent" or "substance with active power", our starting-point is with a large range of examples of the applying of the concept.[46] We

[44] The move from experience to science is a move towards axiomatization according to Plato in *The Republic*, VI and VII, and Aristotle in *Metaphysics*, Alpha, 1.

[45] Duns Scotus holds that "cause" must be univocal to allow reasoning. For his general notion of "univocity", see *Duns Scotus: Philosophical Writings*, pp. 19 f.

[46] Cf. Michael Scriven, 'Explanations, Predictions, and Laws', pp. 181–6.

judge that the bridge was destroyed by a flood, not a bomb and not an excessive load: there was a flood and some people observed the bridge being swept away by it, and there was no bomb and no notable load. We judge that a person was knocked over by a falling rock, not by a gang of attackers. We judge that Smith wrote a letter and posted it, that Jones built and supervised the bonfire wherein he then burnt our favourite cabinet. We judge that Brown spoke certain words. And in these and suchlike ways we apply the concept of "causal agent" to floods, rocks, people, and so on.

It is not then that general metaphysics looks out on the Universe and sees a sea of substances with active powers, typically material substances, and notices the occurrence of personal agency as a local minority phenomenon. Rather, general metaphysics takes cognizance of the concept of "cause" in the same act as it takes cognizance of the concept "substance" or makes use of the term "exist" where what I have called real metaphysical or substantial existence is concerned. But, in taking cognizance of the concept of "cause", it willy-nilly takes cognizance of a concept which is given to it in the various primitive cases of its application in a wide span of types of use.

(This will be found especially relevant to our discussion of the personhood of the First Cause in Chapter VIII. In viewing this wide span of types of use the metaphysician or general philosopher cannot but note, not only the impossibility of reducing the modes of causality to one, but also the relative deficiency of some modes compared with others inasmuch as some agents exercising active power do so in ways which are explanatorily richer in relation to why the effect came about than is the case with some other types of agent with more limited kinds of active power.[47] In this way the agency exercised by lower mammals turns out to be fuller than the agency exercised by inanimate objects, and the agency exercised by human beings fuller still. It may be, and indeed, on examination, it turns out that it is the case that there are certain ways we can appreciate in which human agency is deficient or explanatorily incomplete. It has to be asked whether these identifiable sources of deficiency arise from what it is to be personal as such, or only from

[47] Cf. Chapter III, Section 2(*c*), in regard to the span of types of causation and the impropriety of trying to reduce all the modes of causality to one univocally definable kind such as that exemplified by inanimate substances.

certain aspects of the way in which their personhood is realized in human beings.)

It needs to be realized that temporal causal agency presupposes a nature according to which the thing with active power has continuous existence over time. It does not exercise its active power instantaneously in the sense of at a point of time (that which divides a period without occupying a period). But then it is incoherent to suppose that any temporal causal agent exercises real causal action unless there is an underlying cause maintaining it in existence with its nature over time. This underlying prior cause of continued existence and nature is what gives force to the action of any agent it thus underlies.

From this standpoint it appears that the application of the concept of "cause" which takes us back to a first cause will be presupposed by, and underpin, the causal 'principle' or 'principles' which we rely upon in thinking about mundane objects. It may be alleged that some univocal or, as it were, typically unambiguous general causal principles of a rationalist kind have been relied upon surreptitiously and my present mode of presentation is thus a mere subterfuge. This allegation I refuted at the end of Section 3 of Chapter VI. It has been the philosopher's common instinct to proceed in a way guided by examples, an instinct relied upon in my present mode of presentation, which does provide a genuine alternative to rationalism and to that modern empiricism which is a mere variant of rationalism (imagining being substituted for conceiving). The soundness of this instinct we established in subsection (*a*).

VIII
The Personhood of the First Cause

A proof of the existence of a cause of continuance in existence as such is not yet a proof of the existence of God until it has been shown that such a cause is personal. For this is a key part of the connotation of the word "God", and on our approach to it the way of dealing with all the other connotations not yet touched upon turns.

We have therefore to consider why the concept of "person" should have any particular privilege in metaphysics.

Let me briefly summarize the points dealt with in Chapter VII on which our treatment of personhood will turn.

Firstly we there established that first-order inquiry should always be considered primary, so that second-order inquiry remained parasitic and dependent upon it. At the same time we laid down the requirement that first-order inquiry be continued critically, i.e. in the context of second-order inquiry and supported by it, rather than in isolation.

Secondly we showed that because general knowledge depended upon concepts, many of them polymorphous and not univocal, its primary realization was in judgement about examples, not in assent to general principles, axioms, or laws. This understanding and these judgements in respect of examples provide a basis from which intuitive inductions may be made, and as a result the understanding of a concept may result in an assent to a spread of analogous applications without any global single univocal principle being involved.

In establishing God as cause, we *ipso facto* establish God as causal agent and therefore subject or *suppositum* as well as actuality, but we are still using a concept as extensive as that of "real existence" itself: indeed then we have established a concrete predicate, that of causal agent, in regard to God, but we have not introduced any such exclusive predicate (not applying to all real existents) as the concept of "person".

Against this background we can ask how and whether the notion of "personhood" can be brought into or, as some people would regard it, smuggled into metaphysics. For on the face of it persons are just one particular type of being and God, if personal, is just a particular kind of being. Why, it is asked, should one particular kind of being be elevated within a truly general metaphysics to a unique position? Is it not, the question is voiced in antipathy to what we suggested in Chapter VI, just a confusion when Aristotle identifies the science of being as such with the science of the divine, a confusion begotten of a false synthesis between two different concepts of first philosophy? (The true reason why this synthesis, found in Aristotle only in a rough undeveloped way, is not a fudge but correct will only emerge in the second-order discussion in Section 5 below.)

In Sections 1 and 2 we shall show the richness of the concept of "person", in particular its explanatory richness, and the openness whereby it is not tied to the human case with its deficiencies, the richness and openness which leave room for, and even strongly invite, the presumption that any first cause is personal. But in Section 2, reinforced later in Section 5(*d*), we shall illustrate the vital need to resist the resulting temptations to argue a priori from the concept of "person" alone that a first cause must be personal.

Rather as always the basic argument, as we insisted in Chapter VII, must proceed from actuality. And in Sections 3 and 4 we shall outline the proper lines of such argument and what is required for its fuller development. It will be an argument to the effect that the cause of the continued existence of persons, or of a world containing persons, must be itself personal.

But such first-order causal argument cannot exist in isolation if the approach is to count as 'critical' as required in Chapter VII. And for this the nature of the second-order argument concerned, the argument set forth in Section 5, is of special importance. This is because it shows that the validity of the argument in 3 and 4 precisely turns upon avoiding the Cartesian segregation of mind and personhood from the world: it turns upon insisting that the concepts we have, in particular the concept of "causal agent" which is an element in our concept of "person", are concepts which we in fact acquire and could only acquire in a way appropriate to our being animals set amongst other animals within a biophysical environment, the world.

Once having established that God as subject or *suppositum* is not just 'causal agent' but 'person', we can state the resulting general perspective, the perspective which makes God the intentional agent of all the temporal continuance in the world, immediate to it as an agent is immediate to his action. In this way we give full significance to our remarks in Chapter V, Sections 7 and 8.

In that chapter we sounded a key note for our whole thesis of the reality of time. There we saw that the requirements of avoiding an analysis of action not distorted by dualism have the effect of making the action of the trans-temporal or eternal First Cause an action which takes place in time. That is, it secures that the assertion of God's eternity presupposes the reality of time and the non-existence at any juncture of what is future to that juncture: it is because of this that the future is the area of radical contingency and freedom for God and for man; and it is also because of this that continuance into the future needs a cause, a cause which can only be non-temporal.

The personal character of God is thus secured against any denial of the reality of time. Whether, as I suggest in Chapter IX, it can be secured against restrictive understandings of a person's 'individuality' depends upon the openness of the concept of "person" to which I now turn.

1. THE NATURE OF PERSONHOOD: THE CLOSE INTEGRATION AND INTERDEPENDENCE OF DIFFERENT ELEMENTS IN THE CONCEPT; THE RICHNESS OF THE CONCEPT AND ITS DESCRIPTIVE OPENNESS

Evidently, the concept of "person" marks out the range of significant joint application of such predicates as "know", "understand" and "judge", "desire", "prefer", "rejoice in", "intend" and "choose", "speak" and "hear". But what it is that constitutes the positive content of the notion of "person", so as to bring an object within the range of this spread of predicates, is less clear. The two most common approaches have been to make either intellectuality or freedom of choice the hallmark of personhood. We must see why neither of such approaches is right and discover the richness of the concept of "person", involving an integration of intellect and will and an openness to rapture, creativity, and relationship. Often intellectuality or understanding has been conceived of in terms of the possession of models. These models are supposed to be in some way like or representative of reality. In better accounts of the matter, these models have been conceived of as having some kind of 'intentional' existence as if in some way they established the thinking and knowing subject in an attitude of being mentally directed towards reality, so that it was this reality rather than the models which had real existence and were the objects of knowledge. However, this account still leaves it obscure by what means the intellect knows or judges the models to be appropriate and so gives no account of any relation to reality adequate to secure such knowledge or justify such judgement.[1]

Alternatively, intellectuality has been conceived of in terms of thinking or judging, but this (*a*), leaves it obscure as to how the intellectual subject knows that he has attained the goal of truth in this thought or judgement, i.e. in virtue of what relation to reality he can know this, and (*b*) omits an account of the meaning of the terms or concepts which compose such judgement.

These difficulties about truth lead us on to identify the decisive omission in these accounts. They omit to explain what it is for the intellectual subject to envisage knowledge and truth as goals of intellectual operation in which he, the subject, ought properly to come to rest. Accordingly, the intellectual subject needs to be capable of rejoicing in, or having satisfaction in, these goals and an active relation to the world securing his attainment of these goals; and any theory of models in judgement must be integrated within a recognition of the intellectual subject as a being who recognizes value in such things, either desiring or rejoicing in them.

Moved by these considerations, we may be tempted to give priority in our explanation of the notion of personhood to the ideas of free choice, perhaps *inter alia* exercised in active exploration of the world. However, this reaction away from giving primacy to any supposed theoretical use of the intellect, understood as disengaged from its object, omits to take account of the fact that the much

[1] The argument of this paragraph is directed against any form of 'representationalism', i.e. any theory in which ideas, models, notions, forms, or *species* are conceived of as objects of the mind's attention in the act of knowing or understanding, rather than as the means or principles in virtue of which or according to which the act of knowing or understanding is possible and proceeds, themselves not attended in the act concerned. The rejection of 'representationalism' is to be found in Aquinas, *S. Th.*, Ia, Q. 85, Art. 2, *Comm. De An.*, 718, and more fully in *Q. D. De An.*, e.g. Q. 2, resp. 5, and also in Wittgenstein, *Philosophical Investigations*, I, 73–4, 256–80.

vaunted free choice presupposes deliberation, and deliberation presupposes various judgements about the various alternatives posed to choice. But beyond this it also fails to take an adequate account of the wider aspects of will, appetition, and love presupposed by free choice but also realized in examples of the voluntary which are not examples of choice: voluntariness is exercised in trust and mistrust, hope and despair, boredom and enraptured enjoyment, without these being immediately results of, or expressed in, choice.[2] The point of choice and the making of choice depend on attraction towards, or love of, the goods which give point to the alternatives posed (by "love" here is meant that placing of value on things which is shown as much by gladness or satisfaction as by desire). Man also sets value upon, and loves, things which are not the issues of his choice, both things in nature and things in human life. Always, then, we are meeting with a valuing or love more deeply set than choice, and always this involves both knowledge of its objects and knowledge of their value: it is therefore still intellectual, not just instinctive. Among the objects of this valuing or love are knowledge and truth.

The concept of "person" accordingly involves the bringing together of notions of knowledge and of value or love, in a setting of relationship with objects of this knowledge and value or love.

Deliberating as to what is the case prior to coming to a judgement, and deliberating as to what to do prior to making a choice, presuppose some incompleteness in the cognitive relation to the object known and the absence of what we may call rapture, i.e. some incompleteness in the affective relation with the object loved. The freedom of deliberation on which both human knowledge and real choice (as opposed to chance) are founded presupposes a capacity for self-reflection which allows an unlimited capacity for review and a certain openness or infinity in respect of the possibilities that may come into consideration as a result of such review.

But the point of such freedom, self-reflection, and review is only to bring man yet nearer in more complete cognitive relation to his objects and a more completely engaging enjoyment because a

[2] It will be noted that in Augustine the concept of freedom of will sometimes (e.g. *Epistle 157*) appears broader than the concept of the freedom of choice and that in Aquinas God has a love of Himself which is necessary and therefore the object of will only in a sense broader than choice (e.g. *De Potentia*, Q. 2, Art. 5).

greater enjoyment of the goods attained and an enjoyment greater, not by accident, but because of what makes them 'good'. The aspiration of man to achieve a complete cognitive relation with his objects is seen in the desire to combine sight, which is often almost bare of understanding, with understanding or to realize in understanding a directness of relation and an evidentness of the object to the intellect of which sight is our primitive model. Rapture or enjoyment which is not inspired by such cognitive relation, but lacking integration with it (immoderate, inordinate, or having a measure bearing no relation to the measure and character of the cognitive relation), is disjoint from properly human fulfilment.

The concept of "person" also involves a certain openness in respect of conceiving or imagining different possible sorts of thing and different possible sorts of relationship or connection (the unformalizability of natural language, its incapability of being captured in any mechanically applicable set of rules, is important here), together with an openness in respect of creativity so as to be in some measure unrestricted in what is created. Obviously in these respects also a person may be, even in what pertains to personhood, more or less deficient.

The experience of the actual which, as human beings, we already have suggests (*a*) that the possibility of enrapturement is never securely achieved in the empirically known conditions of human life, and never securely achieved in respect of any human faculty with a cognitive aspect. But it also suggests (*b*) that, nonetheless, it is in some measure achieved (inasmuch as deliberation proceeds without violation of rationality) in relation to Nature and to living beings, but above all to other persons.

Let us develop this second point, (*b*).

The capacity for a person to envisage something as a good (as an element within the total structure of a good, fortunate, or roundedly happy life) is associated, for instance, with his capacity to see appreciative knowledge and creativity as integrated within, or even expressions of, a personal relationship. Accordingly the concept of a person is such as to render defective all realizations of personhood in which cognitive relations and rapture are in identifiable respects lacking in integration with, or incomplete in respect of, relation to other persons (or in which there is no openness to completion, cognitive and motivational, in relationship to other persons). The prominence of other-person-relatedness in the concept of "person" is exhibited in the prominence of the concepts of "speaking" and "hearing", metaphorically if not literally, in our thinking about persons and in the propensity to speak of understanding as having as its objects not natures and situations but persons. For an intellectual and loving or conative being to be closed in on itself is a limitation: personhood as such involves at least the openness to relatedness to other persons, although not the choice of relationship in every instance of relationship offered.

In our experience persons grow from immaturity to maturity and depend for all agency, creativity, knowledge, and communication upon the use of bodily faculties. In their creativity, their freedom is restricted by the nature of the stuff with which they work and the conditions within which they work. Although the billiards player is more largely responsible for the state of the game at any point than are the individual billiard balls (his action is explanatorily richer), he is still not totally responsible: he did not make the billiard table or balls, let alone the stuff of the Universe and the order of nature, upon which his actions and those of his fellow players depend. The deficiencies of human cognition are much adumbrated in philosophical literature. Accordingly, in sum, human beings are indeed persons, but they exercise and realize personhood only in ways that are in many ways limited and deficient, in some cases deficient in respect of the ways in which personal action provides an explanation of its outcome, i.e. deficient in ways pertinent to explanation.

But none of these deficiencies seems to arise from the concept of "person" itself. Therefore, if the First Cause is not thus deficient but is explanatory in more respects than human action could be, this will not debar the First Cause from being personal: the concept of "person" does not tie personhood to the case of agency as limited as the human.

That is, the concept of "person" is 'descriptively' a concept of extraordinary openness. From all this it does not follow that there is no deep conceptual incoherence in the statement that the First Cause is personal in nature, let alone that this states a real possibility. All we have exhibited in respect of openness and possibility is that, at the level of the making and exploring of hypotheses in life or in metaphysics and of story-telling in history and cosmology, to use the concept "person" in regard to the First Cause involves no immediate, prima facie, or 'descriptive' absurdity.

2. FALSE AVENUES OF APPROACH TO THE QUESTION OF THE CENTRALITY OF PERSONHOOD: A PRIORI ARGUMENT FROM THE CONCEPTS OF "BEING" AND "CAUSE OF BEING"

Such then, in broad outline, is the concept of person as our experience of the actual gives it to us. Our second question now confronts us. Why should it be supposed that the causes of continuance in temporal existence of the various kinds of agent with active power with their natures that we know, *viz.* persons in the limited human style we experience, higher animals and other living things, and all the variety of inanimate agents we learn of empirically, should have as their cause something personal? and why, not only this, but something personal in a way which is in no way deficient either *qua* person or *qua* causal agent?—inasmuch as to be deficient *qua* causal agent would be one way of being deficient *qua* person since personal agency is one mode of causal agency, indeed the least deficient we meet with empirically.

The first grand point which needs to be seized, and which it is useful to announce in advance of the consideration of particular fallacious arguments, lest the reader lose sight of the wood for the trees, is this: any argument to God's personhood which proceeds merely from the most general undetermined concepts of "being" or "cause of being" must be fallacious for the following reason.

If God exists and His being or *esse* is identical with His activity of living, knowing, and loving, i.e. "identical" in the way explained in Chapter V, then for Him to understand His *esse* will be the same action as His understanding of His personal life, and the intelligibility to Him of His *esse* as it is in itself will be no less than the intelligibility to Him of being personal in His unrestricted and perfect manner. But the intelligibility of what concepts human beings can form of *esse* is less than the intelligibility of the concepts they form of the personal. Human concepts of the personal have content, as indeed the concept they form of other kinds of causal agent also have content, whereas *esse* (which is what gives actuality to whatever has actuality) is understood by us as a content-less presupposition, unmodelled, rather than conceived in that content which it has in God.

Accordingly, the concepts of "person" and "life" do not enter metaphysics, as an area of human thinking and reflection, merely as unfolding the pre-conditions (known by second-order reflection) of the application of the term "real", nor through any empiricist or rationalist unfolding of the concept "cause of being", since we have already excluded these avenues of approach, in the earlier sections of this chapter. Rather, they can enter metaphysics only as a result of metaphysics being a science or study which is shaped to those most basic aspects of reality which it investigates, and which reflection on reality provides a critical control upon. Accordingly, it can only draw upon those concepts with which human apprehension of reality supplies us.

Therefore, it will remain that some further consideration has to be adduced before it can be concluded that the cause of the continued existence of the stuff and order of the Universe is a cause, not only complete and undeficient in respect of efficient causal explanation, but also in respect of every kind of explanation, or that it must be personal rather than being non-deficient in some quite different and incommensurable way unknown to human thinking.

Let us now review the fallacious arguments concerned in more detail.

(i) Lack of deficiency in the First Cause, *qua* cause, does not imply determinism. The natures of living things may be such as to exclude physical determination. The natures of human beings as persons are certainly such as to exclude any kind of general determination whatsoever. The nature of the physical in its physical aspect as we know it through quantum physics is such that, at the physical level of description at least, chance in some absolute sense[3] is intrinsic to it. Whatever causes the continuance in existence of things with these natures must do so in a way that does not nullify but preserves these kinds of freedom and indetermination. Within these limits we can say that whatever causes the continuance in being of the stuff of the Universe, whether this be matter, energy, animate beings, or whatever, must be explanatorily the complete

[3] By this I mean: not just in the sense of coincidence nor just in the way things which fall out according to some family of laws statistical in character are said to be 'by chance' because they are undetermined relative to the level of description embodied in that family of laws.

cause, deficient in no respect whatsoever pertinent to efficient causality, of the things of nature.

Further, inasmuch as where *esse* is 'what is done' or 'the effect', the *esse* concerned has no existence in any way separate from the activity or action of doing or causing it, we may expect the causal agent concerned to explain the effect 'formally'. But what I mean by this word "formally" here is not what is meant when, for instance, Descartes says that (e.g.) a cause has to possess the properties of its effect formally (i.e. properly or in the same sense) or eminently, but needs quite different explanation. For we are not yet in a position to lay down that the effect must be like its cause (i.e. 'formally' like in the sense in which Descartes would use the word), even if only a deficient likeness. All that we can say is that the cause is or contains a principle (we could call it an Idea or Form) which is directed 'intentionally' towards the effect, intentionally both in some Brentano-like[4] sense of being directed towards an object and in the logicians' sense of not implying the existence of the object (since the causing is not a necessitated act). In this so-called formal causation we have part of the structure of final causation or teleology. However, we do not yet have it that the cause is an intellect which knows, understands, or appreciates its effect or that it not only is directed towards its effect as an end in its action but desires and rejoices in this effect as an end which, being known, is valued for its own sake.

(ii) We might have thought of arguing that if the first cause of the Universe contained Ideas 'intentionally' directed towards its effects in the mundane, temporal world, i.e. principles in virtue of which it was able to give rise to the range of temporal existence that there is, then there must be no distinction relevant to efficient causal explanation between this primary causal agent and these Ideas or principles, i.e. nothing to compromise the incompositeness or radical unity (*simplicitas*) of this prime cause. For it might have seemed that unless these principles were such as to involve a likeness of the effects to the Prime Cause, one would have an accidental coming together of aspects of God.

[4] I say 'Brentano-like' because his 'being mentally directed towards' and 'being teleologically directed towards' are two forms of 'being directed towards an object'. Michael Clark in 'Intentional Objects', pp. 123–60, gives a useful corrective to modern tendencies to use the term "intentional" as a technical term in logic roughly co-extensive with the term "referentially opaque", insisting on Brentano's connecting of it with the mental.

However, the fact that this coming together might be accidental so far as human conceptualisation is concerned does not show that it is accidental absolutely, since there may be conceptual connections and necessities which the human mind has not reached or cannot reach to recognize.[5] And the fact that this coming together might involve some priority of the prime cause to the Ideas from the standpoint of *non-efficient* types of explanation (of which, as we shall mark later, it is not clear that we have a complete tally) likewise does not show that the radical unity of the prime cause in the way relevant to *efficient* causal explanation (explained in Chapter V) is compromised. (It is noteworthy that in Duns Scotus, as in some forms of theistic neo-Platonism, the Ideas are not absolutely identical with God but identical only according to some relation or in a certain respect.[6])

Accordingly, we cannot, by this abstract argument from the unity of the prime cause with the principles in virtue of which it is able to act towards an end, conclude to its being personal.

All the time there is a temptation to leap prematurely to the conclusion that the cause of temporal existence is personal. This temptation must be sturdily resisted, firstly lest we violate the requirements for metaphysics to be critical and lose all right to being heard as well as any colour of plausibility, and secondly lest in our arbitrary and premature move we intrude some (arbitrarily latched onto) extra feature of the supposed personhood and personal agency and erect an idolatry which the theologian does right to spurn.

I say this because I am now going to mention some more attractive arguments which still need to be set aside as non-probative.

(iii) One of the features of the concept of personal agent which is most notable is that the mode of causality which it exemplifies is the richest of the modes of causality of which human experience gives us knowledge, and that it not only exhibits no deficiency relative to the other modes of causality we know of from experience, but also, as I have argued above, contains no root identifiable to us from

[5] Cf. Chapter VII n. 7.

[6] See É. Gilson, *Being and Some Philosophers*, p. 86.

which deficiency might proceed. Its explanatory richness is of a peculiar character inasmuch as it involves in itself, in inseparable union integrated together, formal, efficient, and final causality in relation to its effect, and, as we have seen, if the cause of temporal continuance as such be personal, it would envelop in the same integral whole the causing of material continuance as such as well. But, somewhat as in mathematics, the system of complex numbers, including all algebraic and negative numbers, is closed under a certain set of operations and their converse operations, and in this way complete, so from the standpoint of human understanding the concept of personal first cause of temporal existence achieves a certain closure in respect of all the types of explanatory inquiry known to us, and, in this respect, has a certain maximal completeness.

Therefore, at this point, the temptation to which I referred reaches an extreme point. Yet it is not yet established that a personal being could exist who was free from the limitations we are aware of in the case of human personal existence. Only the pre-critical rationalist can argue directly from the mere absence of formal contradiction between propositions stated in human concepts to real possibility. Nor is it yet established that the cause of temporal continuance, as such, must have the maximally explanatory role in respect of the Universe of temporal things, only that he or it must have a certain completeness in respect of efficient causal explanation.

Further, it is not established that the kinds of explanatory relation appreciated by man constitute a complete tally of kinds of explanatory relation so that reaching beyond the concept of "efficient cause" to that of "personal efficient cause" has not yet been shown to exclude non-deficiency absolutely, but only explanatory non-deficiency cognizable by man. In the Eastern Orthodox theological tradition it has been envisaged that within God there may be more than one type of non-efficient explanatory or 'causal' relation, not comprised within the Aristotelian list (i.e. not precisely material, formal, or final), and that the distinctions between the being or essence (*ousia*) and persons and between the being or essence and the so-called 'energies' of God arise from these.[7] We

[7] See Vladimir Lossky, *The Mystical Theology of the Eastern Church*, ch. 2–4, and Basil Krivosheine, 'The Ascetic and Theological Teaching of Gregory Palamas', ch. III.

may not feel obliged to espouse these Eastern Orthodox conceptions of the 'energies' or of the incomprehensible priority of *ousia* to 'energies', or this account of the incomprehensibility of the priority of the Persons of the Christian Trinity to its *ousia* and of the equal incomprehensibility of the different priority of the *ousia* to the Persons. We can only note with interest the recurrence in this exotic location of the agnosticism in regard to our capacity to appreciate the kinds of 'real' connection (connection involving an *aitia* or 'cause' in the exceedingly generalized Greek sense) occurring within nature as it presents itself to the mathematical physicist.

True, it should now be evident that just as there is no way of being more real than the real (being real in a different respect does not help), and no way of being more than good (but only richer ways of being good), so there is no way of being higher or more than personal, since the personal, as such, in concept, has no deficiencies wherein it might be surpassed: the self-reliance, autarchy, or individuality opted for by some represents a determination and restriction of personal opening out, and the relatedness associated in common opinion with the consummation of personhood is not a restriction unless the objects of relation are restricting. However, it is still not yet established that, although there is no way of surpassing the personal, there may be other ways in which things may be non-deficient *qua* efficient cause, and such that these ways do not involve them in deficiency in their own terms, i.e. *qua* whatever they are.

3. EARLIER APPROACHES TOWARDS A VALID ARGUMENT FROM CREATED PERSONS TO GOD'S PERSONHOOD

The starting point for our further investigations at this juncture must be the consideration that the different modes of agency and life, and among them the different modes of being a person (for in common opinion and in many past and present philosophies the human represents only one way of being a person amongst others), all represent different modes of existing, different ways of being a substance. Accordingly, whatever causes the existence of such things, whether their beginning out of nothing or their continuance, has to cause their existence in every positive respect or aspect in which they do exist. In traditional terms being a causal agent, living, and being a person are 'perfections', not here in the sense of being things of value worth aiming at, but here only in the sense of being ways of being without qualification, positively real. The cause of a thing which is in a certain way perfect (a way picked out by some concept such as "causal agent", "living being", or "person") must be a cause of that thing, in respect of every aspect whereby that thing is thus perfect. In other words, there is not one way of being real, but there are many, and efficient causation has to extend to all of them, not merely that exemplified in inanimate bodies.

It has been held that a cause must have 'formally' (i.e. in the proper or literal sense), or in a more eminent way, all the reality to be ascribed to its effect, or at least this holds where the cause is the complete efficient cause (not needing to rely upon or utilize anything else) of the effect. From this plausible principle, it at once follows that the cause of personal existence, or of the temporal continuance of a Universe including persons, must either be a person or have in a more eminent way all the reality which a person has. But there would seem to be no way in which a cause which was non-deficient from the point of view of efficient causation could be incommensurable with the personal, and yet contain in a more eminent way all the reality of the personal.

However, the principle that a cause must have formally or eminently all the reality of its effect, although indeed very plausible and utilized by Descartes and even by the empiricists (e.g. by Hume, when he considers that every simple idea must derive from an impression), is as yet quite unproven, besides involving an obscure or ill-explained notion of degrees of reality. We must in our next section discover what can be said to justify some such principle, while avoiding such notions as that of "degrees of reality". (It will be noted that, inasmuch as we have observed that it is part of the human concept of "personhood" that it does not exclude but even invites the envisaging of the possibility of a person whose knowledge and love is direct and perfectly engrossing of personal faculties, our argument is reminiscent of causal arguments to God from the existence of the idea of God, the existence of persons with an idea of God, a mystical experience, and the phenomena of conscience and recognition of the 'sublime'—supposing that these last are conceived of as intrinsically religious.)

In sum, although, if anyone is willing to concede such a principle as 'that only the personal can be unconditional or immediate cause of the existing of the personal' as a premiss, we will be able (in the way indicated in the preceding paragraph) to prove to him the existence of God on a basis somewhat wider than that appealed to by Descartes, inasmuch as it will be only to the existence of persons, not to their possession of a clear and distinct idea of God, that appeal will be made, nonetheless until someone makes such a concession we will have no argument.

The plausible causal principle here being appealed to stands, like other plausible causal principles, in need of some underpinning, preferably an underpinning which allows all these causal principles to be seen in some integrated relationship with each other.

4. OUR OWN ARGUMENT AND THE REFUTATION OF THE HUMEAN ANALYSIS OF THE CAUSAL RELATION

In order to identify what is rightly inspired in this last argument, we need to get behind it and consider the structure which it shares with all parallel causal arguments, and the defect in their presentation which leaves them all open to sceptical criticism. We shall at the same time be identifying the key defect in the classical empiricist and sceptical approach to causation itself.

It is natural in human thinking that causal explanation should proceed analytically, in the sense of proceeding by breaking a causal relation down into its supposedly discrete component parts, i.e. by 'analysis'.

That is, so far as possible we look for separate descriptions of the cause, the effect, and any intermediate there may be between cause and effect, and descriptions such as will make at least two of these into logically separate and distinct existences: if the intermediate is conceived of as the activity of causing itself, then an attempt is often made to envisage this in a way which makes it logically separable either from the cause or from the effect. And where we have a case of a productive agent producing another object distinct from itself, or producing a change of state or situation in an object distinct from itself, then the linguistic structure of human thought is such as to enforce the provision of descriptions of the nature of the cause and the nature of the effect which leave nothing impossible in formal logic or in imagination in supposing either one to exist without the other.

Further it comes about that the way the nature of the effect is conceived by human beings is such as to leave no formal or imaginative impossibility in supposing the effect to exist without there being any cause or activity of causing at all, and almost indiscriminate freedom in formal logic or imagination to suppose almost anything to be the cause[8] (the activity or the relation of 'causing' is in any case something stated, rather than sensorily imagined).

However, although our human concepts of "cause" (linguistic or verbalizable concepts) allow this analytical approach to causation to be envisaged and widely utilized, nonetheless it is not the case that the absence of any conceptual impossibility of a kind which can be identified as utilizing a combination of formal logic and considerations as to what is sensorily imaginable, constitutes the absence of any conceptual impossibility whatsoever, nor even the absence of any conceptual impossibility knowable to human beings.[9] The fact that something is 'descriptively possible', i.e. possible at the level of description (e.g. at the level of sciencefiction or fairytale), does not establish that it is conceptually coherent absolutely, or even conceptually coherent within the limits of what human beings cognize. In particular, in the case of the notion of "causation" it is demonstrable that the analytical approach (although it may have some usefulness within certain controlled or contrived contexts within certain carefully circumscribed types of human inquiry) is inappropriate, if the concept of "cause" is to have any application at all—which, as I have indicated, human knowledge of reality is not such as to leave open to real doubt.

Let me now set these considerations forth in greater detail.

In the particular case with which we are concerned, what I call the analytical approach has had extraordinary results.

In the first place, it is supposed that we can conceive of persons

[8] Hume in the *Treatise*, bk. I, pt. III, sect. XV, tells us 'Any thing may produce any thing'.

[9] See G. E. M. Anscombe, ' "Whatever has a Beginning of Existence must have a Cause": Hume's Argument Exposed'.

with temporal existence as existing, and each and every other kind of living and non-living causal agent as existing, without any of them having a cause of the continuance in existence involved in each case, despite the fact that they are all individually and collectively impotent to achieve this continuance in existence. We can readily picture any picturable thing without picturing its cause and there is no formal contradiction in supposing it to continue to exist without a cause, but neither of these facts does much to suggest that reliable continuance in existence can coherently be supposed to be without a cause. And if what is concerned is the existing of a thing, and if in fact this depends on a cause of continued existing over the whole period of continued existing, then this existing has no existence independent of the activity of causing it.

Moreover, in the second place, it is not just, as in other cases, that the activity of causing is accidentally specified with reference to something which can in fact exist even when the activity has ceased, but that there is no specifying of the activity concerned except precisely as the activity of causing this continued existing.

In the third place, we are asked to imagine by some that the activity of causing the continued existence of persons, of living things, and of other causal agents could be, in some way, accidental or external extras—independent of the inner nature of the cause acting or operating in this causal action.

Let us amplify this third observation.

The idea that a purely material entity might (logically might) be the logical subject of the activity of thinking is found in Locke, but seems to be one of the most mad excrescences of human thought. At least more modern thinkers have wished to identify, in the material or physical, structures of function and activity isomorphic with the structures supposed to be involved in thinking; that is, the attribution of thinking to the material is alleged to be grounded in some appropriate structures within the material realizations of the nature of the material, not foreign to it. To suppose that the principles of a thing's activity could be extrinsic to the thing or its nature is to generalize the absurdity of which Locke's *jeu d'esprit* is but a particular example. In general, in respect of each type of example we can appreciate that for something to be an internal cause (material, efficient, or otherwise) of something else it has to have a quality, structure, or constitution formally adapted to or reflected in that of which it is the cause: only when the relationship is purely extrinsic so that it can be some mechanism of Nature which secures the connection is it conceivable that such formal relation is absent; in other cases it is vital if the supposition of causal connection is not to be merely chimerical, an incoherent fantasy.

To suppose that causing or doing (as referred to in the adage 'the agent does what happens') can be accidental to the inner nature of a thing, or (if there is no distinction between a thing and its nature) accidental to a thing as it is in itself, is to assimilate causing or doing to 'being compatible with or allowing to happen by chance': to do this is clearly incoherent inasmuch as 'to be by chance' is precisely to arise within a setting of causal pre-conditions and acting natures (operative, active powers) without being, at least under the description given, caused or done by anything. Free action and natural (necessary) action are each of them expressive of the nature of the thing that acts. We have no basis in example or otherwise for doubting the necessity, if the notion of causal agency is appropriate to a certain being at all, that the action will be expressive of what the thing is in itself. That is, the suggestion that the principles of causal activity might be in any way external to or accidental to the inner being or nature of a thing, and that yet the thing should be described as the cause operative in the case considered, does not seem coherent.

We are now in a position to state how it is that Hume may be considered to have presented a refutation of his own position in respect of causation.

The concept of "cause" which our experience of the actual gives to us in a variety of types of example is such as to involve (i) some unity between a thing or its nature and any active power in which this nature may be realized; (ii) some unity between a thing with its nature and active power and any action in which its nature and active power are expressed or realized; and (iii) some unity between any such action, to which (of course) the agent or thing is internal, and its immediate effect. But this runs contrary to what Hume's arguments suggest.

If we know the concept of "cause" to have application, then we know the validity (within the range of application which second-order considerations, drawing upon our knowledge of the real, may be needed to make clear) of the presuppositions of the applications concerned. And if we know these things, then we know that any argument suggesting that the concept of "cause" lacks application in the commonly received sense must contain a *reductio ad falsum*, so that its premisses or principles must be mistaken or misconceived. Now, we know that the concept of "cause" as understood in the common way has application directly to things in their relations in the world and not merely indirectly in virtue of a relation between our thoughts about things. But Hume's arguments purport to prove the opposite. Accordingly, Hume may appropriately be considered to have provided us with a refutation of the analytical or atomistic approach by exhibiting the absurdities or falsehoods which result from his reliance upon it.

These considerations enable us to give a better statement to the type of argument pointed towards in Section 3—albeit still requiring the possibility of support along such lines as we suggest in Section 5.

In brief, in each case it will have to be admitted that 'there is either in the nature of the cause or in Nature whatever it takes for it to be non-chimerical to suppose the effect to be indeed the effect of the cause concerned': and this involves that, when the cause is internal to what is caused, what is caused must be, in the respect in which it is caused, like its cause. But we have not proceeded, like the mediaevals, by boldly presuming a principle that effects are like their causes. Rather we have explored the consequences of the rejection of the Humean basis of approach, *viz.* the formal adaptation of a thing's nature to its action and to any effect to which it is internal, consequences required if the suppositions that the action is the action of the thing or that the cause is thus internal are not to be merely chimerical.

So, for it to be not merely chimerical to suppose that the incomposite cause of the continuance of temporal things is what gives, unconditionally and unqualifiedly, continuance of existence to personal beings (of which we know the existence of the human), this cause must have as an aspect of its nature the formal character reflected in the existence of these effects, *viz.* personal existence. Otherwise, it is as if persons are given existence, as we say, 'out of a hat': the problem is not their being given existence without reliance upon matter, but their being given existence without there being in the nature of their cause of existence anything connected with 'what they are' in respect of being personal. Such is the principle of my argument.

It will be noted here that my argument depends upon insisting that 'being personal' must have something answering to it in its cause of continuance which cannot be other than this cause's being personal, while rejecting the seemingly parallel argument that 'being human' must have something answering to it in its cause of continuance which cannot be other than this cause's being human: it suffices that the cause of human continuance be personal—it need not be human.

Two things are involved here: (i) the irreducibility of the personal to the non-personal of which I will not give further explanation in this book beyond that in Chapter III; and (ii) the distinction between proper genera of existent such as 'human being' of which the natural sciences give an account and positive trans-generic properties such as 'being personal', of which they do not—but of which, rather, they presuppose an account. 'The personal' is rather a mode of existence than a proper genus of existent.

Because being personal has to do with a distinction in modes of agency or causation and therefore with a distinction in modes of existence, as being human does not as such have, what is involved in continuing persons in existence is as different from what is involved in continuing non-persons in existence as it is from not continuing anything in existence: what natural explanation requires in accounting for genera presents a measurable task whereas differences in modes of existence are without measure or account. (One thing is not more real than another, but different things are real in irreducibly different ways, ways which fall into a hierarchy only in the limited sense explained in Section 5.)

Our earlier remarks enable us to indicate a second and related argument. The principles whereby the first cause of continuance in the world is able to cause existence as such (i.e. to cause continued existence unqualifiedly and unconditionally) and whereby it is able to cause respectively each of the modes of existence of which we know, must be intrinsic to, and therefore identical with, its radically one nature. Within this perspective, we have now to bring back to mind the intimate internality of the causal action of the First Cause whereby it gives continued existence to a nature and whatever it is that is involved in that nature's being actual. A nature or, rather, a thing in its nature does not exist merely instantaneously or, as it were, lie about the world statically, as if its characteristic actions were an accidental excrescence external to its nature's being actual. Rather, the nature of time, and the mode of existence of causal agents within time, is such that the actuality of a nature, far from being a static datum, has its tendency to realization in characteristic action, according to its nature, intrinsic to it. As we expressed it in Chapter VI, the existence which God upholds is not just a state but an actuality or something 'active' (we might say, a 'being active').

Therefore, for the causal action of the First Cause to be internal and intimate to a thing involves its being internal and intimate to the existence of that thing as dynamic, expressed in its characteristic actions. Accordingly, the intrinsicness to the nature of the First Cause of the principles of its causal action, an action which, by its nature, extends to this underpinning of the characteristic actions of temporal natures, therefore involves the intrinsicness to the nature of the First Cause of being personal in order that it should have the capacity to sustain human beings in those of their actions which are personal.

That is, personhood, as what gives their special character to personal temporal actions as against other temporal actions, has to be intrinsic to the nature of the first cause in order for the first cause, as such, whatever it is, to be immediate or immanent to these actions, which it has to be in order to be immanent to the existence and the natures of the temporal persons whose actions they are.

However, these types of argument should not be considered in isolation, as if, were they incapable of being supported in the way suggested in Section 5, they would still be appropriate to the shape of human thinking and experience. Considered in isolation, the arguments we have just presented appear merely as those arguments which are thrown up into a false pre-eminence by the modern preoccupation with the subject of thinking and of experience and with subjectivity. The modern propensity has been to regard as more certain the phenomena or alleged phenomena of self-consciousness, with its supposed reflective, self-critical, mystical, moral, and *praxis*-orientated modes, than the Universe of living and non-living beings within which human subjects or persons exist in relation and which is constitutive of the human mode of personal existence.

That is, the need for argument such as Section 5 presents arises from the need to disambiguate the philosophical conception of how human beings are persons. If a Cartesian standpoint is adopted, the support suggested in Section 5 becomes alien, but then we are left in the situation of uncritical rationalism.

5. SECOND-ORDER SUPPORTIVE ARGUMENT

Earlier in this book, in Chapters I to VI, we presented an argument to the existence of a cause which gave continuance in existence unconditionally and unqualifiedly, subject to the reserve that certain epistemological queries had to be dealt with—although the final drawing together of the argument will only be in our concluding chapter, after our consideration of the oneness of such a being. Then, in Chapter VII, in effect, we reached the position that there was a valid line of first-order argument, but that here, as elsewhere in philosophy, positions neither needed nor ought to be held uncritically. That is, it was of crucial importance that the argument stood in the context of the possibility of certain second-order considerations from the philosophy of language about the spread of uses and interrelations between uses of various causal expressions and between those uses of modal expressions like "possible" and "contingent" which are related to causation.

Our situation in relation to the personhood of such a first or underlying cause is similar. In Sections 3 and 4 we arrived at a statement of what we judged to be a valid first-order argument to the personhood of any first cause. But in this argument we argued from the continuance in existence of certain parts of the Universe, *viz.* human persons, to the personal character of any underlying cause of the whole. And the question arises at a second-order level as to whether we were making a licit use of the concept of "causal agent" in reasoning as we did. Therefore again the first-order argument may indeed stand, but only in the context of some second-order argument from the philosophy of language.

The main second-order argument to be adduced runs upon different lines. We have to ask ourselves how the concept of "causal agent" is given to us. By this I do not mean "how" in the sense of genetically, by what process or kind of learning, to which such answers might be offered as 'from experience' or 'by the exercise of intellectual intuition'. Nor do I mean "how" in the sense of under what epistemological conditions, as it were, through what restricting or distorting spectacles or what avenue of approach to knowledge. Rather I mean: in what shape is the concept given to us? That is, I am in a sense asking simply what the concept given to us is - given to us so that there is indeed no doubt that we have it and no doubt that it (in this shape) does have application (rather than its being any other concept of a more indeterminate kind under the name "cause" that is given to us, in our experience of the actual, as having application).

The shape of this polymorphous concept[10] as it is given to us is such as to incorporate a hierarchy of modes of causality internally related to each other. The question is as to the nature of this hierarchy. Upon this our further argument about the personhood of any first cause will turn.

(a) Preliminary remarks

If we consider not the concept itself, but its application alone, it is evident that we find variety among causes: human persons, a range of kinds of sub-human higher animals, and the vast range of lower animals, plants, fungi, bacteria, viruses, and non-living things. On reflection, it is implausible that there should be any conceptual obstacle to the existence of non-human intellectual and linguistic animals and a wide variety of possible, but not actual, other forms of non-intellectual living and non-living bodily existence.

Leaving aside the case of the existence of God, the subject of concern in this book, it is not certain a priori that it is possible for there to be intellectual beings higher than any linguistic animal, beings like the angels, the devils, and the intelligences supposed to occupy the celestial bodies and move the spheres, which have been at different times supposed to exist. Certainly there is no obstacle from formal logic, or from considerations of what violates the imagination, to conceding the existence of such beings. But apart from argument from actuality (e.g. on the basis of some Revelation) it does not seem that we have a priori knowledge of the absence of just any kind of conceptual incoherence in the supposition of the existence of such beings. (The limited character of our knowledge of conceptual possibility was explained in Chapter VII.) Indeed, the objections made by Aquinas, Ryle, Strawson, and many others against dualism in the case of man would apply to some accounts of

[10] Cf. Chapter III, Section 2, especially (*c*).

the way angels, devils, and celestial intelligences move bodies and know about things in the Universe, and do naturally invite a suspicion that these types of being may be conceptually impossible, a suspicion only removable by argument from actuality.

However, our question is not about what kinds of living and/or intellectual beings exist, or may be supposed in some story of the Universe to exist, but of the shape of the concept of "cause".

(b) *The main clue*

It is noteworthy that physical or material causal agency is integrated within the agency exercised by living, including personal, bodily beings. Likewise, the elements of agency involved in growth and regeneration, exemplified in plants, are exemplified also in animals. Thirdly, the kinds of agency involved in touching and motor response to being touched are exemplified, not only in lower, but in all higher animals. Fourthly, whatever kind of agency is involved in the complex which embraces perception, imagination, and memory arising out of perception, desire, and action, both explorative or perception-seeking or directed to an extraneous end, exhibited in pre-linguistic animals, appears also in man.

However, the higher forms of life, in each case, include aspects which do not, in concept, incorporate the lower aspects. Thus, the knowledge, memory, and desire we attribute to higher animals exemplify this. And, at least in concept, the causal agency of higher animals extends to the causation of some "inner" states of imagining, desire, and enjoyment, e.g. in dreaming and daydreaming, and perhaps this is a necessary pre-condition of play. And in the case of man the examples are multifarious and better known to us. Accordingly, at least as the concepts of "life" and "personhood" are given to us in our actual experience of reality, they are not given to us in such a way that their causal aspect is limited to exercises which incorporate a material or physical aspect.

Therefore, upon reflective examination, we should not conclude that in the way the concept of "cause" is given to us each lower and less rich mode of causality is integral to each higher mode of causality, so that we have a hierarchy closed in within the physical, of which the summit is exemplified in language-using animals such as human beings—plausible though such a suggestion might have seemed and attractive therefore to the materialist sceptic. Rather, what we have discovered is that each lower kind of causality is such that in principle it could be integrated within a higher mode. And we find that we can form no conception of the mode of causal agency of a lowly kind which could not in another instance of the same highly general mode be integrated within a higher type, the material or physical within the living and both within the personal.

This is the origin of the common accusation made by followers of Hume against believers in real causal action on the part of purely material entities that these believers are superstitious animists at heart, ascribing to one billiard ball something analogous to authority over another, and attributing their conformity to physical law to the subservience involved in obeying a human law: 'animism', if not 'mere metaphor', they say.[11] In fact, the use of notions of "cause" and "necessity" in connection with the interactions between inanimate things precisely serves the anti-animist purpose of excluding the supposition that the effect follows or the law is obeyed as an exercise of free will: to say that the effect will certainly follow but 'could not-follow' inevitably imports the connotation of will.[12] And it is simply false, and without a shred of basis in reason, to suppose that causal action as such involves 'animism' or, by yet wilder supposition, personification: these are floundering suggestions, born of the difficulties resulting from the mistake of first swallowing the Humean or any other atomistic analysis of causation. In any case the idea that the exercise of authority by a master over his servants is an example of causal action, let alone a paradigm of it, is far-fetched.

I say that my considerations reveal the origin of this mistake on the part of the followers of Hume because beneath their error and confusion there is, I believe, truth. It is my contention that the notion of a mode of causal agency that could not be subsumed in any instance within an exercise of that mode of causality characteristic of the living and personal, but is, as it were, totally immiscible with it, as if foreign to the same world order, is not coherent. In this way of thinking the different uses of the term

[11] See A. J. Ayer in *The Foundations of Empirical Knowledge*, IV, 18, pp. 183–99, on causation as animistic and involving the attribution of 'authority' to causes.

[12] There is no way of accounting for human knowledge of the future as certain or reliable without this knowledge being in virtue of knowledge of natural necessity or likelihood except in terms of the knowledge human beings have of their own future intentional actions, cf. David Braine, 'Varieties of Necessity', pp. 147–50.

"causal agent", i.e. the different modes of causality, could, as it were, lie together as a heterogeneous heap unrelated to each other in any systematic way: what the term might turn out to be applied to would be indefinite or potentially infinite (because completely unrestricted by ontological, epistemological, or any other considerations, and thus totally undefined). Such a view would be credible (if at all) only if what unified what we call causal agents was the applicability of the word "cause" in some univocal, henomorphous, logically single sense or meaning. But this view of the concept of "cause" we refuted in Chapter III.

We have here identified the root of what has been regarded as a baseless prejudice: that denying life and personhood to the First Cause in itself inescapably involves the distancing of the effect from the cause, forcing upon us a model appropriate only to the material or physical. The view that the impersonal cannot be thought of in an open undetermined way so as to transcend, or be incommensurable with, the personal, but only in a sub-personal physicalistic way, has, we have shown, a proper intellectual rooting, and is not baseless.

Let us now make an ordered statement of the second-order argument to which we have here found the clue.

(c) Analytical statement of supportive second-order argument

Our argument, in effect, proceeds in stages as follows.

I. The concept of "causal agent" as given to man in the act of articulating his experience in speech is a polymorphous concept.

That is,

(i) This concept is not univocal, henomorphous, or definable in terms of any mechanically applicable set of criteria, simple or complex,

and

(ii) Properly or literally (not metaphorically or in a transferred or secondary sense) used of each of the kinds of thing of which it is used.

II. There must be some relationship, hierarchical or otherwise, between the different uses of the concept "causal agent", i.e.

different modes of causal agency, for it to be possible for I(ii) to hold despite I(i).

III. We know of certain modes of causal agency, e.g. material object, plant, animal, and human modes of agency. We are aware that material object causation is an element integrated within some exercises of plant-type agency, plant-type agency integrated in some exercises of animal-type agency, and animal-type agency integrated in some exercises of fully human agency. We are also aware that the life of higher animals includes aspects which, even if they depend in reality upon exercises of less rich modes of causality, include elements which do not, in concept, incorporate these less rich or lower modes of causality. Likewise, human life includes exercises of causality which do not incorporate, in concept, lesser modes of causality. We can therefore give sense to the notion of a lower mode of causal agency which within the life or existence of some being is 'subsumed' within exercises of higher modes, these higher modes having other exercises not involving the lower modes concerned. (This is intended as a contextual definition of the notion of being 'subsumed'.)

IV. Exercises of material, plant, animal, and human causal agency do occur, and we know them as different modes of causal agency in the sense indicated under I. The notion of a mode of causal agency which is so unrelated to lower modes of causal agency that none of these lower modes, not even the personal, can be subsumed within it is not coherent.

As against IV, it might have been supposed that, as material agency is subsumed within plant, plant within animal, animal within human, so human causal agency was imaginably subsumable within some other kind, this other kind within yet some other kind and so on, imaginably unendingly, so that the First Cause exercised a mode of causality within which no lower mode was subsumable.

But apart from Revelation, if Revelation has been given, man has no genuine knowledge of the real possibility or any mode of causality intermediate between the human and whatever mode of causality is exercised by a first cause. And man has no knowledge of the freedom from conceptual incoherence of any fancied concepts of such allegedly "imaginable" intermediate kinds of mode of causal agency. So the conceptual coherence of supposing an infinite regress has not been established: only it has not yet quite been disproved.

However, disproof of this wild supposition lies within our reach.

What we have been offered includes not a shred of constructive argument to the effect that there is or could be an infinite hierarchy of modes of causality above the human, an infinite hierarchy all of which supposedly lies below some transcending mode exercised by a first cause. We are offered not a shred of argument in theology that, e.g. the hierarchy of angels is infinite. And unless a superhuman mode of causality is actual, and known to be such in the context of Revelation or otherwise, there is no constructive knowledge of its conceptual possibility. But, we must say at this point, no use of the concept of "cause" which is not constructively based could be supposed to be grounded in knowledge of reality.

Therefore, matters stand as stated in IV: there is no coherence or content to a use of a concept of a mode of causal agency within which none of the modes of causal agency which we experience is subsumable.

(d) False avenues of approach to the question of the centrality of personhood: no valid a priori argument from the concept of "person"

In order to argue that whatever gives existence or continuance to the personal must be personal or higher than the personal (or, as some would say, more than personal or supra-personal) it might seem attractive to argue from the internal features of the concept of "person" itself. However, I believe such an approach to involve the uncritical rationalism I attacked earlier: it turns out to involve the false move from the mere 'descriptive' openness of the concept of "person" to an unqualified conceptual openness and, with it, a false move from conceptual to real possibility. It is for this reason that I myself adopt the more indirect approach outlined earlier in this section in order to clinch the dispute, an approach more appropriate to human beings whose immediate experience is primarily of a world of human persons, other animals, plants and material things, to pick out some of its discrete elements. I now outline the attractive rationalist-style argument which I eschew and detail why it fails.

According to this latter style of argument, the concept of "personal agency", because of its connections with teleology, intellectuality, and creativity, creativity extending over the whole range of what the intellect conceives of as open to being created, while being compatible with material modes of causation since these are evidently subsumable within the personal, has a peculiar character. It would embody a special openness, so that there would be no restriction or limitation on what a person may be agent of, at the same time as exhibiting a connectedness with the actually experienced modes of causality instanced above. This special openness would preclude personal agency from being subsumed within any higher mode of causal agency. For it to be thus subsumed would be for there to be some exercises of this higher mode of causality which could not be conceived of as exercises of personal agency. But this the concept of personal agency would preclude.

Yet, however attractive this line of thought may seem, nonetheless I would incline to think that this line of argument is a priori in the objectionable sense of not being constructively based on knowledge of reality, but only upon mere supposition or imagination in regard to what so-called 'persons' could do. From the imaginability or supposability of a person's (e.g.) causing unconditionally and unqualifiedly the continuance in existence of other things the genuine conceptual coherence of a person's achieving such causal action indeed does not follow, let alone its real possibility. True, granted that something has achieved and continues to achieve such causing in fact, we are not occupied in inquiring in a void what so-called 'persons' can do, as if we were asking whether a person could do this or that which nothing at all in fact does. Rather, granted that nothing in the concept of "person" excludes, so far as we know, a first cause from being a person or persons (i.e. the concept of "person" is in this way descriptively open), the question is as to whether any first cause is such, and indeed whether it could be anything else.

Nonetheless it remains in my view that the correct basis for concluding this is not from any consideration a priori of the concept of person such as this one, but by an a posteriori approach from the consideration that human beings are persons, the human mode of causality being subsumed within the personal, and from the need to insist that there is no content in the use of the concept of causal agent in regard to a being supposed to exercise a kind of causal agency within which none of the modes of causality experienced is subsumable.

Because Revealed theology proffers to us a system of models for thinking about persons, God, and the world which may seem to have a certain coherence, we are tempted to suppose that in some deep or implicit way the possibility of the instantiation of these models and associated concepts is empirically and/or rationally given even independently of Revelation. Thus tempted, we then read whatever we will into our philosophy and produce all our desired conclusions with a flourish, while in fact merely expressing a surreptitiously introduced uncritical rationalism.

(e) *Informal restatement of our overall argument*

Our considerations in these three Sections, 3–5, give a rationale for the constant impulse to suppose that the act of causing or of maintaining existence, involves such a zest or *élan* as to force upon us the thought that it is expressive of life, and, if of life, then also of personhood, since the life is evidently immaterial. We feel that only in the activities of living beings can the term or end of that which is causally active (i.e. the object or effect which makes the activity to be the kind of activity it is) be itself part of the life or a realization of the being of the cause. In non-living things something can be internal to the realization of the being, or nature, of a cause only by being in some physically defined sense part of it, as emanations may be conceived of as in a way parts of their source. In living things we know that some activities, e.g. imagining and thinking, involve internal realizations of the thing's life. In creation, the being of the creature has to be, from the point of view of efficient causality, utterly distinct from the being of the creator; but the being of the creature has no existence, independent of the causing of its being or of the causing of its temporal continuing, which is expressive of the nature of the creator.

We can form no conceptual model of this causing which does not bring in life and personhood. The considerations I have offered in this section suggest that this is not *faute de mieux*, but entirely justified by a fuller reflection on the notion of cause itself.

This then is in my view the clinching argument which exhibits the introduction of the concept of "personhood" into the centre of general metaphysics and cosmology as fully acceptable in a 'critical' philosophy. Earlier argument had already indicated (we will rehearse it again in our concluding chapter) that the cause of existence as such, i.e. of temporal continuance as such, must be incomposite, underivative, and immediate to the temporal existents to which it gives existence; there was no room for intermediates or regresses, let alone infinite regresses. We have now shown in this chapter why such a cause must be personal in nature.

IX

Aspects of God's Power and Personhood: The Frontiers of Metaphysics

What we have established is the existence of a First Cause, the well-spring of all positive existence, whose own existence and uniqueness are intrinsic to Him in the sense that there is no explanation for them outside Himself and no room for such explanation. Therewith we established the negative and relational attributes associated with these properties, and whatever belongs to God as being personal, e.g. life, intellect, and will, along with the intentional character of His exercise of causal agency in giving existence.

But the connotations of the word "God", and therefore what is required in a 'proof' of God's existence, go far beyond this.

In the first place, we need to give the conceptions of God's power and personhood further positive content, most clearly in respect of omnipotence and perfections of intellect and goodness, if we are to be able to regard Him as holy and indeed alone proper object of unqualified worship and abasement.

In the second place, we need to do this in such a way as to avoid the anthropomorphism invited by thinking of God as personal. For, although we have built into our approach entirely adequate safeguards against God's being conceived to exist in the way that His creatures exist by means of our exposition of His self-existence and incompositeness, it is not clear that we yet have any adequate safeguards against His being conceived as a person in the same way as we are conceived of as persons. As a result, it might seem that, through the concept of "person", a backdoor has been created for the introduction of finite models, since in all our thinking we are apt to model God's intellect on our intellect and God's goodness on our goodness. Until the problem of giving further positive content to our conception of God's personhood and yet avoiding anthropomorphism is resolved, the conception of God's personhood remains, indeed with the openness we have vaunted, but in almost total indetermination.

We shall discover that a careful philosophical approach, marked by the a posteriori character invited by the consideration of God as cause of existence and the cause of the character of what exists (Section 1) and by the two aspects of infinity, one negative but the other positive, implied by the concept of God as "incomposite" (Section 2), is indispensable if one is to avoid a rampant anthropomorphism. That is, it is necessary in order to prevent any Revealed theology from degenerating into idolatry.

Within this framework we shall indicate how the concepts of God's holiness (Section 3) and goodness (Section 4) are to be approached, opening the way to showing how the insistence on an a posteriori approach in philosophy to determining the content of the concepts of "person" and "good" as applied to God marries very well with the requirement of Revealed theology that their ultimate lineaments be set by Revelation (Section 5). In the consideration of the mode of God's goodness all the problems of anthropomorphism are brought together with those of, on the one hand, the problem of evil and, on the other hand, the status of the doctrine of the Trinity or other highly 'personalized' presentations of God, such as the Jewish or the Muslim.

But it remains that any development whatsoever of the concepts of "personhood" and "goodness" in regard to God, even within Revealed theology, has to remain true to the insight that God alone exists in such a way as to be able to communicate existence so as to be thus the very well-spring of existence. And at the same time it must remain within the constraints set by this unique mode of existence—that is by His incompositeness—to escape the pit and folly of anthropomorphism. These key metaphysical elements are internal to theology itself, even if the orders of presentation, knowledge, and understanding may differ as between philosophy and Revealed theology (Section 6).

In this chapter my design is to indicate schematically how various problem areas left by the earlier discussions in this book should be approached. The conception of God as the well-spring of existence provides, I believe, the key to the solution of all these problems. But it is not for this book to deal with all the problems of the conceptual analysis of activity (with its 'object-directedness' and its 'intensity'), of the theories of explanation and evaluation, of phenomenology and cross-cultural studies, and of Revealed theology that open out—but only briefly and provisionally to indicate how a path is to be cut through the maze.

1. THE BASIS FOR AN A POSTERIORI PHILOSOPHICAL APPROACH TO ANY FURTHER POSITIVE ATTRIBUTIONS TO GOD

(a) *The problematic of any metaphysics proposing further positive content to the concept of God*

We have reached a posteriori the conception of God as the well-spring of existence, the cause at each juncture of the continuance in existence of things beyond that juncture, and therefore, if there was a first juncture or set of junctures, the cause of the inception of temporal existence from nothing. We have considered what is involved in different modes of causal agency and existence, and have concluded that this well-spring of existence must be personal, and therewith living and exercising knowledge and intention in His giving of continued existence, not just cause but author of temporal existence as it unfolds.

In all this we have been careful to proceed in a way which is 'critical' as well as a posteriori, exhibiting the second-order controls that come into play in regard to the application of the concept "cause" in Chapter VI, 3, and Chapter VII, 1(*e*) and 2, and in regard to that of the concept of "person" in Chapter VIII, 5. That is, we have avoided at any stage latching onto some positive concept of God and then proceeding uncritically and a priori to 'develop' this concept, 'proving' new positive notes of God's nature on the basis of the earlier note.

(b) *False or fragmentary approaches*

What we have to avoid is a cheat in which under the cover of introducing (*a*) negative qualifications on any suggested attribute, and negative attributes, cunningly working on the connotations of words like "perfect" and "infinite", and (*b*) causal arguments requiring the pre-possession in God of whatever it takes to support the excellences to be found within creation, including presumptions based on the realization that this 'support' is intentional, it is presumed to have been proved that God Himself possesses these excellences in some positive way of which the created cases provide a model so that words like "infinite" and "perfect" come to imply something positive about God's way of exemplifying these excellences. For instance, the absence of qualification in the way in which words like "know" and "good" apply to God is construed as implying that, to use metaphors derived from the senses, He has perfect 'sight' and 'grasp' of Himself and that, on the presumption that His personhood is adequately like the human, He has perfect 'charity' and 'justice'.

It has not yet been shown either that these excellences of creatures are of an unlimitedly trans-generic character, free of any internal feature debarring them from applying to the infinite, or even that at least they must be specific forms of such trans-generic properties and that the relevant trans-generic properties are indeed 'positive', and not merely formal or negative: only if they are at the same time both 'positive' and thus general can they properly be called 'perfections' in the restricted technical sense. As we have presented the problematic of a metaphysical approach to theism and the conditions of its being critical, it is essential that what 'positive' attributes are exemplified in God is something to be determined a posteriori upon reflection on actual experience: for each attribute which introduces fresh or richer content, we need some further consideration, reflection, or argument.

It will be helpful to illustrate these remarks in detail, striving to reach as far as we may on the basis already stated in order thus to exhibit the limitations of any approach proceeding by what I have called 'development'.

It is important in the first place to note that in our earlier presentation, in Chapter VIII, Section 1, the concept of "person" was a unitary integrated one, involving the notes of "life", "intellect" and "will", with intellect involving both self-knowledge and knowledge of action. It was not a concept based solely on some one of these features so that the others readied by 'development'. Therefore, if self-knowledge and the integration of self-reflection and self-knowledge with activity involving enjoyment and/or desire is integral to the concept of "person", then something about God's self-comprehension would indeed seem to follow. And, since reliance upon the metaphors of 'sight' and 'grasp' seems indispensable to the description of the ground and fruits of thought or judgement in the human case, it is unsurprising that to refuse to use these metaphors of God would be to deny (not just to qualify) the positive content of the attribution of intellect.

But this type of example exhibits how we have to proceed warily.

An account of the incomprehensibility of God to any finite

intellect, and at the same time for some account at least of some negative aspects of God's own self-comprehension, waits upon some general consideration of the notion of "comprehension". But the groundwork for such an account should be evident from what we have already said. Part of the function of using the metaphor of sight in regard to God's self-comprehension is the negative one of excluding reduction to anything verbalizable, i.e. of insisting upon His ineffability. Since God's upholding of things is intentional, we know that what is involved in His understanding Himself is richer than what is involved in understanding the workings of Nature and their basis in the constitution of things and understanding this through and through, but we can form no proper model either of the content or of the mode either of His self-knowledge and self-understanding or of His knowledge and understanding of the world.

A more general basis for a proof of this incomprehensibility of God is provided by the consideration that God is incomposite.

Obviously, since anything incomposite must be non-bodily, non-spatial, and non-temporal, it must be unimaginable and it must be impossible to form any model of its mode or modes of existence and action. But, since its nature is nothing other than itself and so is an individual actuality rather than anything quasi-abstract, the barrier to comprehension is yet more deeply set: any mode of comprehension of which we can form a model involves representing what is modelled or conceptualised in such a way as to carry the implication that what is modelled is something of a nature which may not be exemplified, i.e. is of a nature which is 'quasi-abstract' or general.

However, it does not belong to this book to develop a theory of comprehension, either by the human intellect or of any more general kind, but only to uncover the point of take-off for any positive attribution to God, clearing away false approaches.

True, the divine intellect will be free from the deficiencies of any particular finite way of being intellectual, as well as free of any deficiencies or limitations in its own way of being intellectual. True also, it can be argued, it will be free of any weakness generative of deficiency within creation. For, although individuals may fall short of the good which belongs to their specific natures, this is not a falling short of the general good of Nature inasmuch as it may fall within the general system 'intended' by Nature, so that there is no obviously cogent contrary argument. And, at least if we are willing to prescind from certain difficulties about omnipotence connected with the distinction between creating from nothing and upholding in existence, it would seem evident that, since all nature is within its sphere of action, there is no way error can enter into its knowledge or action: therefore what it thus gives existence to must be good in all its positive or real aspects because of its intention or will. However, all this does little to ground any positive attribution of comprehension, let alone comprehension in unqualified completeness or infinity, to God.

If God's creative and upholding activity because it is intentional has to proceed from His own self-understanding, this does not show anything richer in Him than lies within the scope of possibilities (known a posteriori) within creation unless something more is involved in being the cause, and having the power to cause, existence as an actuality. But that this involves anything more has not yet been demonstrated. As we shall discover later, this problem is brought to the fore even more by the consideration of the way in which God is good than by the consideration of how He is intellectual. That is, it has not been at this stage by this kind of argument shown that God has knowledge and understanding or has goodness in a style incommensurate with the styles exemplified in created persons.

Or, to state the matter in a way more exactly to the present point, we have not yet established that, because the mode of God's existence is different from that of creatures, the mode of His personhood and the modes of its realization must be different also and in a related way.

But, in order to unpick the problems, we need to approach matters in order, beginning by exploring further the significance of God's role as the well-spring of existence.

(c) Our approach to the problem: the clue provided by the concept of God as the cause of positive existence

The concept of God as the well-spring of existence, doing nothing except intentionally, is a fertile one. And this is the key to the resolution of the problems just raised, along with a further exploration of the significance of God's 'infinity'.

The strategy of arguing, as in Section 4 of Chapter VIII, that the

First Cause needs to have a certain character (*viz.* personhood) in order to maintain things of that character (i.e. persons) in active existence can be generalized to apply to other cases of the maintaining of things of a certain character in existence. In brief, as suggested in Chapter VI, the First Cause is immediate to the causation of all positive existing whatsoever and from the character of what exists in its positive aspects'inference may be made, with appropriate provisos, to the character of the First Cause. We need now to consider in general how such inference is possible, leaving the full consideration of 'infinity' to Section 2.

In considering the case of personhood in Section 4, we took special care to examine the primary case of an inference from the mundane to properties of the First Cause more contentful in concept than that of real existence. There we showed that, because personhood was not a limiting property tied to the members of any proper genus such as those the natural sciences take cognizance of, but rather encapsulated the positive distinguishing aspect of a mode of existence in such a way as to be open to application to something transcending the temporal or, in common terms, not finite in nature, the upholding of persons in existence, that is dynamic or 'active' existence, required nothing less or other than something personal. We took this special care, drawing in our argument upon the feature that personhood picks out a mode of substantial existence, because of the strategic importance (*a*) of personhood itself in any development of a metaphysical approach to theism and (*b*) of establishing that something beyond substance and causality was accessible to such an approach.

Because the case of personhood was of such strategic importance, we not only restricted the basis of our argument so that we could draw upon the very special logical situation of modes of substantial existence, but also in Section 5 of Chapter VIII took peculiar pains to exhibit the critical character of our approach. In now attempting to widen the basis of our approach in order to establish other attributes of God, it does not belong to this book to proceed with the same detailed care and precision, but only to indicate the schema or general shape of the approach required.

In order to establish the basis for such a more general type of argument of which we can view this case of personhood as a special case, we need to consider what it is for a mode of existence as a substance, for the possession of a quality, for the existence of a relation, or for an exercise of causality or passivity (undergoing or experience) to be something 'positive'.

In Chapter V, we came to the view that what it is for an 'attribute' to be 'positive' was for the having of the attribute to be an example of 'positive existence', and that was for it to be an example or facet of existing as an act or actuality. It is, then, *qua* way of existing that an attribute counts as 'positive' at all and it would seem necessary that it be always something pertaining to it *qua* way of existing that makes it positive, rather than something limitedly generic. So, what makes a more specific attribute 'positive' must be its being a specific form of a trans-generic property which is positive, whose 'positiveness' consists in its being as such a respect in which things exist more richly, a facet or aspect of richness in existence. Therefore, for instance, what makes the excellences of the various genera and species of being count as 'positive' in the sense with which we are concerned is not their being evaluated as 'excellences'—such evaluation, *qua* evaluation, is consequential—but their being specific forms of trans-generic positive attributes.

It is at this stage that we introduce an argument of the same pattern that we relied upon in Section 4 of Chapter VIII and urge that, in order for the First Cause to be immediate and immanent to things whose existence, nature, and actions are marked with a positive character of this kind, and immediate to them precisely as unconditional cause of continuance, this First Cause must have this character intrinsic to its own nature. The continuance of things, not only in the natures they have because they are of the natural kinds they are, but also in their secondary properties, e.g., in the case of human beings, their virtues and vices, areas of knowledge and ignorance, etc., requires the maintenance of these states not just as static but as dynamic. That is, it is not only substantial existence which requires maintenance not just as a fact or state but as an activity or actuality, as we argued in Chapter VI, but also existence in respect of so-called 'accidents' such as these attributes, when positive, of mind and will. God does not in general perform our episodic acts, but He remains as immediate upholding cause of all positive dispositions expressed in them (where there is no positive disposition there is the absence of it, some privation or negative state; God does not need separately to uphold privations or negative states since it suffices that He upholds their contexts, defined by more limited positive dispositions).

In brief, what makes the having of an attribute positive is its being an act of existing. Accordingly, what is required in its cause or causes is required in them *qua* causes of existence. The cause of existence has to have such a character in order to maintain things in existence with such a character. It must have in its nature what is required to give existence of this kind: but there is no way of its containing the richness or active power necessary for it to be non-chimerical for it to be the basis of the existence of things with the attribute X other than by being X, because of X's trans-generic character not limited to the finite. That which possesses existence unqualifiedly and is the well-spring of existence to other things must be the archetype not only of every mode of existence but also of every way of existing more fully in order for these things to be designated 'positive'. Or, one could express it, the first cause must have it in its character, by being itself X, in order to maintain existence of things as being X because there is no way of being richer in reality which is pertinent to causing X than by being X. This would seem to follow *a fortiori* from the causation concerned being intentional and proceeding upon a background of the cause's own self-understanding.

Schematically, on this basis the argument will, in the case of each attribute introducing a new positive note, open out as follows.

(i) Whenever, proceeding a posteriori, we discover among the excellences which it belongs to some particular specific or generic nature to be realized in or to be open to and hope for[1] a property which marks a fresh richer way of existing, we reflect upon what makes this thus 'positive' or involving of a 'fresh richness'. We then seek out the trans-generic character not descriptively limited to the finite which is exemplified in this excellence and which, in the case concerned, is that in the character of the way of existing involved which renders it positive. Finally, we reason in the way outlined above that this trans-generic character, because actually within the

[1] If it be alleged that there is some excellence of intellect or will which can be conceived on the basis of our knowledge of human beings, but which is not actually realized among them, it remains that, if it is within the scope of human nature even to hope for it, the rest of our argument will follow.

scope of some temporal beings (known a posteriori to be so), must be exemplified pre-eminently in the First Cause.

There is no question of these attributes being exemplified separately, one in one 'first cause' and another in another 'first cause', since it is only as modes of existing or ways of existing more fully that it is being argued that the First Cause of existence *qua* First Cause of existence must have these attributes in its character. We learn a posteriori the respects in which there can be 'positive' existence, i.e. in which existing can be 'richer' or more full, and from this we learn the notionally different aspects of God's existing as the actuality which immediately underlies, as upholding cause, all positive temporal existence—aspects integrated together in the 'inner unity' or incompositeness of God as First Cause. More precisely, the incompositeness explains how all these things are welded together in one subject while the fact that there cannot be more than one incomposite thing means that there cannot be separate subjects for these different attributes.

And, because there is in the First Cause no nature as a quasi-abstract object to limit the mode, style, or way of existing of the First Cause, and because of what we may call the simplicity or undeterminedness (i.e. not restricted by any nature) of these trans-generic attributes, which is precisely what allows them to belong to the First Cause (i.e. makes them 'not restricted to the finite'), they must be possessed by the First Cause in an unlimited, perfect, and unsurpassable way.

As to real possibility, these attributes must be possessed in an unsurpassable way because there is no way anything else could possess them except by being maintained in existence or deriving existence from the First Cause, and therefore possessing them in a way determined and limited by a nature, a nature given in the giving of determinate existence. God alone is self-existent and intrinsically underivative, and there is no way in which any thing can arise except from Him and His self-understanding, and no way in which from Him there could arise anything which equalled, let alone surpassed, Him. As to our knowledge, we saw that we can say for certain—there is no epistemic possibility otherwise—that there can be only one 'incomposite' thing: and that will imply only one thing with no 'quasi-abstract' nature limiting its way of existing, and therefore, even conceivably, only one thing as such the well-spring of all positive existence because (by this *reductio ad absurdum* argument) there could not be two things available to be such.

However, as will become plain in our next section, it is not just that, for the extrinsic reason of the lack of an available cause, there is no real possibility of there being anything which equals or surpasses God, but that any such thing is intrinsically impossible. As will become plain when we distinguish 'sortal' and 'intensive' attributes, nothing which exists and has any positive attribute whatsoever in the mode of the derivative can equal or surpass in respect of that attribute that which exists and has attributes in the mode of that which is intrinsically underivative, i.e. the incomposite: for anything composite has a nature which limits the way in which any attribute is realized.

(ii) From this it can be seen that any such positive non-limiting trans-generic attribute exemplified in the world or within the scope of mundane natures, whether we know of it philosophically or not, and even if we are supposed only to know of it in God's making Himself known in Revelation, must be pre-possessed by the First Cause as a condition of its existence or real possibility in the world. But in order to make clear what is involved in 'infinity' and 'unsurpassability' we have to distinguish between 'sortal' or 'adjectival' attributes of God, which belong to Him 'non-intensively' as agent, person, subject, or *suppositum*, and attributes which belong to Him as 'actuality', i.e. attributes which we might call 'intensive', 'active', or 'verbal'. That is, we have to distinguish between personhood, being good, being omniscient, and suchlike, and, on the other hand, existence, activity, dynamism, vigour, life, comprehension, and love. This distinction is brought to the fore by a consideration of the power of God and His omnipotence.

2. THE SIGNIFICANCE OF INCOMPOSITENESS: THE INFINITY OF GOD NOT ONLY AS SUBJECT BUT ALSO AS ACTUALITY: THE NATURE OF OMNIPOTENCE

(a) *Problems of God's omnipotence or infinity of active power*

It might seem that each task takes only as much power as is defined by the type of task concerned, a power being defined by the object which it is directed towards, and therefore a finite effect only requiring a finite power to produce. Therefore, it would seem that an a posteriori argument to the power of God could only reach to showing that, since it is by God's power that nature has what powers and dynamic character it has, God's power reaches to the sorts of thing which are within the scope of natural causation, and even that God's power may be limited to what He can produce by natural causes.

True, God is not limited by any conditions set by an environment, or by being bodily and restricted by the character of the bodily means of action available to Him, and indeed He is not limited by any 'nature' given as a 'quasi-abstract' datum prior to and limiting His possible modes of action. But the inference from this fact to the suggested conclusion that He is so unlimited in His power as to be able to produce just any intrinsically possible thing may be resisted on the ground that a nature and properties do not have the role of restricting their subject's behaviour but of opening it out and making it possible: any nature and properties may seem to be what, *qua* positive, put things within the reach of an agent's power. Any suggested inference from the infinity of a thing in being and nature to its infinity in respect of the objects of its power might be objected to on the ground that it involves an illicit transition in thought from the 'magnitude' of a thing with its power *qua* thing to the magnitude of the 'intentional' or 'formal' objects of the power, the things which the power is 'directed towards' and so in terms of which, *qua* power, it is defined or specified: for a power to be unlimited in respect of its being would not seem to be for it to be unlimited in respect of its objects. So there seems no inference from the fact that no external or prior thing limits God's power to this power's being unlimited in respect of its objects.

Two inadequate responses must be set aside.

Firstly, it might be argued that, since there is no way for any being to exist except derivatively by the Divine causal action, there is no real possibility of anything existing except by His power and, therefore, every really possible thing is in God's power. However, this does not show that those intrinsically possible things which are never actualized are within God's power, and we cannot abandon the notion of intrinsic possibility (that in regard to which there are no intrinsic reasons making it impossible) because it belongs with our notion of the incomposite (that in regard to which intrinsic reasons leave no room for the possibility of its non-existence), and both notions seem well-conceived and not mere rationalist inventions.

Secondly, it might be argued that, since God as First Cause is personal and perfect in intellect, He can conceive any intrinsically possible thing (which, *qua* being, will be like God in this or that respect and, if it exists, will exist in virtue of His intentional causal action), and that, since God as First Cause is unrestricted in power, His power will reach to every such intrinsically possible thing. But this argument does not deal with the problem of inferring from the unrestrictedness of God's power *qua* thing to the unrestrictedness of this power in respect of its objects. The only purpose this argument serves is not to prove God's omnipotence from His personhood, nor even to indicate how, conversely, His personhood might be inferred from His omnipotence (on the ground that we lack any model of how a power might be unlimited in its objects other than that provided by the idea of an intellect open to all things), but to show, once granted that the First Cause has unlimited power, the way in which the having of this power is realized: acting out of an understanding which can reach to conceiving through and through each and every intrinsically possible nature is the way in which unlimited power is exercised.

(b) *The resolution of the problems of omnipotence*

Accordingly, our response to these problems should proceed along quite different lines.

Firstly, for the proper handling of the problems here concerned, it would seem that our starting-point should be the distinction between, on the one hand, things marking God as subject or *suppositum*, i.e. things said 'sortally' (most so-called 'concrete nouns' and many so-called 'adjectives'), and, on the other hand, things marking God as actuality, i.e. attributions of activity (one might say, things said using activity verbs). Upon examination, one will find that the norm is for the concrete nouns and adjectives concerned either to apply without qualification or with qualification or in a secondary way or not to apply: that is, except derivatively in virtue of relation to activity, there is no question of degree or 'intensive magnitude'. By contrast, the norm is for attributions of activity 'directed towards objects' to have an 'intensive magnitude' defined by reference to features of the spread and character of these objects. The word "object" is to be understood here in a very broad sense to cover whatever an activity is directed towards which need not be a 'thing' but may be any kind of 'effect', 'appearance', or state of possibility or tendency: i.e. whatever in the logicians' sense is comprised in the 'formal object' of the activity.

At the most primitive level, this is exemplified in the case of light, which we conceive as an activity defined not in terms of something contained within the light-source but in terms of what it is directed towards, that which is present at the place where the perceiver is, what we might describe as the brightness available to be experienced or to have its appropriate effects. In the case of love, it might be objected that the intensity of love was not mainly to be measured or assessed in terms of the 'extensiveness' of the things loved but the force or *élan* in the lover with which its objects were loved: love, it is rightly said, is a relation signifying something real in the lover. In this it is no different from light and even from physical force or active power since we conceive both of these as 'real' in what gives light or has the physical power. But in all these cases it remains that there is nothing 'real' in the love, light, or physical power in that which relates to objects or possible objects except in so far as these are understood in terms of such relation: love as actualized has its force or *élan* in its exercise precisely as a relation, i.e. precisely as other-directed, and this force is present as something available through being known to inspire an answering love; it is only in physical cases that the 'force' in the exercisers of active power can be assessed (though not, *qua* force, defined) from a consideration of their constitution, rather than only through a consideration of its possible effects or the appropriate response to it.

It is not for this book to resolve all the detailed problems which arise. Obviously, we would have to make a careful distinction between intimately related sortal or adjectival and active verbal terms. We would have to distinguish between being an agent, a living thing, a light, or a brave person, sortally, and exercising agency (i.e. acting), living, shining, or being brave in the sense of exercising bravery, as activities. We would have to note that sorts of things are typically graded into sub-sorts according to the intensity of their activity: brave people into the more or less brave according to the intensity of their bravery in action and lights according to their luminosity. And we would have to distinguish the case of where a power or disposition and its exercise are conceived of as inseparable from the case where the power or disposition can rest for a large part inoperative.

Secondly we need to note that the exercise of active power is in its concept the exercise of something 'directed towards' objects. But it is in terms of active power and kinds of active power that we have explained or illustrated the content of the notions of 'real' existence and of modes and degrees of richness of this 'real' existence. Therefore, to describe a thing as infinite or unlimited in respect of its existence or active power or in respect of things with the logic of these (dynamism and, as commonly thought of, vigour, strength, life, and suchlike) is precisely not to describe it as infinite in respect of something partly or wholly specifiable apart from 'directedness' towards 'objects' but in respect of something with no content except for one consisting in such directedness. In the case of a bodily action, a material nature limits the objects and modes of possible action, and, in the case of non-bodily action, if the agent's nature is a 'given' prior to its existence, this nature will likewise determine or limit the objects and modes of possible action: but, in the case of God, because of His incompositeness there is no such nature as a prior 'quasi-abstract' given to limit the objects or modes of action, and there is no content to descriptions of His existence, active power, and suchlike except in terms whose content is to imply some mode of object-directedness.

Thirdly, whereas we have in effect argued that the First Cause must be personal because some of the things upheld immediately in existence are persons, we have gone further than to argue that the First Cause must be an agent (a first substance, subject, or *suppositum* as an exerciser of active power), since we have represented the First Cause as not only *suppositum* but also as actuality—although, as I insist again below, the precise statement of this point turns upon finding a consistent exposition of God's incompositeness. It is not just that God possesses active power because He causes the continuance in existence of created things with active power, but that He is the active power, activity, or actuality which, in the sense explained in Chapter VI, grounds or energises all created actuality or activity. To express the point with a flourish of rhetoric, we might say that God is the actuality which gives actuality and the character of being an actuality to every other actuality, or, as some with less discipline have said (presuming that life and love belong to God), God is existence, activity, vigour, life, and love itself.

Therefore, whereas each new note which we prove of God as *suppositum* or subject, e.g. that He is agent, that He is person, or that He is 'personally good' (see Section 4 below), has to be, according to my approach, proved of God by fresh or amplified a posteriori consideration, the situation with the concepts we apply to God in describing Him as actuality is different. We learn a posteriori of each new aspect of God as actuality, adding to the aspect of active power inseparable from activity the aspects of life, love, and comprehension, aspects whose bringing in is marked by the sortal concept "person": but we do not need fresh a posteriori argument to establish each new degree of intensity in these aspects, assessed as it is in terms of its object-directedness.

These clues enable us to do justice to one of the insights we met with in Chapter VI, namely the suggestion of Aquinas that, whereas to bring one being to be from another by some transformation within the context of presupposed natures required only a finite power, to bring that being to be from unqualified not-being (i.e. to make a thing stand outside nothingness without anything being causally presupposed) requires an infinite power: infinite in respect of its objects. In place of the way of thinking whereby a nature and properties (so-called 'second natures' or dispositions) are what, as it were, liberate or give power to a thing we need to substitute the conception of Chapter VI that anything whose existence and activity is pre-set by a nature will be limited or, as it were, imprisoned by that nature and therefore above all impotent in respect of the giving of existence. But, in order to refute the conception of natures as giving power, we have to recognize that being of a certain nature or having a certain property are mere facts or aspects of actualities propositionally picked out: it is the *suppositum* which is the logical subject of the power, nature, or property, and which exercises it; and it is an actuality of some kind which grounds the power, nature or property, or has these as aspects.

In our general strategy we argue from the need that temporal continuance have a cause and from the fact that only something incomposite, in the case of which there was no real distinction between subject and actuality, could serve as such a cause to the conclusion that such an incomposite thing exists. Thus we do indeed argue a posteriori to the existence of an incomposite actuality causally immediate to all other actualities, but what we establish is the existence of an incomposite actuality *simpliciter* upon the basis in reasoning that it is needed as a cause. The notions of activity and active power are included in the notion of this incomposite as an actuality and therewith is also included whatever belongs to the unlimitedness of this actuality as activity and active power. That is, we do not argue to the existence of an incomposite thing and then register as established by successive a posteriori considerations first its being a cause and then separately each aspect of its exercise of causal power.

Accordingly, the presumption that what can cause continuance in existence unconditionally and unqualifiedly is infinite in power, that is with a power extending to all intrinsically possible things, i.e. omnipotent, proves to be correct.

(c) *An aside: the character of our proof that there is only one God*

I have said that what we prove is the existence of an incomposite actuality *simpliciter*. This has significance in another way.

I have wished it to appear that in my approach to the question of whether there is only one God I avoided the approaches of either pre-critical empiricism or pre-critical rationalism.

That is, firstly, I did not argue from the unity of our Universe to the unity of God because that kind of argument taken in isolation would leave it open that there be other gods who created other universes or no universe at all, besides leaving a puzzle as to why there might not be a company of gods co-willing our Universe.

Yet this type of argument is very common and very attractive. Note is taken of the grand unity which excites contemporary cosmologists and physicists. Note is taken of the marvellous contriving of things considered absurd by the biologist, the biochemist, and the neurologist. And sometimes nowadays note is made of how conveniently the universal constants of nature as well as the unusual conditions provided by the earth seem selected to allow, as it were, by an almost infinitesimal chance, the existence of stable conditions suitable for life to develop, while the discordant and ugly features, 'nature, red in tooth and claw', 'life, nasty, brutish and short', and the scale of human greed, folly, and wickedness, are disregarded, perhaps reasonably, but reasonably only if they can be exhibited as merely accidental excrescences on a nature or a system of nature in itself good.

But those who argue in this way proceed as if the primary function, work, or *ergon* of God, and indeed of any Divine being, would be the causing, contriving, and maintaining of the Universe: Gilson, in *God and Philosophy*,[2] notes how often rationalism decays towards the same conception of God as mere Demiurge. As against this one must concede that, if to be Divine is to be transcendent, so that creation is therefore free and unnecessitated, then any god's primary work or activity must be in the exercise and/or enjoyment of his own life, and it is not clear why there might not be lots of gods who have not troubled to create anything or who have created universes in no relationship to the Universe within which we live and move.

Nor, secondly, did I rely in a rationalist way on the principle of the Identity of Indiscernibles in regard to 'intelligible' or immaterial objects, arguing from the singleness of a concept to the singleness of the instantiations of the concept in every case of that whose essence is immaterial. This is the principle traditionally drawn upon in order to prove that each angel must be of a species or essence peculiar to itself. God is, in this way of thinking, as it were, a single unmultipliable Idea or Form.

We find this not only within all forms of Platonism and in Aristotle's conception that forms must be identical if they are not individualized by matter, but also in Duns Scotus and later in Leibniz. And, peculiarly, we find the same rationalism straggling on in quasi-empiricist guise in applications of Ockham's razor, whereby it is alleged (to mention a common presumption in Duns Scotus or Ockham) that we should regard the falsehood of what it is pointless or inconvenient to suppose as probable: the rationalist and aesthetician's predilection for simplicity here reappears in a so-called empiricism.

In Aquinas we find the same way of thinking. He presumes that immaterial entities have to be discernible (i.e. different) in nature, not just in accidental properties: human souls could not be distinct unless they came into existence in matter; and angels have to be already distinct by nature from each other before they make any choices, e.g. between good and evil, between dependence on God and self-reliance, or become in any other way differentiated in respect of accidental states or actions. There is, as it were, no room in logical space for two identical immaterial entities, any more than there is any room in ordinary space for two bodies in the same place.

[2] Étienne Gilson, *God and Philosophy*, ch. III.

In expressing this thought, Aquinas tells us that, in an angel, its nature is identical with the angel itself. Thus, the nature of the angel, in such a case, is conceived of as if it were a particular primary substance or *suppositum*, and as what exists rather than as the angel's existing or any fact presupposing this existence. Yet to have a nature which is, *qua* nature, an object of understanding, a particular or identical to a particular (I mean by "particular" something designated by a singular term and which has real or substantial existence attributed to it) is odd in the context of human discourse, and it is yet more strange for such an entity (called a nature) to be conceived of as in a way abstract, in as much as it is a nature which could perfectly well not have been instantiated. Whereas when we speak of an angel as an individual or under his proper name we 'presuppose' his existence, by contrast, when we speak of his nature, we do not in a technical sense 'presuppose' its existence or instantiation. Moreover, it is odd to suppose that a nature in a particular case is something open to being instantiated either in one or in zero instances but something not open to being instantiated in any different number of instances, e.g. two.

A Thomist such as Jaques Maritain may attempt to alleviate the apparent mess by saying that the individual exercises existence whereas the nature receives existence, and this may be useful in discussing how one person, namely Jesus, may exercise one act of existence, although for his human nature to receive existence is not the same as for his divine nature to have existence.[3] But this piece of scholastic jargon-inventing betrays the fact that we have here within Aquinas's own thought an undigested remnant of an alien Platonism: the conception of natures or essences which are quasi-individuals. The alien character of this nugget of Platonism is apparent when Aquinas discusses the Trinity: he has to explain that in God there are three primary logical subjects (or 'primary substances' in the sense of Aristotle's *Categories*, translated '*subsistentia*' to distinguish it from substance in other senses, *supposita* in Aquinas's own technical vocabulary) with only one actuality or act of existence (what he calls esse in the sense of *actus essendi*); and it is this *esse*, not the three *supposita* or Persons, which is identical in the Leibnizian sense with the *essentia* or nature—i.e. the concept of *essentia* has had to be separated from the concept of primary substance, *suppositum*, or individual in that sense.

Aquinas appears to treat the identity of God with His *essentia* as the root of His unity, which would seem to put God's uniqueness in the same position as that of the angels, and to make the argument for it depend upon the Platonism I have described.[4] But if the *essentia* or nature as an object of understanding is what is concerned, then what makes the case of God significantly different is not a supposed real identity between *essentia* and

[3] *The Degrees of Knowledge*, appendix IV.

[4] In *S. Th.* Ia, Q. 3, Art. 3, Aquinas treats the identity of God with His *essentia* as a consequence of His not being material, rather than as a consequence of the unity of *esse* and *essentia* in God (*S. Th.* Ia, Q. 3, Art. 4), although in *S. Th.* Ia, Q. 11, Art. 3 he seems to treat this root of unity as belonging to God alone, despite its seeming to be something which in Aquinas's view God shares with the angels.

suppositum, nature and the individual of the nature, but a strict identity between *essentia* and *esse*. In this way, the very *esse* which the Person or Persons of the Godhead exercise or share, itself a 'particular' (in no way even 'quasi-abstract'), is intrinsically, *qua* object of understanding, unmultipliable—the only case of an object of understanding to be intrinsically in no way a 'universal', a 'possible', or 'quasi-abstract'. Therefore, Aquinas does not need even this residual nugget of Platonism.

These two approaches, then, I say that I avoid, the approach of uncritical empiricists and this last approach inspired by Platonism.

Yet it might seem that I had established the existence of a cause for each bit of temporal continuance that went on and then presumed on the basis of the uniqueness of the incomposite cause that all these separate bits of continuance were caused by one thing. Thus these separate notes, of being cause (*a*) of this bit of continuance and (*b*) of that bit of continuance, would follow a priori, (*b*) from (*a*), by the concept of "incomposite cause"—albeit not according to the principle mentioned above. We would then seem to have two arguments for there being one cause of our Universe, one being of an a posteriori kind from the inseparability of the different bits of temporal continuance in our world to the singleness of the cause of this Universe, and the other argument of a more a priori kind from the concept of an incomposite actuality. And I might seem to have drawn upon the pre-critical empiricist gambit in response to a feeling of disquiet about the apparent smell of rationalism arising when separate notes are proved to belong to God by a priori argument. It is indeed always suspicious in philosophy when we have two quite different arguments of discrepant natures for the same conclusion.

However, I do not proceed in this way.

We do not prove first the instantiation of the concept of a cause of the continuance of this bit of existing and then prove the instantiation of the concept of the cause of that bit of continuance and so forth and then by a piece of a priori reflection prove that the causes of all these separate bits of continuance are one. Rather in each case we prove at once the existence of just one cause of all the continuance there is. That is, in each case we prove the existence of an incomposite actuality, whose character as incomposite (i) excludes its being situational or dispositional towards activity so that its actuality is an activity, an activity marked by the active power to cause, a power all exercises of which presuppose the giving of existence, and (ii) implies its uniqueness. True, anyway, we know these pieces of continuance, at least in our Universe, are not discrete from one another so the question so far as our Universe is concerned could only be of the cause of continuance of the whole Universe, not of individual snippets of continuance; but this is not precisely to the point; rather each bit of continuance is known as caused by something incomposite and alone incomposite in the same act as it is known as caused at all; i.e. in the same act as we know it as caused we know it as caused by the same individual actuality as is the cause of every other bit of continuance.

When we recognize any one bit of continuance in the Universe as caused we *ipso facto* recognize it as caused by the same individual actuality as is the cause of every other bit of continuance[5]. We do not prove the instantiation of the concept "cause of the continuance of such and such" as a general concept and then at a second stage show that this cause is incomposite and unique, but show the existence of this incomposite unique cause (causing all continuance whatsoever) in the act of showing the need of a cause at all.

This way of stating the matter needs a more accurate formulation, but the main lines or general direction of the solution should now be clear.

(d) Summary: the significance of incompositeness: two aspects of infinity

The concept of "incompositeness", as applied to God, whereby He is at the same time subject and actuality or whereby there is no real distinction in His case between the subject which exists and the actuality of its existing, has two aspects.

On the one hand, it is the basis of the negative aspect of all that we say about God, His not being limited in being, nature, or mode of personhood in this way or that way or in any way at all: this exhibits His infinity as a negative concept.

On the other hand, it is the basis of our understanding that where concepts signifying actuality or activity in the sense we have indicated are concerned, they apply to Him without limit in 'intensive magnitude', 'intensive magnitude' assessed in respect of the objects which the actuality or activity *qua* activity is directed

[5] The Revealed theologian may make a comparison with the case of our only recognizing the divinity of Jesus, as having the same character of divinity as the Father, in the act of also recognizing Jesus' unity with the Father.

towards: in this way, God's active power, and with it (if they have been proved to belong to God in some way) His life, His comprehension, and His love, are not limited to those modes which we can find exemplified in Nature and they are not limited to what is within the scope of Nature's possibilities; and, in this way, the concept of God's infinity is not merely negative in content but such as to take us in the direction of the 'more positive', 'fuller', and 'more eminent', and this without limit.

3. THE PRESUPPOSITIONS OF HOLINESS: PROBLEMS IN THE PHENOMENOLOGY OF THE HOLY

We have outlined the basis of the approach which the strategy of this book would suggest for establishing by a posteriori argument the possession by the First Cause of all those trans-generic attributes which mark aspects in virtue of which the existence of temporal things is positive. In the use of the word "positive" and the theory of evaluation which I adopt this word "positive" is not itself an evaluative term. But, on the approach which I would take, it is precisely such positive trans-generic attributes that make things right, good, beautiful, great, or objects of any other such general pro-evaluation. Therefore, it will follow from what we have established that the First Cause is the primordial archetype of whatever in the world is the object of any such pro-evaluation in respect of what makes it thus to be valued. Thus there is nothing positive in creation which makes anything to be valued which is not exemplified pre-eminently in the First Cause. And all such attributes are possessed by the First Cause in an unlimited, perfect, and unsurpassable way.

Argument along these lines provides a prolegomenon to a consideration of the worshipfulness and holiness of God.

But speaking of God as "holy" involves two further features: (*a*) the consistent development of a negative theology of God's 'Otherness' and incomprehensibility; and (*b*) a proper phenomenology of the Holy.

As to (*a*), we have already in our account of incompositeness and 'nature' in God laid the groundwork of an account of how God is 'Wholly Other' in the sense that in any respect where there is likeness there is also greater unlikeness so that He is whatever He is in a way that is special to Himself. Whatever in creation exemplifies some positive ground of value has this because in the respect which makes it valuable it reflects the same character pre-possessed in God, but this character will be possessed in a way which is not only perfect and unsurpassable but also utterly unlike.

As to (*b*), whereas with other evaluative or quasi-evaluative terms such as "great" and "good" it appears that the description of an object may suffice to justify the application of the terms, it does not seem that the presentation of God as an object of thought, even as an object of thought known to have actual existence, is enough to ground the use of the word "holy". That is, it might be held, it is only when God becomes for man an object of involvement,[6] so that the knowledge of His nearness is not (to use Newman's terms) notional but real,[7] that the word "holy" can be applied without idolatry: only God Himself, not just our ideas of Him, suffices for religious involvement to be non-idolatrous. Or so it might seem: but, apart from the limited comments in Sections 4 and 6 on how any 'presentation' of God so as to be known as holy precludes any rival in value so that holiness has implications for both goodness and uniqueness, these issues of phenomenology are obviously outside the scope of this book to consider.

Yet we have spoken of positive attributes in God such as would ground the most general pro-evaluations, making Him primordial archetype of whatever is object of value or wonder in the world. Yet the interpretation of such evaluative or quasi-evaluative terms as "great" and "good" in regard to God, at the same time as involving a whole theory of evaluation and part of the background of our use of the word "positive", provides the ultimate test of the

[6] For the understanding of this notion of "involvement" some consideration of the concept of "*die Sorge*" in Heidegger may be helpful as well as the more obvious issues raised by religious attention in faith and in theistic mystical experience and by participation in sacred history (e.g. in the goings-on reported in Scriptures) and in liturgy. The distinction between being an object of 'involvement' and being an object merely of thought is represented also in the differences between remembering people and events and remembering facts.

[7] In the rhetoric of Scriptures and much other theistic discourse, it might be said that God is known as holy, known in His holiness, only as He actually 'draws near' and makes Himself 'present'. But, since He is near and indeed present, the immediate energizing upholding cause of the continued dynamic existence of each thing, it is better to say, not that He draws near as if He had been absent, but that He makes His presence known in a different way—and use this as a way of interpreting the metaphors of 'drawing near' and 'becoming present'.

possibility of theism without anthropomorphism. To the questions here involved we now turn.

4. THE 'GREATNESS' AND 'GOODNESS' OF GOD: SOME PROBLEMS IN THE THEORY OF EVALUATION

In considering what is 'positive' and developing a theory of 'evaluation', we must distinguish between 'real' attributes, qualities, or properties and 'evaluative' ones, which will be 'consequential' upon the 'real' ones, avoiding common mistakes in making or in applying the distinction. It cannot belong to this book to offer full examination and justification of the required non-conventionalist and objectivist theory of evaluation, a theory always grounding any evaluation not in the structures or formal characteristics of a human evaluative or choice situation but on the 'natural' or, more generally, the 'real' characteristics of the objects valued. But what our argument in this book does require is that we illustrate how these problems arise in connection with the 'greatness' and 'goodness' of God.

At a second stage, a theory of evaluation has to make a distinction among 'evaluative' and 'consequential' attributes between, on the one hand, those whose application does not connote being an object of 'involvement' or 'experience' and, on the other hand, those whose application does connote this. It will, for instance, be a matter for inquiry as to which attitudes or emotions—e.g. sorrow, guilt and unworthiness, fear and terror, horror and dread, awe and wonder, delight and enrapturement, regard, respect, admiration, and worship—connote 'involvement' with their objects and which do not connote more than a factual knowledge of them. Then, it might seem that to call something 'right' or 'good' did not connote its being an object of 'involvement' but that to declare it 'holy' did this. But it would remain that the application of any such 'involvement-laden' evaluative term and the continuance of the associated attitude or emotion has preconditions: in particular, for something to be 'holy' there must be no way in which it can arise that something stands beside it as a rival locus of value, e.g. a rival locus of greatness or goodness. With these remarks I return to a consideration of greatness and goodness.

The term "great", so highlighted in the writings of Augustine and

Anselm, is the most general pre-evaluative ('descriptive') evaluation-geared term in our language. God is great in respect of His being and nature, in respect of His active power (His omnipotence), in respect of His intellect, and, in whatever way He is good, in respect of His goodness: in all such respects, He is not only 'great' but 'unequalled' and, as we have said, 'unsurpassed' and 'unsurpassable'.

The background of this extraordinarily general term (a thing may be described as of great fineness of detail or of great minuteness) is strangely limited, arising by metaphorical extension from the spatial case. In origin, "great" is a special kind of relative term like the term "large", differing only in that, as it were by design, the English language has the two terms and refuses to "large" the same metaphorical extension. We compare things as being 'larger than' or 'smaller than' each other, and for some purpose or from some standpoint group those which are a certain span of sizes together as 'large':[8] that purposes and standpoints may differ is evident in that what counts as a large animal may differ in different contexts, and the same feature is more important in the evaluative case; but beyond these remarks the spatial origin of our usage of the term "great" is of no present relevance.

We have sufficiently discussed what belongs to God's greatness in respect of being and nature, active power and intellect, but some further comment is needed in respect of His goodness.

It has been established above that there is no way in which it can arise that there is in creation some object which has value in a way limited to itself so as to constitute a rival type or source of value. Therefore, if that 'personal goodness' which when attributed to God is in danger of generating anthropomorphism does indeed represent a positive ground of value, then we cannot but conclude that 'personal goodness' in the sense concerned is to be attributed to God. In this way the problems of avoiding anthropomorphism and at the same time establishing theism turn upon a consideration of this case.

It may well be argued that the conception of "personal

[8] In a logically parallel way we distinguish among past events, ordering them by the relations 'earlier than' and 'later than', and group for some purpose a certain 'span' of them as 'early', what counts as early being different according to whether (e.g.) one is treating geological time, general human history, or the history of England.

goodness" as involving a certain love, abandon, and generosity in relationships with other beings met with in Jewish and Christian tradition, and perhaps also present in the conception of Allah as Merciful and Compassionate, is a special one, notably different from (say) Greek and (say) Buddhist conceptions of the goodness to be valued in human beings. Whether or not this conception is thus special, it has become the predominant Western conception of what it is for a human being to be a good man or woman.

The conception concerned, like other global conceptions of human goodness, has, of course, the other most general virtues built into it. There can be no love or generosity which shows respect and regard for other people which does not first recognize and give them their due, i.e. this 'personal goodness' presupposes justice. There can be no soundness in relationships with others without soundness in relation to oneself, i.e. this 'personal goodness' presupposes a healthy attitude to one's own good, excluding self-hatred and injustice to self, healthy but balanced, not anxious to give priority to self or over-protective against risk. The exercise of this goodness requires, like the exercise of any virtue, practical wisdom (*phronēsis* or *prudentia*), fortitude (courage and perseverance), and steadiness of mind (*sōphrosynē* or temperance, freedom from scatter produced by variable desires). But it is characteristic of the Jewish and Christian conception of goodness (the Hebrew *hesed*, the Greek *eleos*, and Latin *misericordia*, "mercy" or "loving kindness", as realized in *agapē* or *caritas*, the friendship of God towards man and responded to by man in his *agapē* or *caritas* towards God and towards other men[9]) of which we are thinking that it is without desirable measure, so that, as it were, the more of it the better. That is, whereas Aristotle expects a measured love of others as oneself in the contexts of friendship and community, the ideal of Jewish and Christian goodness is to go to the extremes of abandon and generosity in living out this love of others. The 'good' shepherd, we are led to understand, lays down his life for his sheep, and leaves the ninety and nine in order to seek out the one that is lost. It is not surprising that the word for 'good' here is not *agathos* (the usual word for 'good') but *kalos* (elsewhere translated as

[9] The same idea, together with a making explicit of the strange conception of men and women sharing this characteristic with God, seems to be expressed in St Matthew's 'Be ye perfect (*teleios*) as your heavenly Father is perfect' (5: 48), and in St Luke's 'Be compassionate (*oiktirmōn*) as your Father is compassionate' (6: 36).

'beautiful'): it is not clear that the shepherd with prowess as a shepherd and who is in a common Greek sense *agathos* would do these things. Doubtless to be a 'good man or woman' in the Greek sense is a global characteristic involving justice to others, as well as generosity emulating that of Aristotle's 'great-souled' man, but this might be because prowess in being a human being involved whatever belongs to the noble and honourable, not because of the internal connection between justice in relationship and generosity in relationships of love. Likewise, the kind of goodness exemplified in the Buddha or the Bodhisattva is also a global virtue embracing the others mentioned in the last paragraph, but rather than making the basis of virtue in other-relatedness the centre of what it is to be good, it appears to make it into a by-product of freedom from self-centredness and the reign of desire, a freedom which is the fruit of 'enlightenment'.

What matters for our argument is (*a*) whether such a state of character can be conceived of, (*b*) whether thus conceived it is an excellence which it is part of human nature to be open to, (*c*) whether it would be itself a positive attribute, a facet of the richness of human existing, or only a consequential evaluative property ascribed in consequence of some logically prior 'real' attribute, and (*d*) whether, if it would be itself a positive attribute, it would be trans-generic in character and open to the infinite.

It is unimportant for our argument whether or not the concept of "personal goodness" picked out is special to Semitic-rooted or Western ways of thinking: there seems no plausible doubt that it can be conceived of as a 'descriptively' possible state of character, i.e. no plausible doubt as to (*a*).

The more difficult question is (*b*). Firstly, we have to ask whether the existence or full flowering of this state of character depends upon a fullness of abandon in generosity which can only be in accord with 'reason', 'soundness', or 'balance' in a person's life, if he can have a hope in Providence in respect of the openness of possibilities in his own and others' lives which is made possible by faith: this would make a difference to whether this so-called 'personal goodness', conceived of as explained above, can consistently be conceived of as a 'virtue' or good characteristic to have, rather than violating the requirements of truth and truthfulness. It might not seem that (*b*) should be in any doubt on this ground, although the extent to which this view is 'philosophical' rather than in some sense 'empirical' and the extent to which it is independent of the 'experience of Revelation' and of Revealed theology are obscure. Secondly, viewing this state of character as an excellence or perfection of human beings depends upon accepting a perspective in which the absence of personal relations is seen as a privation of a person in respect of what belongs or is to be hoped for as a fulfilment of personhood. That is, it involves rejecting the perspective which would regard the Greek autarchy or self-sufficiency as the ideal (perhaps regarding this as the Satanic temptation), and therewith rejecting any supposed Eastern conception of 'nirvana' as a grace-independent and non-relational state.

As to (*c*), the point turns, not only on how the kind of goodness concerned is conceived, but also upon the theories of explanation and of evaluation to be adopted.

It is, I believe, vital to realise take the view that so-called 'personal goodness' is a 'real' attribute alongside other 'excellences' such as knowledge or understanding, even though to mention it is to praise or 'commend' its possessor and to express or imply an evaluation. We should do this because, in the explanation of the behaviour of human beings, the full explanation of any properly human kind of action is always, I believe, holistic in form. By this I mean that, for the purposes of understanding and explanation, human beings are to be treated as psychosomatic wholes, with 'veridical' perception and effective intentional action being regarded as the paradigms of perception and action, and excellences of understanding and will being treated as unities, integrated by their 'ends' as directions of the human *élan* concerned. Thus, for instance, knowledge as a 'property' mentioned in an explanation of a piece of behaviour is not to be viewed as a complex of sure belief, well-groundedness, and truth, the three incidentally co-related in such a way as to be non-accidental in their co-existence and origin: rather knowledge is an identifiable whole, the achievement of which marks a stage in the dynamic of the unfolding of a human life, as the arrival at a stepping stone, the striking of a new note, or the firm capture of new strong points can be a stage towards the crossing of a river, the resolution of a chord, or the firm capture of a new terrain. The grasp of the unity of the intermediate step, in its value, is an element in the explanation and understanding of the larger advance, its value and role being inseparable. Hence, mention of such excellences, both the more global and the more particular, must enter into the causal account of the upshot of human action. This is true even, and perhaps especially, of the more global excellences since, for instance, as a human being develops, his or her 'goodness' becomes less and less an accidental agglomeration of less global virtues and virtuous habits, and more and more an integrated seamless whole, none of whose aspects are accidentaly related in his or her life to the others.

If we were to consider global concepts of human goodness in general, (*c*) and (*d*) would need to be considered together. Thus, for instance, if it had been the supposed Greek conception of being a good human being or even an excellent one (*agathos* or *spoudaios*), it might be arguable that the adjective concerned considered by itself was applied merely consequentially and did not signify a real attribute, although it would be indeed trans-generic (the word being, on this view, used as we use it in speaking of good farmers, good actions, good essays, good parties, good screwdrivers, and so forth), while taken with its noun it would not be trans-generic. Or there might be obscurities in regard to the supposed Buddhist conception of human goodness as to whether, in the first place, the Buddhist conception of individuality involved being tied to desires associated with a particular historical condition so that 'goodness' involved 'non-individuality' and, in the second place, as to whether their ideal of goodness did not involve an independence of personal relations: according to the perspective mentioned above the second feature would represent a privation so that, even if such a state of character were trans-generic, it would not be 'open to the infinite'. But we do not have to explore these dark questions turning on the interpretation of many concepts in each of the two cultures concerned since our concern is limited to "personal goodness" as explained.

The question (*d*) is only of the 'descriptive' openness of this particular concept of goodness such as human beings are supposed to exemplify: if it is 'descriptively' thus open, then so long as the requirements (*a*), (*b*), and (*c*) are satisfied the rest of the a posteriori argument to God's personal goodness will follow. That is, it will straightforwardly follow, if 'personal goodness' is within the scope of creaturely perfection, that it must be a perfection pre-possessed by God. And, if there is value not only in the works of human virtue but also in being the sort of person who is personally good, then personal goodness must have a value in itself which makes it above all necessary that God be personally good if He is indeed personal, not impersonal, and the ultimate object of awe and abasement.

But is the concept of "personal goodness" even descriptively open to the infinite? Let us now consider the difficulties of this question. We shall discover that these ultimately turn on whether this 'personal goodness' is indeed a perfection even in the human case, the question raised under (*b*) above.

5. THE PROBLEMATIC SET BY THE NEED TO AVOID ANTHROPOMORPHISM

(*a*) *The general problem set forth*

In conceiving of God as the personal well-spring of existence we may, following the suggestions of Revealed Religion, be attracted towards going far further in a way which, *if taken outside further content supplied by the religions of Revelation*, might seem merely romantic and sentimental.

In this further development, God will be conceived not only as First Cause and person, but as a Father, all of whose work in creation is the work of Love, so that beneath all is not only a mind but a heart. It is God from whom all good things come, and they come to us as human beings as gifts to us as persons, and in this way He exercises His Fatherhood towards us even in creation (and not only, as might be suggested, in grace and redemption). But, as we think of human gardeners as tending their plants with fatherly care, so God's Fatherhood extends to the lilies and the sparrows. In this way of thinking, each created thing, each snowflake, each plant, each animal, each human being, is watched over and cared for with an infinitely greater wisdom and care than any it may have for itself. In exercising this care for each thing within the context of care for the whole, fostering it within the limits set by care for the whole and by its own nature, this Father makes everything else in Nature a gift to it, the means of its fostering.

Any such 'highly personalised' way of thinking of God brings to the fore the problem of evil since, at least on a superficial view, there seem to be many things in the world not very evidently in accord with the unqualified goodwill of an omnipotent person for other persons. Further, since God is supposed to be self-sufficient and creation a free overflow of the Love which He has in Himself without need of expression outside Himself, if perfect goodness and love of their nature involve personal relations, it seems that God's fatherly love must have some expression within God Himself and not be expressed solely in love towards creation: but in this case we would seem to have laid the basis for the Christian doctrine of the Trinity within philosophy, by proving within philosophy that there is more than one Person within God,[10] e.g. a Son beside a Father, antecedently to any creation.

But, in any case, such apparent modelling of Divine personhood upon human personhood seems to generate folly.

Firstly and most obviously, the way of thinking concerned infects our conception of the relations of God to creation. Thus, it invites vain speculations as to what better worlds God could have created, as if we like God could conceive and evaluate a 'whole' which comprises a succession of stages of development, each stage including open potentialities[11] relevant to evaluation not only in virtue of what they are potentialities for but also in virtue of their openness, and as if we like God could conceive every intrinsically possible thing or actuality rather than just those within the scope of natural power. But who are we to judge of the mysterious or hidden purposes of God as if we were in His position of independence, freedom, and omnipotence and His position to judge of the potentialities of creation? Again, in this way of thinking, God, imagined as if He made plans and acted in the manner of a finite person, must either 'by decree' predestinate all things antecedently to creation so that the ultimate root of evil seems to lie in His arbitrary will, or else be temporal and deficient both in knowledge and power in respect of the future: we seem to be forced into a choice between 'strong predestinationism' and a coeval interdependent demiurge.

In these and other ways the models which thinking of God as personal entices us to embrace generate conclusions which run contrary to the understanding of God's trans-temporal character already gained from a consideration of the significance of His being cause of temporal continuance. This involved His being immedi-

[10] See P. T. Geach, *The Virtues*, pp. 71–80.

[11] In terms of the mathematical theory of the mathematical intuitionists, the world is a 'whole' (never a completed totality) more analogous to a 'spread' than a 'species', cf. A. Heyting, *Intuitionism: An Introduction*.

ately present in the totality of His incomposite life to the existence of things as this existence successively unfolds, but precisely in such a way as to preserve the character of the future as in no way yet existing so that, as we expressed it, the past and present constitute what God has done (declaring His will by His act) whereas the future and His future action do not yet exist and in regard to it His will remains free, unbound by His action. For God also the present is a 'dramatic now' (Chapter II, Section 1 (*d*) and (*e*)), the take-off point for His action, because it is as a point where He is still free that He has constituted it by the very way He constitutes Time.

This group of problems, concerning the relations of God as personal to creation, is thus easily dealt with in the light of the approach we have adopted in this book which leads to this position announced in Chapter II and summarized in Chapter V, Section 8. Our approach has the negative effect of excluding any such interpretations of personhood as would make God have an existence external and parallel to the Universe, as it were alongside it, interpretations associated with anthropomorphism and with deism. For these arise, as we saw in Chapter V, from combining essentially temporally tied models with a dualistic analysis of agency, both incompatible with the immediacy of God as cause of continuance to the existence He upholds: the incompositeness on which the power to cause continuance depends at once excludes mediation and requires a gulf of nature, as between the infinite and the finite.

Secondly, the way of thinking concerned affects one's understanding of how God is in Himself, in particular the problem of how the notion of 'individuality' applies in connection with God. Let me develop this point separately.

(*b*) *The problem of individuality*

The notion of "personal goodness" as I have explained it seems to turn any doctrine whereby there is more than one Person in God into a belief in more than one God. The doctrine of the Trinity thus interpreted makes God into a 'Divine Society' of three good individuals, being and doing good in regard to each other. That is, whereas the explanation of Chapter V of the unity of God's actuality might seem formally to salvage the unity of three *supposita* in one *esse*, the application of the concept of "personal goodness" might seem to reimport an idea of 'individuality' incompatible with this explanation.

This issue of 'individuality' is more difficult to deal with and represents the crux of the problem of the relations between so-called 'natural theology' and Revealed theology. It also presents the crux of the problems raised by the different conceptions of 'mysticism' in different religious traditions.

Let me explain this latter point.

In Chapter V we made it plain that the 'nature' of God as that in virtue of which counting was possible to this extent, that we could say that there is only one God, was nothing 'quasi-abstract' but identical to the individual incomposite actuality, God's existing itself. It follows from this that, once it is established that He is personal in nature, His self-understanding will have as its object nothing general or 'quasi-abstract' but only what is singular.

In some developments or unfoldings of what is claimed to be Revealed theology, this is what makes it appropriate to speak of the nature or the being of God as an object of vision. The intellectual act of understanding a nature is something we normally presume to have as its object something in some way or other general, abstract, or potential, or to particular things only in respect of aspects which are thus in some way general, abstract, or potential. By contrast, in the case of God, the Jewish or Sufi mystic or the Christian are not supposed to have transmuted a general concept or the meaning of some general term in use among human beings, by some Platonic sleight of hand or misconception, into some existing, living, individual, vibrant actuality. Rather, God's actuality is as it exists in itself intelligible through and through, and it is something both (*a*) unitary or inwardly unified or integrated, and (*b*) unique and unmultipliable.

It is therefore something which God Himself comprehends in understanding Himself and which, if we see, we see (albeit without grasping or comprehending in any vision we may be brought to) in a vision which reflection shows us as desiring and as incapable of satisfaction in anything short of it. (I have in mind here what St Augustine expressed more elegantly when he spoke of our hearts being forever restless until they 'find their rest in Thee'.) The possession of God which in the so-called Beatific Vision men are understood to enjoy is not a grasp, or comprehension, or enclosing within the survey of a finite intellect, of that which is unfathomable, but an enjoyment wherein man is enraptured and rather captured by, than capturing, the One who enraptures him.

In any further development of this way of thinking, the distinction between understanding natures (the ways things act and the principles of their action and the constitution which may underly these ways of acting) and understanding persons will need to be marked. In attending to creatures the understanding of natures always takes us to something general or in some way potential, capable of zero, one, or many instantiations, whereas in the case of God it takes us neither to something general nor to an individual *suppositum* or primary substance but to a certain unique activity, energy, or actuality. By contrast, in both the case of creatures and the case of God, our understanding of persons is always directed ultimately towards an individual *suppositum* or individual *supposita*, an individual subject or individual subjects, and this is as much true of Islamic and other unitarian religions as of Christian.

The main religious question has been as to whether the understanding of persons has any fulfilment in respect of God or the ultimate at all—or whether, as many Hindus have conceived, the ultimate is impersonal and attention to a person or persons is ultimately a form of attachment, confining the human spirit, so that man's own ego must seek rest or submergence in understanding of or union with some primal energy conceived of impersonally. Once it is conceded on either philosophical or religious grounds, or both, that the ultimate is properly conceived of in personal terms, it becomes plain that the Ultimate Object of understanding and love in respect of persons must be a *suppositum* or *supposita*, a Subject or Subjects, while the theoretical intellect in a certain way will reach its goal in the understanding or contemplation of the actuality, activity, or life wherein the existence of this Subject or these Subjects consists. But for this to be granted the concept of God's personhood must be purged of all those implications of 'individuality' to which the Hindu and Buddhist rightly object.

(c) *Our treatment of the problem: the dovetailing of an a posteriori approach in philosophy with the requirements of Revealed theology*

That is, a key requirement at the fourth stage (*d*) of argument in the strategy laid down in Section 4 would seem to be to demonstrate that attributing personal goodness to God, as something whose exercise is antecedent to any creation, does not import any such undesired 'individuality'. But it might seem that such 'demonstration' is to be had, if at all, only through a posteriori reflection (i) on the situation in the human case of personal relationship in respect of this 'individuality' as exhibited and (ii) on the expositions of personal relationship, e.g. between God the Father and Jesus as God the Son, in Revealed theology: and Revealed theology would suggest that the full understanding of personal relationship in the human case is only to be gained in considering man's relationship with God. In advance of such reflection the situation of 'descriptive openness' seems to many to be irredeemably obscure: but this would seem to be because they hanker for an imaginative model in a context in which no imaginative model is adequate to the purpose. However, all that we have required in the parallel case of the concept of "person" is the absence of logical incompatibility, together with the recognizability of personhood as having a special kind of perfection as a mode of causal agency: in the case of "personal goodness" again we have the absence of logical incompatibility, and we also have a certain kind of perfection as a mode of exercise of personhood (that is, *if and only if* the perspective suggested under (*b*) in the strategy outlined in Section 4 above is correct).

If this broad plan of approach is acceptable, then it allows an integration between the adoption of an a posteriori approach to theism from within metaphysics with Revealed theology.

Once it is realized that God is personal, acting intentionally in His exercise of causal agency, each new positive note that we discover a posteriori as an aspect of His personhood exhibited in His operation throws what is going on in the whole of His creative activity into a new light. Just as reasoning to His personal character from His upholding created persons in dynamic existence establishes not just the personal character of this particular upholding action but the personal character of all His operations, so discovering His goodness in some way or respect in upholding created persons in dynamic existence precisely as good in this way or respect establishes His goodness in the way or respect concerned not just in this but in all His operations, and indeed in His overall design for Nature as a whole. If there be some kind or facet of goodness in creation known only through Revelation, then from this would follow the presence of this in God also as archetype. It might be that "personal goodness" as cêntred on other-relatedness and described below is in practice dimly known only under the incipient influence of Revelation, so that the full application of Semitic and now common Western conceptions of personhood to God would be based a posteriori on this fragment of 'Revealed anthropology' if it had not been already established in a Revealed theology of personal relations in the Trinity.[12]

The situation is then this: part of creation has become, as the human species has emerged, reflective, reflective upon the world, upon itself, and upon God. Within this context, each fresh aspect of personhood as it comes to light in human thought and awareness in one act shows something of man and something of God, something of the workings of this creation and something of the character of God's dispositions for creation as a whole. Man's recognition of God is like one person uncertain as to whether he recognizes another, initially even indeed whether he is having to do with another person or 'thou' at all, and who, stage by stage, gets to know the other in new aspects and, as the development proceeds, gets to know himself in new aspects in the same acts.

(d) The indispensability of the concept of incompositeness in any theology

In all this the general metaphysical approach which we have adopted serves an essential negative role within Revealed theology itself.

This appears firstly in regard to the problem of 'individuality'.

For, if God is personal and good, then His mode of being a person must be His mode of being and vice versa, and His mode of being good must be His mode of being and vice versa, in virtue of His incompositeness. There is not one actuality, His existing, which underlies another actuality which is His being a person; nor one actuality, His existing or existing as personal, which underlies another actuality which is His being personally good or exercising personal goodness. Rather all these constitute one actuality picked out by description under different notional aspects. If His being

[12] P. T. Geach, *The Virtues*, p. 75 f., suggests that the theological usage of the term "person" is the root of the now normal modern use of the term.

personally good involves internal other-relatedness then His being involves this. But it follows from this that His way of comprising other-relatedness within Himself will be in accord with His incomposite way of being. From this it follows that there could be no such 'individuality' in different proposed Persons within God as could imply three Gods or three subjects the roots of whose personhood had any distinct or independent standing as if They first stood as Persons and then related to each other. And They will have no ontological relations prior to Their personal relations: as actualities Their ontological relations will be Their personal relations.[13]

We have already observed how the immediacy of God to the existence He causes, immediate because any hypothetical intermediate would be impotent in the communication of existence in the way concerned, secures the rejection of deistic, 'strong predestinationist', and Platonist models of eternity and personhood. It does this in the act of securing the rejection of a dualistic analysis of action. But the notion of God's incompositeness secures a rejection of any finitary model of personhood whatsoever.

6. CONTRASTS IN THE ORDERS OF PRESENTATION, KNOWLEDGE, AND UNDERSTANDING BETWEEN PHILOSOPHICAL AND REVEALED THEOLOGY

In considering the relationship between metaphysical approaches to theism and Revealed theology, confusion has resulted from the mixing up of three very different things. For we have to distinguish between the order of knowledge in the following out of some avenue of approach to knowledge (heuristics), the order of presentation in some system of exposition dictated by the need of ordered pedagogy or the conveniences of elegant axiomatization (rhetoric), and the order of understanding or 'being' ("science" in an antique sense).

As to questions of presentation, a cavalier exaggeration of their significance has contributed much to polemics but very little either to the issues of substance or to the interpretation of the texts or

[13] I owe the formulation of the point in this sentence to conversation with Mr Keith Whitefield, although the far-reaching development of this thought which he would propose comes to me as quite fresh.

authors who may fall subject to *opprobrium*.[14] Our need is to avoid being distracted by them.

What is of somewhat more importance and interest is the question of the order of knowledge. Plainly, within metaphysics, one is envisaged as progressing from a knowledge of the existence of temporal creatures to the existence of God as a self-existent first cause. One is envisaged as capable of advancing to a knowledge a posteriori of His eternity and His radical unity or simplicity and, from the intrinsic character of His existence, to a knowledge that there is only one God. Nothing in this argumentation turns upon God's being living or personal.

But thirdly, as to the order of 'being', whereby one thing is understood as prior to another as its 'root' or 'why', it has seemed that the whole order in which conceptions about God are unfolded within a metaphysically based theology is alien to the order proposed to us in Revealed theology.

In metaphysics, we may seem to proceed from the concept of God as intrinsically underivative cause to such concepts as those of His radical unity or simplicity, His infinity and completeness (perfection in a metaphysical, not a moral sense), His unity, and His eternity as a feature of His own life or mode of being; and we see as consequent on these all the aspects of His relation to creation, such as His omnipresence in time and space and His power and omnipotence. The concepts of "life", "personhood", "intellect"

[14] e.g. so far as questions of presentation are concerned, the facts that in the *Summa Theologica* Aquinas treats the non-personal attributes of God before the personal, the things pertaining to God's one nature and one *esse* or act of being before anything pertaining to His subjecthood or to there being a distinction of Persons or Subjects within Him, and that in discussing God's he simplicity has an article on the unity of *essentia* and *esse* in God previous to his article on God's unity, are no more significant than his proving that God is not a body before proving anything else in respect of His simplicity. It is a matter of rhetoric and techniques of pedagogy only, and it turns out on examination that, in the First Part of the *Summa Theologica*, one meets with distinctions and arguments upon which nothing turns until some much later section of the First Part. For instance the distinction in Q. 13 between the thing signified and the way of signifying affects the interpretation of identity propositions in Q. 3 but nothing turns upon this until the discussion of the Trinity, e.g. QQ. 39–40, and other early discussions are affected by the discussion as late as Q. 93, Art. 4, of how man is made in the image of God (so that it is only in relation to understanding and loving God or the aptitude or disposition to it that man is properly in the 'image' and not just the 'likeness' of God) and the discussion of God's governance of creation in QQ. 103 ff.

and "will", "knowledge", and "loveliness", which are indispensable to His being an object of worship, supremely and only to be adored with religious awe and abasement, making the questioning of His goodness and of the rightness of obeying Him seem both mad and impious, do not seem to have a look in until a later stage or stages.

By contrast, in Jewish, Christian, and Muslim theology, the basic datum is that of God as primary personal agent alone to be worshipped.[15] He is in an ineffable (this means unspeakable) way primary in respect of His own life so as to be self-existent, i.e. to be *a se* and radically one (i.e. incomposite or simple). He is primary in relation to all else that exists, so that all else exists only in accord with His wisdom, and is omnipotent and subject to no control in His own workings in respect of all other things. And His primacy is such that His love, goodness, and justice are subject to no external law or criticism. His being, love, and wisdom are such as to enrapture the creature which is true to its nature and open to knowledge and love, and drive into fear and dismay any intellectual creature who refuses the dependence involved in such personal relation and prefers self-reliance to the consummation of his own personhood. The Holy is fascinating and fearful.

It is not for this book to expound either Jewish or Christian or Muslim or any other claimant to the status of Revealed religion so as to indicate in what way, if any, Revelation gives such contentful knowledge. Suffice only to illustrate the difference between the

[15] This perspective I have found most powerfully presented by Kornelius Miskotte in *When the Gods are Silent* and in Karl Barth's *Church Dogmatics*, vol. ii, pt. 1, ch. VI, sect. 28. I note the association made by Barth between God's aseity and His freedom (pp. 257–321), an association met with in Greek tradition although there always qualified by the distinction made by Athanasius between what was of the nature of God (e.g. the generation of the eternal Son) and what is by His free will, or *boulē* (*viz.* creation). The activity which underlies the world is indeed, not anything physical or fixed by nature, but a pure delighting and love, certainly an act of will, but not 'voluntary' in our common meaning of the word (cf. Aquinas, *De Potentia*, Q. 2, Art. 3).

The intimate connections between aseity, freedom, and omnipotence are connections into which the philosopher has a posteriori insight as I try to indicate in Section 2 above. Barth's objection to speaking of God as necessary (*Church Dogmatics*, p. 307) are in part merely linguistic, i.e. not taking into account that in the peculiar idiom of the logician 'necessity' signifies nothing more than the denial of a possibility of not being (a denial with which Barth would agree, e.g. p. 306). But in part Barth appears to feel that to speak of intrinsic necessity in God presupposes some ontologism of prior essences making things necessary, the ontologism which Chapter V, Section 11 was primarily designed to refute.

approach made in Revealed religion and anything suggested within philosophy by citing one example: in St John's Gospel, the unity (this is the first-level kind of unity) of God, whereby He has not given His glory to another, is explained by the personal rootedness of Jesus, the Son and Word (logos) of God, in the Father, and the origin of the Spirit in the 'breathing' of the Father and of the Son, so that the openness of the concept of 'person' is made the vehicle of an indication of how in God's revealing Himself as within Himself containing relations involves nothing alien or contrary to Jewish monolatry (the worship of One alone); there is nothing like this in philosophy, and nothing arising from it about God as in Himself Merciful and Compassionate. Beyond this we need only to add that even Revelation has never claimed to give comprehension of God, and in Jewish, Christian, and like traditions the content seems to consist in the formation and orientation of the person intellectually so as to be directed in loving unconceptualised knowledge towards a Person or Persons.

The picture which we sketched of the Fatherhood of God as exercised in fatherly love towards creation represents only a fragment of the perspectives suggested within Christian theology, perspectives of which philosophy shows no knowledge, let alone any different (albeit more limited) avenue of understanding.[16] Within this theology, God's Fatherhood's primary exercise is in eternity within the Godhead towards His Son, and His supreme exercise of love towards creatures is in making possible to created persons an adopted sonship[17] so as to be sharers with this eternal Son in the Divine Life which the divine Persons share, in virtue of the indwelling of the Holy Spirit. Further, the dispositions of this Father's Heart towards creation include at the root of His whole plan for creation a willingness that this Son become man in order to accomplish this sharing and further that this Son having become man should die for the sake of the redemption of the Universe from frustration.

[16] This perspective, whereby the God who creates is a 'God of the Covenant' whose Father's Heart, 'with reasons which Reason does not know' (Pascal), cannot deny His very Self in the Son and is Himself Gift, the uncreated gift of Love (the Holy Spirit), is nicely introduced by Karol Wojtyla in *Sign of Contradiction*, III (i) concluding paragraphs, VI (i) and VII (i).

[17] It is with reference to this alone that, it turns out, Man is to be described as made in the image and not just the likeness of God (see n. 15 above), and it is said to be by the Holy Spirit that human beings are thus able to address God the Father as 'Father'.

Of these aspects of the Divine Love and Unity philosophy gives no knowledge or hint, and of the understanding they provide it furnishes no trace. However, it remains that always philosophy supplies the essential qualifying negative elements we have mentioned.

In respect of the existence of God and there being only one God, Revealed theology provides no more understanding than philosophy, since there is nothing to be said as to the root or the why of these things. But even here the mode of knowledge is different because it is not by indirect (i.e. *reductio ad absurdum*) proof, e.g. that of Chapter V, but part of the phenomenology of the holy, whereby God cannot 'draw nigh'[18] as an object of 'involvement'[19] without being known as One beside whom no other can stand. As to the existence of God, philosophy knows this purely a posteriori from God's effects, these effects being known without God's being known in them as an object of involvement: but, if Revelation is given, then God is known in the revealing effects as a present object of involvement, somewhat as people and living things are known immediately in their actions.

[18] See n. 7 above.
[19] See n. 6 above.

X

The Inescapable Conclusion: That One Living God Exists

I have completed a programme of inquiry into the notions of time and causation, substance and existence, explanation, understanding, and science, and into the relationship between first-order inquiries—directly occupied with questions of fact and reality—and second-order epistemological and methodological questions. And I have examined critically those elements in the history of metaphysics (Chapter VI) and epistemology (Chapter VII) that have a bearing on the shape of causal argument from the existence of mundane things, as such, to the existence of God, as immediate upholding cause of all things and (Chapter VIII) personal in nature.

Against this background we shall in this chapter pick out the pivotal features of our resulting argument. And we shall see the metaphysical perspective which I present, marked by its a posteriori approach which considers the conditions of temporal existence itself and not the conditions of rational thinking, as internal to any proper theology. It will be internal to any such theology in the account it gives of the self-existence, transcendence in nature, immanence, and eternity of God, and internal in the negative constraints it sets up excluding any such mere demiurge as some contemporary philosophers have proposed to worship. And in Chapter IX we have already discussed its integration with the perspectives of Revealed theology on the primacy of the personhood and love of God.

1. THE LINEAMENTS OF OUR MAIN ARGUMENT

What we have done is explore the significance of the proposition "There exist temporal beings", thereby in Chapters II to V drawing out a certain metaphysical perspective on the notions of "time", "cause", and "existence", and their interrelation. Then, against this background, we have drawn out a particular argument to the existence of God as the cause of temporal continuance.

What we must now do is summarize this metaphysical perspective and set forth the lineaments of my main argument.

(a) *Underlying metaphysical perspective*

By the 'reality of time' we meant this: that at any juncture the future relative to that juncture has a real contingency because it still simply does not exist and is in no way a 'given'. And we found that any coherent account of time as real was inseparable from the acceptance of the reality of causal agency, involving a whole structure of causal agents with active powers, with natures determining their spread of active powers and modes of interaction. The peculiarity of the present was not primarily cognitive or psychological but dramatic: the present was the 'dramatic now', the stage now reached in the unfolding of a causal or dramatic history, the 'now' relevant to what is next to happen and therefore to action. But we established not just the inseparability of a realist view of time and a realist view of causation but the need to accept both.

We then proceeded to review analytically the full spread of types of existential statement and demonstrated the primacy (contrary to current philosophical orthodoxy) of what we called singular existential statements. Reflection on these singular existential statements calls for a picking out of a use of the word "exists"to signify what we called metaphysical reality, belonging primarily to primary substances ('subjects' in a logically and metaphysically primary sense, for which we adopted the technical term *supposita*). We were forced into the recognition that this primacy could only belong to these 'primary substances' or *supposita* by considerations of the philosophy of language. Yet always the argument that a certain structure of thought has actual valid applications proceeds not a priori (e.g. from the conceptual analysis which goes on in the philosophy of language) but a posteriori from the fact of the structure's being actually exemplified—in the case concerned, actually exemplified in the world of temporal agents considered earlier.

The consideration of existential statements exhibited a twofold aspect of causality as always involving a causal agent but yet as always also involving a causal agent as something in its nature active, i.e. as involving in its existence what we called act or activity. As to effects, always causation was of an actuality or of an aspect of it (a fact), never of a thing as such (primary substance or *suppositum*). Accordingly, effects were always what we called composite: it was not excluded that causes be incomposite, but even in that case they would have to be capable of being spoken of in distinguishable ways, spoken of as *supposita* using 'concrete' modes of speech, and spoken of as actualities in 'abstract' modes of speech—using the words "concrete" and "abstract" in a quasi-grammatical sense, as in Chapter V, Section 10, against the setting of Section 3.

This whole perspective is the one forced upon us by acceptance of the reality of time and the things that go with it, *viz.* the openness of the future associated with real contingency and the existence of real causal action (the exercise of active power by causal agents).

(b) *The basis of our metaphysical approach to theism*

Realization of the reality of time grounds not only the general metaphysical perspective just outlined but also the acceptance of the first two stages of our main argument. At the first stage, it is consideration of the real contingency of the future that makes us register the impotence of the temporal to cause the continuance of things into the future. At the second stage, the real contingency of the future is the background which makes us concede that this continuance needs a cause.

To state the matter informally: for things with natures to exist at all they have to exist not just for an instant but through time. Nature reliably continues, and causal action could not proceed at all unless substances with their natures reliably persisted in activity in, through, and during their causal action, a continuance which Nature and temporal things are impotent to effect. The application of the concept of "natural contingency" presupposes an applicability of a wider, metaphysical, concept of "real contingency", the real contingency exemplified by the future not explained in the natural sciences, which only makes sense against a wider causal background than the natural: in brief, a non-temporal cause has to underly Nature in order for the future to be possible. Things have to be upheld in existence by such a non-temporal cause precisely as things whose natures require non-instantaneous existence or existence through time, in order for this existence to have the necessary reliability.

This is the nub of the argument developed in Chapters I, VI (in Sections 2 and 3), and VII (in Section 1(*e*)). In our presentation, we avoided any appeal to the principle of sufficient reason or any similarly general causal principle which might be suggested as a norm or ideal of reason or rationality because we regarded such universal univocal principles as having no place in ordinary causal thinking besides being unnecessarily strong for the purpose in hand. Rather, we represented what is involved as the acceptance of two interdependent and diverse analogous applications of the concept of "cause", the acceptance in judgement of applications of such trans-generic concepts being prior to the formulation of any analogously understood principle embracing them all.

What was crucial to our presentation was the interdependence of the applications concerned. There was no secondary causation without the 'non-chance' continuance of things, and no 'natural', 'empirical', or 'scientific' real possibility or contingency except within the context of the underlying metaphysical possibility and contingency. Without this latter what we speak of as 'causation' or 'possibility' and 'contingency' might be merely matters of regularity or limits to it—not involving real causal action, freedom, or chance at all—because not involving any real openness of the future. Indeterminacy in physics, as a limit in that in virtue of which there is predictability, implies no such real openness unless understood within the metaphysical context I have portrayed.

Thus, at every point we see our argument dependent upon the metaphysical perspective described earlier, the perspective which involves an acceptance of the real contingency of the future and the existence of causal agents with natures and active powers exercised in real causal action.

This then is the basis of our main argument: for some reason yet to be commented on, the temporal is impotent to cause the continuance of the temporal; but temporal continuance does need such a cause. It will be noted that recognition of the reality of time, so that the future at any juncture lacks any real existence, prevents the Universe as a supra-temporal whole from being treated as a genuine object. And raising the question of the cause of the continuance of things instantly side-steps the objection that there is no respect in which one can raise causal questions about the Universe as a whole without relying upon some such principle as that of sufficient reason, e.g. without having to argue that for any set one can ask an explanation of the whole set as well as of its individual members. It will also be noted that, by reason of the unity of Nature, there can be nothing that continues things in existence in their natures which does not continue all things in Nature according to nature: it is an aspect of the natures of particular things that they dovetail with the natures of others within the context of underlying structures in Nature, including for instance the general character of their parts or 'matter'.

The questions as to how this cause can be shown to have all the attributes, 'metaphysical' and otherwise, required for it to be properly called God and as to how the whole approach can be accepted within a 'critical' philosophy and, at the same time, integrate with any claimed Revealed theology will be summarized as we proceed. But first we have to attend to the second pivot in our whole course of reasoning. The first was the reality of time, with the impotence of the temporal to ground temporal existence and the need of such a ground which went with it. The second is the distinction between the composite and the incomposite.

(c) *The role of the notions of compositeness and incompositeness*

The notion of "compositeness" is pivotal because it is the very same property, namely this that we label 'compositeness', which renders temporal things impotent to cause temporal continuance and at one and the same time debars a thing from being intrinsically underivative. As to the converse property, "incompositeness", we can say that, if anything is 'incomposite', it therein has the internal character which makes it intrinsically underivative, and also that it will follow that there is only one such thing. That is, "composite" is the middle term we introduce into the structure presented at the start of Chapter VI so that, formally, our argument is this: once it is granted (I) that, if temporal continuance has a cause, it is not composite, and (II) that this continuance has a cause, it follows (III) that this cause is not composite and therefore (IV*a*) incomposite, intrinsically underivative (self-existent or *a se*), immediate to its effects, and unique and (IV*b*), since some of the temporal things upheld are personal, itself personal and possessed of all the attributes required for it to be appropriately called God.

The inspiration behind the notions of compositeness and incompositeness as presented in this book is easy to state. It seems obvious that the temporal is impotent in the way explained to cause continuance in existence, and obvious that anything at all like mundane or ordinary things will be equally impotent: only something radically different in character could be free of such impotence. It might then be hypothesised that the cause of temporal continuance was itself caused by something else, i.e. was itself derivative, and that this further 'something else' might itself be caused by yet another thing, so that either there would be no first cause or, if there was a first cause, there would be one or more intermediates between it and the continuance of temporal things (with the result that it might be thought to work through deputies or ministers). However, as against any such hypotheses, it would seem an obvious suggestion that anything which was itself derivative would be *in the relevant way* like the temporal and no less impotent to give existence or cause its continuance unconditionally and unqualifiedly. On this suggestion, only that which is such as to be incapable of having a cause, i.e. intrinsically underivative, can thus give or sustain existence. On this suggestion, only the intrinsically underivative can cause temporal continuance and there is no room for a regress even beyond one cause, no room for even one intermediate, let alone for an infinite regress in causes. The First Cause will be immediate to the continuing existing of temporal things as its direct cause.

My study of compositeness and incompositeness represents an attempt to establish this plausible suggestion on a firm basis. To be composite is, I argue, to be *in the relevant way* like the temporal and therefore like the temporal in being impotent to cause temporal continuance (Chapter VI, Section 1). But to be composite is, on my explanation of the term, precisely to lack that which it takes internally for it to be intrinsically underivative (such by its internal character as to be incapable of being caused), and thus such as to be intrinsically necessary in its existence and uniqueness (Chapter V, Sections 9 and 11). Alternative proposals as to how compositeness should be explained and proposals of alternative candidates for the role of the middle term "*M*" may represent alternative attempts to establish the plausible suggestion mentioned on a firm basis.

We showed that the compositeness of ordinary things is the immediate consequence of the proposition that time is real: the distinction between activity and situation goes with the reality of time and involves compositeness.

In brief, the reality of time involves the compositeness of things in such a way that the future or what is later *always* has the status of a distinct existence (a new contingent continuance) in relation to what preceded it. I say 'always' to mark that, besides the more obvious connotations of this statement, it has the more general effect of limiting what can be legitimately insinuated into various forms of holism. For, if we take an overview of the whole spectrum of possible philosophical opinions, we can see this statement as preventing any conception of natures (whether of individual things or of Nature or the Universe globally) as wholes from raising wholes to the status of absolutes. Within any holistic view of individuals or the Universe it will remain that, if time is real, then, at every juncture, what happens next and even whether there will be a 'next' is open. (By itself, the indeterminacy discovered in modern physics shows little because it does not exclude a holism within which what is undetermined at one level of explanation is discovered to be determined at a higher level of explanation. Rather, what is needed is the truncation of any holism whatsoever in the name of the openness of the future.)

My account of compositeness and incompositeness rested upon the same metaphysical perspective in regard to what is involved in real causal action that I indicated earlier.

The existence of the composite as the possible object of real causal action involves a distinction between that which exists and its existence, subject *(suppositum)* and actuality: that which exists without intrinsic necessity is of a nature such that it might not have been instantiated, and it is only this existence or actuality, not the subject or its nature, which is caused. Thus, anything caused, any effect, is as such composite.

By contrast, a cause is not as such composite. Under the aspect of causal agent, it is a subject or *suppositum* which causes; but without activity there is no action, and, under the aspect of activity, it is an actuality which causes. Both subject and actuality are always involved in a causal activity so that not only is one not compelled to say that it is one rather than the other that causes, but also there is no general compulsion upon one, i.e. no compulsion arising from the character of efficient causation as such, to say that there is a real distinction between subject as cause and actuality as cause. Subject and actuality always differ in our modes of speech referring to them, so that what is referred to as subject is spoken of as causal agent and what is referred to as actuality is spoken of as activity, but, if the incomposite exists, do not always differ in pertinence to efficient causation: their difference would then affect how we describe causal relations, but not what there might be room for efficient causal explanation of or what its explanation might be.

The only reason we have to suppose that the incomposite exists is that its existence is a causal condition of temporal continuance. If anything incomposite exists, it will possess its existence in such a different way from ordinary mundane or temporal things, so as not to be just another thing alongside ordinary things, itself wanting a cause, but rather to be intrinsically underivative. Such a thing, as cause of temporal continuance, thing communicating existence to other things, will be so different in character from anything it communicates existence to that these will be relatively all alike, and all alike in being impotent to communicate existence: therefore, the giver of existence will not be able to deputise the power to give existence: in giving or upholding existence it will only be able to operate unmediatedly or directly, immediate as agent of existence to all else that exists.

(d) *The rest of our strategy*

As to the rest of our strategy, it flows naturally from the starting-points already identified.

Firstly, in regard to the modes of God's being *qua* subject or *suppositum*, we showed in Chapters VIII and IX these might be established a posteriori on the basis of His role as First Cause: e.g. His personhood was inferred from the character of His effects. We do not need to resume those arguments and how we safeguarded their critical character.

Secondly, in regard to aspects of God's being *qua* actuality, we penetrated in stages into some of the significance of His incompositeness.

In itself, as we showed in Chapter V, Part II, incompositeness is a form of unity or integratedness, primarily the unity of God's actuality whereby from the standpoint of efficient causation there is no real distinction between its focus or focuses (its subject or subjects) and itself, so that He exists *a se* or is self-existent. If He is living, then eternity as the enjoyment in one act of the whole of His perfect life is an aspect of incompositeness. From the absence of any distinction in Him relevant to explaining His existence, it followed that there is no standpoint which would allow of a nature which might or might not be instantiated or might be instantiated a different number of times, so that the necessity of the existence and uniqueness of God follow from His existence and incompositeness. And since His nature is nothing other than His existing, i.e. it is this incomposite actuality itself, it follows, as we realized in Chapter IX, in knowing this actuality as cause of the mundane continuance in existence of which we have experience, we know of it as implicitly cause of any possible continuance in existence and any possible derivative existence whatsoever—His uniqueness is a corollary of His type or mode of actuality. (It will be remembered that we carefully explained that God's uniqueness and the necessity of His existence were logical corollaries, not ontological consequences, of His existence and incompositeness.)

However, returning to our first insight, our apprehension of the reality of time, we have to recognize a third stage in reflection upon its significance. At the first stage we register the impotence of the temporal to cause, unconditionally, temporal continuance—indeed, as it were, the irrelevance of their powers and their constitution to the achieving of any such end. At the second stage we realize that the very reality of mundane causal power, presupposing as a 'given' a certain continuance of agents in their natures, means that this continuance must nonetheless have a cause. But now at the third stage we are forced to an apprehension that the power of the incomposite cause which can ground the continued existence of the temporal unconditionally, relying upon nothing outside itself, making the temporal 'stand outside nothingness' simply of its own power, must be unlimited, open, uncontrolled, and untamed, in a peculiar way. This insight is what I explored in Chapter IX, Section 2, noting how, in the case of things generically defined in a way not restricting them to material operations (e.g. 'persons' and 'causes'), nature and power are defined and measured only in object-related ways, so that the unlimitedness of God's power *qua* actuality might imply omnipotence.

In that same context, I indicated how the concept of incompositeness, as applied to God's being *qua* actuality and activity, meant that each new note in our understanding of God as subject carried with it infinity or unlimitedness in those aspects of His being (*esse*), life, and activity in terms of which it was understood, as personhood and being living are understood in terms of the activities of understanding, love, and life.

As to more detailed points there is no need for me to resume the discussions of Chapter IX. My care in maintaining an a posteriori and critical approach was evident there, but its successive facets are recapitulated in my next section (part (*c*)) when I explain how this care dovetails with the requirements of any postulated Revelation.

2. THE INDISPENSABILITY OF THIS METAPHYSICAL APPROACH TO ANY THEOLOGY

(*a*) *The role of metaphysics in any 'positive' theology*

The argument presented in the book shows that the Universe has a personal first cause which is incomposite in the sense explained and unique. Chapter IX explored the limitations of the argument at this stage and what more was required to show that this First Cause is indeed what men mean by "God".

But let us now explain the vital function of the points we have picked out in this chapter within any theology whatsoever.

In the first place, an approach such as we have proposed provides the key to explaining such indispensable terms in regard to God as those of "self-existence" or "aseity", "infinity", "eternity" and "omnipresence", "transcendence", and "immanence".

In the second place, all these concepts have to be understood in such a way as not to undermine the reality of created substances and the causal action and freedom of created agents, an understanding which goes with a grasp of the reality of time. And this grasp of the reality of time, along with the reality of causal action and of freedom within creation will belong primarily to metaphysics, not to theology. If theology is represented as the sole true source of such understanding, it will be found that what is presented is a construction of the mind in which neither God's action, eternity and presence nor creaturely freedom and reality are safeguarded.

Let us begin by considering the key terms I mentioned in regard to God.

Revelation needs the concept of a being which is underivative, but whose underivativeness is not just *de facto* but intrinsic, arising from a difference in the way this being possesses existence, a difference setting it apart from all creatures and rendering it incapable of having a cause: it has to exist 'of itself' without causing its own existence. Revelation needs to understand this being as having an existence in no way limited by a nature as if by something quasi-abstract and prior, i.e. as infinite in being. That is, this being has to be not only transcendent in identity (it is distinct and causally independent and prior) but also 'transcendent in nature', so that there is an 'infinite' gulf of nature between it and creatures.

Yet it also needs a non-metaphorical explanation of what it is for God to be not only transcendent but also immanent, i.e. 'present in' things 'immediately' and indeed 'omnipresent' throughout time and space.

Commonly, the use of these terms proceeds without explanation as if relying upon mere rhetoric. Without such explanations as the approach we have adopted makes possible, it is not clear that any coherent sense can be given to the terms concerned.

It would seem that the notion of a self-existent being, i.e. of a being whose underivativeness is not just *de facto* but intrinsic, arising from this being's very mode of existing, could only be known to make sense because it had to be conceded that it had actual application. This has to be conceded because created beings can be perceived to need just such a cause, since any other kind of being would be as impotent as they are to ground temporal existence. We do not have insight into the conception of "self-existence" so as thus to know its coherence, but know its coherence in virtue of knowing its actualization, which we can know only a posteriori from the existence of creatures. In the same act of understanding whereby we perceive that only the self-existent can ground that which is not self-existent, we also perceive both the gulf of nature whereby the self-existent transcends in nature (not just in identity) the non-self-existent, but also the immediacy of the self-existent to all non-self-existent, ordinary, temporal beings.

Or, to re-express the same point: until the reason why temporal things are impotent to give existence unconditionally is realized to be some property (say, compositeness) which would be *ipso facto* shared by anything not intrinsically derivative, and yet that the temporal must have such a cause of continuance which therefore must be opposite in nature (say, incomposite), and because of this intrinsically underivative, it is not clear that the notion of the 'intrinsically' underivative or the 'self'-existent is meaningful. At the same time, once it is realized that only that which is thus opposite in nature (incomposite and thereby intrinsically underivative) even might be capable of communicating existence to other things, it becomes evident that there can be no mediation between the cause of temporal existence and the temporal existence caused. Thus, the notions of God's transcendence in nature and God's immediacy or immanence are explained in one and the same act and explained in the same act as establishing that these concepts have application, i.e. exhibiting temporal existence as having a cause which can only be of this kind.

Moreover, our explanations of these notions of transcendence in nature and immediacy have been in an adequate way non-metaphorical. We have not merely spoken of a 'gulf' of nature but have explained the respect in which there is this 'gulf'. And, because we have made it plain that God is present to what goes on in the world in the same way as an agent is present in his actions (logically immediate as subject doing the actions and causally immediate as unmediated), we have by implication made it plain that He is present 'in' the world in no more metaphorical a sense than the sense in which personal agents in general are present in the world. As Chapter V, Section 7, made it plain, being present goes with being the subject of propositions of certain kinds, not with being 'contained'.

The God who is thus 'immediate' and 'omnipresent' by being unmediated cause of continuing existence to whatever exists has to be not just everlasting but 'eternal' in the sense of 'non-temporal' in order to be able to communicate existence into the future. But our basing of this understanding upon a perception of the reality of time and the real contingency which arises from the absolute non-existence at any juncture of what is still future to that juncture excludes any interpretation of 'eternity' which would undermine this contingency and non-existence: the 'dramatic now' of Chapter II is a dramatic now even for God—indeed He concreates Time with the world by at each juncture constituting the present in being.

Our explanation of what is involved in a non-dualist account of action precisely enabled us to see how these conceptions are to be reconciled: the primary expression of God's will is in His action, not in some everlasting pre-existing decree, and this action is in time. The past and present are fixed because in them God's will has been declared in act; the future is open because in respect of it the will neither of God nor of any creature has been declared in act—in regard to the future, God and man remain free. God as cause, agent, and actuality is non-temporal: His action is temporal.

Thus, in this second respect, the metaphysical perspective which we opened out in regard to the reality of time and the character of real causal action is vital in theology. For it is the perspective to which it is vital for a theology to hold if it is not to yield to the temptation to ascribe God so total a role as to deny nature and freedom. Yet the explanation and justification of this perspective belongs, as we remarked, primarily to metaphysics.

Thus it might seem that the unfolding of a metaphysical approach to God was not only compatible with Revealed theology, but also positively demanded by it. It is not only grace, but also nature, man's humanity itself, which is in this view violated in idolatry and even in the worship of nature or the things of nature, and violated again in any pretence that the things of nature could stand without a ground, this ever immediate yet self-existing cause.

'The heavens declare the glory of God', the Psalmist tells us, and, if St Paul is right, 'His invisible nature, namely His eternal power and deity, has been clearly perceived in the things that have been made' (Rom. l: 20). On the face of it, it is part of the character of Semitic religion, Jewish, Christian and Muslim, to suppose that the idolater who worships things of human making and the less corrupt human being who worships the wonderful things of nature are both guilty of some human intellectual weakness, failing to see that there must be some underlying source of the continued existence and order of these things, the source of their wonder. In this way Chapter 13 of the Book of Wisdom in the Greek Old Testament merely makes explicit an underlying presumption.

(b) The indispensable negative role of metaphysics in any theology

As to the indispensable negative role of the metaphysical approach we have put forward, it is a matter only of reminding the reader of earlier argument.

Firstly and most obviously, we again and again found ourselves involved in the rejection of Platonist models, whereby the comparison of God to an Idea or Form was used to give some positive content to the doctrines of His incompositeness (unity as a first-level property), His eternity, or His uniqueness (unity as a second-level property). As to His uniqueness, it should not be regarded as either a negative or a positive property, but our explanation of how it is that the term "God" serves both as a singular term and as a general term (both as a proper name and as the name of a nature) dispenses with the need of any resort to Platonism in this regard.

Secondly and most generally, our explanations have made plain the negative purport of most of the terms we use of God. The concept of "incompositeness" (or first-level unity) provides only the most notable case of this. God is one in the sense that there are no distinctions relevant in a certain way to efficient causation in Him, and not in a sense excluding many other sorts of distinction, above all, as we saw in Chapter V, those between Divine Persons: this 'unity', 'incompositeness', or *simplicitas* has nothing to do with the inner cohesion of an Idea. In key cases it is important to note carefully what is denied. God is incomprehensible but not to Himself: He remains personal and therefore intellectual in being. But, more particularly, God is non-temporal and 'eternal' but, as we have seen, not so as to exclude His actions and presence in time and, above all, not so as to exclude life and activity in Himself. Eternity is not simple non-temporality, but is 'the possession, perfect and the whole at once [in one act], of life without terminus': that is, it is a negatively defined form of integratedness of life, and as such excludes rather than implies stasis, and, so far as terms for the passions do not imply alteration in Him from one state to another or episodes in an incomplete possession or enjoyment of life, they still apply to Him.

Thirdly and most importantly, we explained how the notion of incompositeness was necessary in order to prune theology of false finitary models of personhood. The problem of whether 'individuality' had a place in God was resolved into the question of the nature of the activity which God's being consists in, and here the conception of God as incomposite indeed seemed to impose the right constraint on the mode of realization of the concept of "person" in God. Meantime, models of the relation of God and the world derived from post-Cartesian understandings whereby mind and body stood in parallelism or interactive relationships had invited an anthropomorphic modelling of God's relation to the World either upon an interactionist basis or upon a deistic or 'strong predestinationist' picture within which God's decrees established everything at the beginning. But such parallelism is at once excluded by the non-finite or incomparable nature of God involved in His incompositeness: His immediacy is not that of a coeval co-finite being but that of a root of temporal continuance transcending time.

The theology of many contemporary thinkers incorporates the defects of their metaphysics. They postulate a personal God, existing in a Cartesian parallelism alongside the Universe, without any means within their system of thought whereby they can explain either His infinity or His eternity. Influenced by Hume, they hold active power (even the active power to create) in only loose connection with nature, and require no particular immediacy of cause to effect. For these thinkers, the continuance of things in existence does not particularly need a cause, or at least not a cause of any special kind, since—did not Hume teach us—any thing may cause any thing. And, even if in fact—as the theologians tell us—God is the cause both of the coming into existence from nothing and of the continuance of things in existence, there is no particular reason except His choice why He should not depute the power to do these things to something else. Granted that God has not thus deputed the job of giving existence to another, it remains that in this way of thinking He does not need to possess existence in any peculiar way (e.g. intrinsically underivatively) any more than such a supposed deputy would. Thus, the opening which we have suggested enables us to explain in one move both God's immediacy to created things and how His different way of possessing existence and nature is closed to them.

For these thinkers, indeed, God has a nature and this does not limit His actions towards creation and perhaps, the theologians may tell us, does not limit His actions at all. But there is no particular reason, unless it lie either in what the theologians tell us or in some Ockham-inspired economy of supposition, for supposing God to be incomplex, non-temporal, infinite except in certain powers, unqualifiedly omnipotent, or unique. We indicated the predicament of these 'Newtonian' theologians in Chapter II, Section 1(*e*). And even if, as the physicists tell us, Time cannot be conceived as 'prior' to the Universe so that if God created the Universe He must have created our Time, it may be wildly proposed that He has His own 'Time' (possibly called 'Eternity'), from which He may exercise Lordship over our Time.

Any theology which spurns metaphysics is without adequate remedy in dealing with the Demiurge. There is only the possibility of resort to rhetoric and metaphor. And if the object is particularly great our theologians may avoid the term "metaphor" and speak of '*analogia fidei*' or 'symbol'. For One who is great enough there is no term of honour which, if Hobbes is right, they should hold back from using and it would follow that, within a certain consistency, there should be no limits to a theologian's rhetoric.

Against such anarchy in theology and impiety in multiplying words before God, we have to set the barrier of the *via negativa*. We have to deny the likeness to creatures and comparability with creatures upon which the position of these theologians depends. For this it is absolutely necessary that God should be differently related to His effects than any other possible cause to its effects and that He should possess His nature in a different way from any other possible being. The concept of "incompositeness" enables us to secure the assertion of these things. But this is a concept whose coherence is only known a posteriori from the need of the temporal to have a cause, a cause which can only be incomposite.

Thus, the main weapons of any negative theology can never be any concepts whose applicability is given a priori. Positive or cataphatic theology and negative or apophatic theology are never given or developed in separation from one another—as if, for instance, one might speak of some theology as purely apophatic. True, the path of the *via negativa* has a strategic role in growth in religious devotion, related to the key place of incomprehensibility, unimaginability, and unspeakability in the knowledge of the Holy, but to develop this is not our concern. Our concern is rather to represent the indispensability of the *via negativa* to any theology, and its dependence on the metaphysical perspective we have opened out, a dependence in some respects more absolute than the dependence of the positive predications upon this perspective for reasons indicated at the end of Chapter IX.

(c) The basis for an integration of natural knowledge with Revealed theology

What we achieved in Chapters VI and VII included two things.

First, we laid the basis for the adoption of an a posteriori approach to theism within metaphysics which would, in the way indicated in Chapter IX, dovetail with what might be given in Revelation. In connection with the latter it is vital that, in our a posteriori approach, whenever, in some case or type of case, it is suggested that a cause is needed, it is a first-order question as to whether this is so. It is a vital point of philosophical method that the answer to such first-order questions about the conditions of existence should not be made to turn upon the answer to second-order questions about the conditions of rationality in thinking about the first-order questions (so we argued in Chapter VI, Section 4, and Chapter VII, Section 1 (*d*)). In this way, while we rightly allow a posteriori considerations of the conditions of existence an internal place in any theology, we do not allow any system of epistemology, imposed a priori or *ab extra*, any rule in theology any more than in philosophy.

Secondly, we gave an account of our understanding of general concepts which would allow, and indeed sometimes require, their analogous application both to creatures and to God: a foundation for the possibility of discourse in theology. For, as we have observed, we noted in the course of our attack on purely a priori approaches in philosophy that any appeal to the principle of sufficient reason or any similarly general causal principle which might be suggested as a norm or ideal of reason or rationality was otiose: such universal univocal principles had no place in ordinary causal thinking besides being unnecessarily strong for the purpose in hand of arguing God's existence; and in Chapter VII, Section 1, we represented them as incapable of standing up in any critical philosophy. Rather, we represented what is involved as the acceptance of two interdependent and diverse analogous applications of the concept of "cause", showing the propriety of thus applying concepts analogously in Chapter VII, Section 2.

Yet it is still necessary that the approach be critical, and I show how this is possible in Chapter VI, Section 3, Chapter VII, Section 1(*e*), and Chapter VIII, Section 5. The whole of Chapter IX may be regarded as an examination of the conditions of maintaining a critically acceptable method in any further development of a metaphysical theism, including aspects to do with the omnipotence of God, the unity of the Universe, and God's goodness and holiness. And the last stages of this inquiry, into goodness and holiness, were precisely calculated to include a consideration of the conditions of the relation between metaphysics and Revealed theology being critically approvable.

3. DISMANTLING THE SUBSTRUCTURE OF SCEPTICISM

Such then is my argument. It takes account of the need to examine all the questions I have reviewed in regard to the methodology and epistemology of science. It takes account of the need to exercise careful criticism and watchfulness in regard to use of such notions as "existence" and "act of existence". It takes account of the need to distinguish the goodness which consists in the absence of metaphysical defect and the goodness which is explained in terms also of the possession or exercise or being in a state of personal relationship. It takes account also of all the epistemological worries and enquiries raised by philosophers in modern times, in response to the seemingly humane scepticism expressed in Montaigne and Locke, in the writings of Descartes, Hume, and Kant, and derivatively their successors. In this last context what was of crucial importance was our making clear how it is always illegitimate to give epistemological primacy to second-order considerations and judgements, and also vital in consequence of this to reject arguments for God's existence *qua* cause of rational order when these arguments depend on making the second-order primary. Nor may idealism, surreptitiously introduced on epistemological grounds, be made an excuse for regarding the First Cause as personal rather than impersonal.

At many points, it is vital to note that a proposition in philosophy can force itself upon our acceptance and indeed gain the status of knowledge in advance of our having identified and established its implications or presuppositions—presuppositions, it may be, of varied kinds. Thus we establish the reality of time in advance of attaining any appreciation of whether or not an argument for God's existence will stem from it. We attain conclusions in the philosophy of language in relation to the polymorphous character of the more general first-order and second-order terms, and in relation to the hierarchical shape of the concept of causal agent, in advance of developing a theory of the human understanding such as these things may require. And we develop these conclusions in the philosophy of language in advance of any explicit development of the metaphysics they may be supposed to imply or presuppose, *qua* metaphysics, and certainly entirely in advance of any such further or more distant conclusions in metaphysics as may belong to cosmology or theology. Likewise, in epistemology, we establish the existence of a capacity to generalize, not drawing on prior general judgements or axioms, but utilizing some powers of understanding concepts, in advance of developing any appropriate global theory of the human understanding. To develop such a theory is part of a general philosophical anthropology and does not belong to this book. Of course, if it had been the case, or if it were to prove to be the case, that the conclusions in one area of philosophy turned out to be incompatible with their presuppositions, then both would have to be regarded as together in the melting-pot. But such incoherence, although often alleged, has, I believe, never been proved, nor its existence rendered even remotely probable. The history of philosophy is full of over-simplifications (the bequest of rationalism and the doctrine of clear and distinct concepts) generating such allegations, generating a dust so that no man may see, and offering no plausible way out of the difficulties imagined. In my discussions of epistemology and the role of the philosophy of language, I have indicated some of the ways in which these forms of scepticism are to be avoided and the depth of the incoherence of their premisses appreciated.

GENERAL CONCLUSION

Our argument, therefore, takes as its starting-point the reality of time. That is, it rejects the positivist or descriptivist orientation of some modern conceptions of physics and of human apprehension of the real, in favour of a non-subjectivist and realist account of time and causality as intertwined and directly apprehended, implicitly or explicitly, within every direct apprehension, or as its pre-conditions. And, in this, it takes as its starting point that which is a pre-condition of all real natural causation and of human freedom of action and disposition.

Thereby, it involves a rejection of any conception, deist, rationalist, or predestinationist, of creation as something accomplished in a single act in relation to a unitary object, named the universe. Rather, its starting point, namely the conception of causality and freedom as both equally involving an open future to be determined by the co-action of free and unfree causal agents, enforces a realist view of creation orientated towards an historical perspective, as something not accomplished once and for all in a single act, but unfolding, each stage of its unfolding being fresh and an expression of Divine freedom, pre-condition and context of real human freedom. It does not compel, but leaves the way open for the suggestions Revealed theology may make that the whole cosmos falls within the ambit of a single History, unfolding and at no stage determined before the end.

Therein, also, it gives an appreciation of the genuineness of the tensing (present-tensing) of all statements of God's existence, here and now, and of the immediacy of the whole furnace of the Divine life to each acting and living thing, and to every expression or realization of active causal power or of life. Thereby it indicates how an entirely non-abstract sense is to be given to the doctrine of God's eternity.

By the very structure of the approach I have explained, it becomes apparent that for human beings and human understanding, including the understanding claimed to be of faith, the transcendence and the immediacy of the divine are inseparable. That is, there can be no true assertion of God's transcendence in nature (that whereby He is unlike creatures, or as Rudolf Otto would have it "Wholly Other"), which does not in the same stroke of intellectual assent, perception, or insight make clear the non-metaphorical sense of God's immediacy—the immanence so emphasized in Indian-influenced religion.

This perception shows how it is that the eternal God, whose life within Himself is so unified that it permits no causal unfolding or waiting for development in stages or any kind of temporality, is, at the same time, the One whose breath (*spiritus*, *pneuma*, or *ruah*) enlivens, energises, and gives vigour and existence to all things, immediate within them. What we see or empirically know is, in a certain sense, in its positive aspects, His breathing.

Finally and above all, our argument takes account of the need to give a true or proper reason for holding the First Cause to be personal and living, a reason appropriate to man's situation as a bodily member of the physical Universe, and not proceeding from any overweening intellectualism or surreptitious idealism, making the reality of the bodily and material a construction of the human mind. For it is only with this acknowledgement, not just of God's power (His status as first and master of all) and divinity (His transcendence, uniqueness, and incomprehensibility), but also His personhood, livingness, and wonder or holiness, that we reach the summit pointed to by St Paul in the passage we cited, 'His invisible nature, namely His eternal power and deity, has been clearly perceived in the things that have been made' (Rom. 1: 20). The argument and conclusion of this book—reaching from the reality of time and the impotence of mundane things to perpetuate their own existence, all of them in this respect immediately dependent on God's action or, as it were, on His breathing and His energizing all—is most aptly stated in the words of the Book of Wisdom, 11: 25–12: 1. This would have been a part of the backcloth of St Paul's own thought, one who knew the books of the Greek-speaking Jews with the Septuagint. This passage of Wisdom reads:

> Yes, You love all that exists,
> You hold nothing of what You have made in abhorrence,
> For had You hated anything, You would not have formed it.
> And how, had You not willed it, could a thing persist?
> How be conserved if not called forth by You?
> You spare all things because all things are Yours,
> Lord, lover of life,
> You whose imperishable spirit is in all.

Postscript: The Unity of Philosophy and the Impossibility of a Non-philosophical Theology

It has become clear that philosophy is constituted like a seamless web or tapestry within which it is contrived that every strand is interwoven in complex ways with every other strand.

There is no possibility of severing one's account of causation from one's account of time, nor any possibility of giving an account of either which does not depend upon, while yet being integral to, one's account of substance and of real existence. There is no possibility of giving a satisfactory account of the variety of modes of causality and modes of existence which does not also include a right account of human and animal action. Nor is there any possibility of giving any account of the uses of the word "exist" which does not incorporate the insights and precision made possible by the development of logic, and several quite distinct ways in which existence stands as a presupposition or pre-condition of every other statement, fact, or actuality, rather than as just another of any of these. Yet there is no possibility of giving a satisfactory account of any of these notions if there is not also a satisfactory development of the philosophy of language, indicating how the most general words and structures within speech and writing may be used in a disciplined way (there being no ambiguity created between the speaker and hearer) while yet without being used mechanically or according to formalizable rules. And any such unfolding of the philosophy of language will be rendered impossible from the start by sundry common defective accounts of the operation of human understanding. In this way, we are brought to the borderland between first-order and second-order inquiry.

We further need to recognize the effect of the views as to the primacy of second-order inquiry over first-order inquiry which have become standard among philosophers and some popularizers of science, as well as some theologians. These views are present by implication when materialists and others give a merely conventionalist account of the status of what they deign to call 'basic' propositions and when pragmatists and Marxists in very different ways erect *praxis* into a criterion of truth or substitute for theory. The effect of such views is to subvert the possibility of access of each human being, not only to knowledge of God, but also to knowledge of other human beings, and even before that to knowledge of any public world whatsoever, and indeed, in the end, even to knowledge of self.

Finally, in this connection, we need to note that there is no stage of any inquiry at which it is not presupposed that truth and knowledge are goals, goods for human beings, and no stage at which the hope of satisfaction and enjoyment of the fruit of inquiry is altogether absent. The good and the pleasing or beautiful, and the theories of them, stalk philosophical inquiry, and haunt man's thought as well as his life in a way that cannot be exorcized.

Let there be any strategic over-simplification in the account given of time or of cause, of existence or of generality, of substance or of actuality and relation, of the use of language or of the structure of understanding, of the variety of modes of causality, freedom, and modality (possibility and necessity), of knowledge and the conditions of knowledge, of value and aspects of value, and inexorably an impasse will arise in connection, not only with the account given of each and every one of the others, but also with any theology whatsoever.

In theology there will be a threefold impasse.

In the first place, in any theology whatsoever, a false conception in regard to any of the matters I have instanced will generate a distortion in the human appropriation of the religion concerned. It is useful to give several instances of this.

To take a first instance, the dualism that infects the philosopher in his thinking about man will infect the theologian, not only in his thinking about man, but also in his thinking about God. Thus the doctrine of the eternity of God, whereby it should be understood that the whole furnace of the divine life is immediately present within every expression of positive existence in creation, since the primary expression of will is in intentional action (the action itself) not in anything merely mental, and the intentional action is here the giving, or immediately upholding of, existence to the creaturely thing, is turned on its head. It is made now instead to signify that some God away in His eternity has executed an act of will, an eternal decree, internal to His own mind, indifferently regarded as before, during, and after what goes on in creation, so that creation is an area of mere play within which God's will is expressed. In this way dualism has engendered deism and also the most mis-shapen versions of predestinationism. The externality of the cause to its effect, which is vicious in any theory of human action, is entirely destructive of any theology whatsoever, because destructive of any doctrine of creation and therewith, within theology, of any account of human freedom and any acceptance of the reality of time.

Or again, to take a second instance, the atomistic approach to word-meaning, whereby each word is understood to be used woodenly according to just one or some small number of rigid patterns of use listed in a lexicon, even if the lexicon be imagined to be the gift of heaven dominating all other language-use, emasculates theology. It removes from theology, on the one hand, the capacity to understand either its own inner structure or its development, which involve respectively an indefinitely variegated set of uses for any key word at any one time, and a development in patterns of use over time. And, on the other hand, it also precludes any true integration of theological discourse with the rest of human discourse, so that the theology becomes alien to man and sterile, ultimately incapable of human appropriation except by a kind of ritual obedience; and that wealth of means to understanding provided by the bringing of the almost infinitely faceted riches of empirical experience into the construction of the theological temple—the wealth made use of at every point from the highest to the lowest in the scriptures sacred to Jew, Christian, or Muslim—is spurned.

Or, to take a third instance, we observed how in philosophy the elevation of the second-order above the first-order led to an ever more empty subjectivism or even solipsism, or else to mere conventionalism. The same epistemology in theology leads to the erection of the principle that what is given to be believed by authority is to be received with an obedient and immovable adherence by faith into a position of isolated pre-eminence, with the same subjectivism and arbitrariness resulting. The tradition that the act of faith involves God's exercise of grace, that it is voluntary as other acts of trust are voluntary, that it is certain (firm, and intellectually unwavering), and yet that it is at the same time reasonable is in this approach undermined. Grace, voluntariness, and firmness have been preserved but the character of faith as involving an intellectual aspect, an assent also of the mind, with a certainty in that respect, and its reasonableness have been emptied out, all intellectual content and all reasonableness being removed by a sundering of links between faith and the rest of human reflection and experience.

One has often met in philosophy the syndrome whereby attempt is made to select out one basic idea, one single foundational or structuring principle, and represent the whole of human knowledge as a single edifice founded upon it, or determined according to it: this is often regarded as the hallmark of Cartesian rationalism. What we find in theology is often the same Promethean mistake, the epistemological error of erecting one human idea or principle into isolated primacy. Sometimes it is the idea of the infallibility of some scripture, some theological or church authority, or of conscience. Sometimes it is some particular concept in regard to God or to Christ—"unity", "Trinity", "love", "Word", or "Revelation"—the centrality of one out of a range of concepts of the person or of the reality being substituted for the centrality of the person or the reality, Himself or Itself.

Or, a final instance, the presumption in philosophy that facts or reality have a value-free nature and specification, so that there is an "ought"-free "is" from which no "ought" can be derived and so that what implies value cannot have the status of fact, is destructive of the possibility of ethical objectivism, removes wonder from the heavens and from anything physical, and makes the value of truth or knowledge questionable. The same philosophical presumption brought into theology makes any mention of the goodness, wonder, and worshipability of God incoherent, and reduces theological ethics to an ethics of activist obedience to commands, whether the commands seem to concern ritual, or whether they seem to concern justice and beneficence.

These then are just four instances out of a far larger number, of the first kind of impasse generated within any theology whatsoever if it incorporates false philosophical ideas. And there is no way of developing any theology whatsoever which does not incorporate any philosophical ideas. Either we develop our theology thinkingly, and that will mean philosophically, in the sense that it will involve structures of thought about language, time, causation, the nature of man, etc.; or we do not develop or have any theology at all, but only statements of faith, into whose content and unfolding the understanding has been permitted no inquiry and no penetration. The question is, therefore, if we are able to have a theology at all, only this, as to whether we are to be self-conscious, self-critical, self-reflective in regard to the philosophical ideas we make use of; or whether, while pretending to be theologians or biblical critics free of and above all philosophy, we are merely to fall prey uncritically to whatever philosophical ideas are fashionable in our milieu.

The impasse to which I have just alluded amounts to this, any strategic philosophical mistake will generate distortion and internal incoherence in theology, so that there is no safety in a non-philosophical theology but only uncontrolled peril, including the incoherence resulting from a simplistic philosophy of language. However, this last not only generates an impasse internal to theology itself, whereby it is rendered inaccessible to the human understanding as such (even that of the saint), but also an impasse in the relation of theology to each and every other area of human discourse, whether history or literary studies, natural or human science, ethics, politics, or philosophy. In particular the ground is removed, not only from the possibility of any disciplined use of the most general concepts of all ("exists", "lives", "knows", "acts", "love", "good", "true", and suchlike) within theology itself, but also from any shared use of these terms as between religions or theological discourse and the rest of human discourse. Yet it is upon the shared use of these concepts that the use of these and any other concepts in theology depends. It is, of course, this use upon which even the consideration (let alone the acceptance) of arguments to God's existence and nature depends. But what even the theologians who execrate natural theology should be brought to recognize is that their discourse depends on the use of these shared concepts, common property independently of the question of the status of natural theology.

However, the argument of this book not only shows the possibility and indeed the actuality of a cogent and valid metaphysical approach to certain basic truths in respect of God, but also indicates that acceptance of the existence of such a metaphysical approach may be integral to Revealed theology itself in several respects.

Firstly, it will be integral to the capacity of any Revealed theology to admit both the transcendence and the immanence of God, both the gulf of nature (His self-existence) and His immediacy, each in a real and not a metaphorical sense without the assertion of such combination resting at the level of mere rhetoric, possibly disguising nonsense. Likewise with the concepts of the divine unity and God's personhood. Knowledge of coherence stems from knowledge of actuality, and the knowledge that faith gives of actuality in these respects is such as to demand metaphysical exploration if false and idolatrous models, e.g. of a Platonic kind, are to be escaped.

But, secondly, it will also be integral to the capacity of Revealed theology to secure the recognition of the reality of time, and the reality of natural causation and human freedom, against the tendencies of over-simple models of the dominance of God and of God's grace to sweep nature away. Grace does not destroy nature or finitude: finitude and sin must not be confused as in Hegelian tradition. It is the realization that all secondary causation is immediately underpinned by the primary causation of God that saves the assertion of the force of mundane natures (their real active power) from being undermined by metaphysical (not just epistemological) argument.

Bibliography

Standard works, e.g. Plato, Aristotle, Augustine, Descartes, Locke and Berkeley, and works of Scripture (e.g. the Book of Wisdom in the so-called Apocrypha) are not listed where no particular edition or version is referred to.

ANSCOMBE, G. E. M., *Intention* (Blackwell, Oxford, 1958).

—— *An Introduction to Wittgenstein's Tractatus* (Hutchinson, London, 1959).

—— ' "Whatever has a Beginning of Existence must have a Cause": Hume's Argument Exposed', in *Collected Philosophical Papers*, vol. i (Blackwell, Oxford, 1981).

—— 'Causality and Determination', in *Collected Philosophical Papers*, vol. ii (Blackwell, Oxford, 1981).

ANSELM, *De Veritate*; English trans. in *Truth, Freedom and Evil: Three Philosophical Dialogues*, tr. J. Hopkins and H. Richardson (Harper and Row, New York, 1967).

ATHANASIUS, *Contra Gentes*; English trans. in *Nicene and Post-Nicene Fathers*, 2nd ser., vol. vi (Eerdmans, Grand Rapids, 1957).

AUGUSTINE, *Epistle 157*; English trans. in *Nicene and Post-Nicene Fathers*, 1st ser., vol. i (Eerdmans, Grand Rapids, 1956).

—— *On Free-will*; English trans. in *Augustine: Earlier Writings*, ed. J. H. S. Burleigh (SCM, London, 1953).

AUSTIN, J. L., *Sense and Sensibilia* (OUP, Oxford, 1962).

—— *How to Do Things with Words* (OUP, Oxford, 1962).

AYER, A. J., *The Foundations of Empirical Knowledge* (Macmillan, London, 1964).

—— *Language, Truth and Logic* (Macmillan, London, 1936).

BANEZ, D., *Scholastica commentaria in primum partem angelici doctoris ad sexagesima quartem questionem* (1584); English trans., B. S. Llamzon, *The Primacy of Existence in Thomas Aquinas* (H. Regnery and Co., Chicago, 1966; Library of Congress Catalog. Card No.66-26972).

BARTH, K. A., *Church Dogmatics* (T. and T. Clark, Edinburgh, 1957).

BEDNAROWSKI, W., 'The Riddle of Existence', *Proc. Arist. Soc.*, suppl. vol. 50 (1976).

BETH, E. W., *Foundations of Mathematics* (North Holland, Amsterdam, 1959).

BOHM, D., *Causality and Chance in Modern Physics* (RKP, London, 1957).

BRADLEY, F. H., *Essays on Truth and Reality* (OUP, Oxford, 1914).

BRAINE, D. D. C., 'Explanation and Modality', B.Phil. thesis (Oxford, 1965).

—— 'Two Kinds of Critical Philosophy', unpublished paper delivered before the Société Belge de Logique et de la Philosophie de Science in 1972.

—— 'Varieties of Necessity', *Proc. Arist. Soc.*, suppl. vol. 46 (1972).

—— 'The Nature of Knowledge', *Proc. Arist. Soc.*, 72 (1971/2).

CLARK, M., 'Intentional Objects', *Analysis*, 25 (1965).

DENZINGER, H., *Enchiridion symbolorum definitionum et declarationum de rebus fidei et morum*, ed. A. Schoenmetzer (32nd edn.; Herder, Freiburg, 1963).

DIAMOND, C., 'The Interchangeability of Machines', in *The Business of Reason*, eds. J. J. MacIntosh and S. Coval (RKP, London, 1969).

DUMMETT, M., *Truth and Other Enigmas* (Duckworth, London, 1978).

FLEW, A., *God and Philosophy* (Hutchinson, London, 1966).

FRAENKEL, A. A., BAR HILLEL, Y., and LEVY, A., *The Foundations of Set Theory* (North Holland, Amsterdam, 1958).

FREGE, G., 'The Thought', tr. A.Quinton in *Philosophical Logic*, ed. P.F.Strawson (OUP, Oxford, 1967).

GEACH, P. T., *Logic Matters* (Blackwell, Oxford, 1972).

—— *Three Philosophers* (Blackwell, Oxford, 1961).

—— *God and Soul* (RKP, London, 1969).

—— *The Virtues* (Cambridge UP, Cambridge, 1977).

—— *Truth, Love and Immortality* (California UP, Berkeley, 1979).

—— *Reference and Generality* (3rd edn.; Cornell UP, London, 1980).

GEIGER, L.-B., *La Participation dans la philosophie de S. Thomas d'Aquin* (Vrin, Paris, 1942).

GEROCH, R., 'Space–Time Structure from a Global Viewpoint', in *Proceedings of the International School of Physics 'Enrico Fermi'*, ed. R. K. Sachs (Academic Press, New York, 1971).

GILSON, É., *The Unity of Philosophical Experience* (Scribner's, New York, 1937).

—— *Thomist Realism and the Critique of Knowledge* (1939), tr. M. A. Wauck (Ignatius, San Fransisco, 1986).

—— *The Philosophy of St Bonaventure* (Sheed and Ward, London, 1940).

—— *God and Philosophy* (Yale UP, New Haven, 1941).

—— *Le Thomisme* (5th edn.; Vrin, Paris, 1945).

—— *Being and Some Philosophers* (Pontifical Institue of Mediaeval Studies, Toronto, 1952).

—— *Jean Duns Scot* (Vrin, Paris, 1952).

GRISEZ, G., *Beyond the New Theism* (Notre Dame UP, Notre Dame, 1975).

HAMLYN, D., *Schopenhauer* (RKP, London, 1980).
—— Introductory notes to and trans. of Aristotle's *De Anima* (OUP, Oxford, 1968).
HARRÉ , R. and MADDEN, E. H., *Causal Powers: A Theory of Natural Necessity* (Blackwell, Oxford, 1975).
HEGEL, G. W. F., *The Science of Logic*, tr. W. Wallace (2nd edn.; OUP, Oxford, 1892).
HESSE, M., *Forces and Fields: The concept of Action at a Distance in the history of physics* (Nelson, London, 1961).
HEYTING, A., *Intuitionism: An Introduction* (North Holland, Amsterdam, 1956).
HUME, D., *A Treatise of Human Nature*, ed. L. A. Selby-Bigge (OUP, Oxford, 1888).
—— *Dialogues Concerning Natural Religion*, ed. H. D. Aiken (Hafner, New York, 1948).
JACKSON, R., 'Locke's Distinction between Primary and Secondary Qualities', *Mind*, 38 (1929).
JOHNSON, W. E., *Logic*, vol. ii (Cambridge UP, Cambridge, 1922).
KANT, I., *Critique of Pure Reason*, tr. N. Kemp Smith (Macmillan, London, 1961).
—— *Prolegomena*, tr. P. G. Lucas (Manchester UP, Manchester, 1953).
KENNY, A., *The Five Ways* (RKP, London, 1969).
KIERKEGAARD, S., *Training in Christianity*, tr. W. Lowrie (Princeton UP, Princeton, 1941).
KLEENE, S. C., *Introduction to Metamathematics* (North Holland, Amsterdam, 1952).
KNEALE, W., *Probability and Induction* (OUP, Oxford, 1949).
—— and KNEALE, M., *The Development of Logic* (OUP, Oxford, 1962).
KRIVOSHEINE, B., 'The Ascetic and Theological Teaching of Gregory Palamas', *The Eastern Churches Quarterly*, 3 (1938).
LEIBNIZ, G. W., *Philosophical Papers and Letters*, tr. and ed. L. E. Loemker (Reidel, The Hague, 1969).
LONERGAN, B., *Insight* (Longmans, New York, 1957).
—— *Verbum* (Darton, Longman, and Todd, London, 1967).
—— *Method in Theology* (Darton, Longman, and Todd, London, 1972).
LOSSKY, V., *The Mystical Theology of the Eastern Church* (J. Clarke, London, 1957).
MACKIE, J. L., *The Cement of the Universe* (OUP, Oxford, 1974).
MACROBERT, A., 'Beyond the Big Bang', *Sky and Telescope* (March 1983).
MARÉCHAL, J., *Le Point de départ de la métaphysique*, vol. v: *Le Thomisme devant la philosophie critique* (F. Alcan, Paris, 1926).
MARITAIN, J., *The Degrees of Knowledge*, tr. G. B. Phelan (Geoffrey Bles, London, 1959).

MASCALL, E. L., *The Openness of Being* (Darton, Longman, and Todd, London, 1971).
—— *Existence and Analogy* (Longmans, London, 1949).
MELLOR, D. H., *Real Time* (Cambridge UP, Cambridge, 1981).
MERLEAU-PONTY, M., *Phenomenology of Perception*, tr. C. Smith (RKP, London, 1962).
MEYNELL, H., 'The Intelligibility of the Universe', paper delivered at a conference on Philosophy of Religion at Lancaster, 1976.
MITCHELL, B., *The Justification of Religious Belief* (Macmillan, London, 1973).
MISKOTTE, K. H., *When the Gods are Silent* (Collins, London, 1967).
MOORE, G. E., 'Is Existence a Predicate?' (*Proc. Arist. Soc.*, suppl. vol. 15, 1933), reprinted in, *G. E. Moore: Philosophical Papers* (Allen and Unwin, 1959).
—— *Some Main Problems of Philosophy* (Allen and Unwin, London, 1953).
OWEN, G. E. L., '*Tithenai ta phainomena*', reprinted in *Aristotle: A Collection of Critical Essays*, ed. J. M. E. Moravcsik (Macmillan, London, 1968).
PASCAL, B., *Pensées*, tr. M. Turnell (Harvill Press, London, 1962).
PASSMORE, J., *Hume's Intentions* (Duckworth, London, 1968).
PENROSE, R., *Techniques of Differential Topology in Relativity*, Conference Board on Mathematical Studies, 7 (Society for Industrial and Applied Mathematics, Philadelphia, 1972).
—— and KRONHEIMER, E. H., 'On the Structure of Causal Spaces', *Proc. Camb. Phil. Soc.*, 63 (1967).
PIKE, N., *God and Timelessness* (RKP, London, 1970).
POLKINGHORN, J. C., *The Particle Play* (W. H. Freeman and Co., Oxford, 1979).
PRIOR, A. N., *Time and Tense* (OUP, Oxford, 1968).
QUINE, W. van O., 'Existence and Quantification' in *Ontological Relativity and Other Essays* (Columbia UP, New York, 1969).
—— 'Two Dogmas of Empiricism' in *From a Logical Point of View* (Harvard UP, Cambridge Mass., 1953).
RAMSEY, F. P., 'Universals' in *Foundations of Mathematics and Other Logical Essays* (RKP, London, 1931).
RAHNER, K. *Spirit in the World* (Herder and Herder, New York, 1968).
—— *Theological Investigations*, vol. i (Darton, Longman, and Todd, London, 1961).
RUSSELL, B., *Principles of Mathematics* (Cambridge UP, Cambridge, 1903).
—— 'On Denoting' (*Mind*, 1905), reprinted in *Logic and Knowledge*, ed. R. C. Marsh (Allen and Unwin, London, 1956).

—— and WHITEHEAD, A. N., *Principia Mathematica* (Cambridge UP, London, 1910).

RYLE, G., 'Thinking and Language' reprinted in *Collected Papers*, vol. ii (Hutchinson, London, 1971).

—— 'Categories' (1938), reprinted in the same volume.

—— 'Use, Usage and Meaning' (1961), reprinted in the same volume.

SCHILPP, P. A. (ed.), *The Philosophy of Rudolph Carnap* (La Salle, Illinois, 1963).

SCOTUS, J. D., *Duns Scotus: Philosophical Writings, a Selection*, ed. and tr. A. Wolter (Nelson, London, 1962).

SCRIVEN, M., 'Explanations, Predictions, and Laws' in *Minnesota Studies in the Philosophy of Science*, vol. iii, ed. H. Feigl and G. Maxwell (Minnesota UP, Minneapolis, 1962).

SHOEMAKER, S., *Self-Knowledge and Self-Identity* (Cornell UP, Ithaca, 1963).

SPINOZA, B. DE, *The Chief Works of Spinoza*, Vol. II, tr. R. H. M. Elwes (G. Bell and Sons, London, 1883).

STRAWSON, P. F., *Individuals* (Methuen, London, 1959).

—— 'On Referring' (*Mind*, 1950), reprinted in P. F. Strawson *Logico-linguistic Papers* (Methuen, London, 1971).

SUPPES, P., *Axiomatic Set Theory* (van Nostrand, Amsterdam, 1960).

SUTHERLAND, S., 'God, Time and Eternity', *Proc. Arist. Soc.*, 79 (1978/9).

SWINBURNE, R., *The Coherence of Theism* (OUP, Oxford, 1977).

—— *The Existence of God* (OUP, Oxford, 1979).

THOMAS AQUINAS, *Summa Theologica*, Prima Pars (cited as *S. Th.* Ia); English trans. English Dominican Fathers (2nd edn.; Burns and Oates, London, 1941), vols. i–v; Parallel Latin and English edition, English Dominican Fathers (Blackfriars, London, 1961–70), vols. i–xv.

—— *Summa Contra Gentes* (cited as *SCG*); tr. Fathers of the English Dominican Province (Burns and Oates, London, 1929), and more recently, tr. A. C. Pegis (Notre Dame UP, Notre Dame, 1975), 5 vols.

—— *Commentary on Aristotle's Metaphysics* (cited as *Comm. Met.*); English trans., *On the Metaphysics*, tr. J. P. Rowan (Henry Regnery, Chicago, 1961).

—— *Commentary on Aristotle's De Anima* (cited as *Comm. De An.*); English trans. in *Aristotle's De Anima and the Commentary of St. Thomas Aquinas*, tr. K. Foster and S. Humphries (RKP, London, 1951).

—— *Questiones Disputatae De Anima* (cited as *Q. D. De An.*); English trans., *The Soul*, tr. J. P. Rowan (Herder and Herder, New York, 1949).

—— *Questiones Disputatae De Potentia Dei* (cited as *De Potentia*); English trans., *On the Power of God*, tr. English Dominican Fathers (Burns and Oates, London, 1932–4).

—— *Quaestiones Quodlibetales* (cited as *Quodlibet.*), ed. R. M. Spiazzi (Marietti, Turin, 1949).

THOMSON, J., 'Is Existence a Predicate?' in *Philosophical Logic*, ed. P. F. Strawson (OUP, Oxford, 1967).

VENDLER, Z., 'Effects, Results and Consequences' in *Analytical Philosophy*, 1st ser., ed. R. J. Butler (Blackwell, Oxford, 1962).

VERNEAUX, R., 'Note sur le principe de raison suffisante' in *La Crise de la raison dans la Pensée contemporaine* (Desclée de Brouwer, Paris, 1960).

WILLIAMS, C. J. F., *What is Existence?* (OUP, Oxford, 1981).

WITTGENSTEIN, L., *Tractatus Logico-Philosophicus*, tr. D. F. Pears and B. F. McGuinness (RKP, London, 1961).

—— *Philosophical Investigations*, tr. G. E. M. Anscombe (2nd edn.; Blackwell, Oxford, 1958).

WOJTYLA, K., *Sign of Contradiction*, tr. M. Smith (St Paul Publ., Slough, 1979).

Index

The indication, e.g., 1 n., may signify both page and footnote, or just the footnote. Other key entries are listed under the entry 'main argument'.